The Organic Farming Manual

The Organic Farming Manual

A Comprehensive Guide
to Starting and Running
a Certified Organic Farm

Ann Larkin Hansen

Storey Publishing

The mission of Storey Publishing is to serve our customers by publishing practical information that encourages personal independence in harmony with the environment.

Edited by Nancy Ringer, Deborah Burns, and Sarah Guare
Art direction and book design by Dan O. Williams
Text production by Patrick Barber/McGuire Barber Design

Photography by © Jason Houston
Illustrations by Bethany Caskey
Map on page 130 adapted from GeoNova Pub Inc./www.stockmaps.com

Expert reviews by Harriet Behar, Organic Specialist, MOSES (Midwest Organic and Sustainable Education Service) and Ralph Moore, Market Farm Independent (Equipment chapter)
Indexed by Eileen Clawson

Storey books are available for special premium and promotional uses and for customized editions. For further information, please call 1-800-793-9396.

Storey Publishing
210 MASS MoCA Way
North Adams, MA 01247
www.storey.com

Printed in the United States by McNaughton & Gunn, Inc.
10 9 8 7 6 5 4 3 2

Library of Congress Cataloging-in-Publication Data

Hansen, Ann Larkin.
The organic farming manual / Ann Larkin Hansen.
   p. cm.
  Includes index.
   ISBN 978-1-60342-479-0 (pbk. : alk. paper)
   ISBN 978-1-60342-480-6 (hardcover : alk. paper)
   1. Organic farming. I. Title.
S605.5.H355 2010
635'.0484—dc22
                    2009049116

To all organic farmers — those who blazed the trail, those who are making it happen now, and all those to come

Contents

Foreword

Organic farming is a vibrant and growing sector of agriculture in the United States and across the world. This comprehensive manual covers many aspects of farming, from vegetable crops to many types of livestock. This book contains practical "how-to" options to help the reader decide how to best use their agricultural land, as well as historical background on how and why various practices and equipment were developed. I wish I had this book 30 years ago when I started my lifelong adventure in organic agriculture!

There are few activities in life that offer the same sense of accomplishment as farming, accompanied by the spiritual fulfillment gained by understanding, co-operating, and being a part of the natural world. Organic farming, with its reliance on natural systems rather than synthetic purchased inputs, is attractive to young adults seeking to make their place in the world, people in middle age who want a change from their current path, and older folks who are semiretired and want to pursue an agrarian lifestyle. For those wishing to manage their land and produce a wide variety of foods in harmony with nature, *The Organic Farming Manual* offers the background and options necessary to start a wide variety of farming systems.

From what to look for when buying land, to deciding what to do with that land once you have it, this book offers clear guidance for those who wish to farm organically, whether solely for their own food production or as a business venture. The United States has laws governing the use of the word *organic* on agriculturally produced products. Ann Hansen has incorporated the regulations of the USDA's National Organic Program into these pages, allowing the reader to have a better understanding of what they would need to do in order to access the dynamic organic marketplace.

Organic farming is not difficult, but it does have its own unique set of challenges. *The Organic Farming Manual* helps the reader navigate the various challenges and possible approaches to these issues. Even after a lifetime working with organic farmers, I found so much useful information in this book. Whether for providing a basic overview or explaining nitty-gritty details, this book should be kept as a reference through the years, as one explores and experiences the joys and rewards of organic farming.

Faye Jones
Executive Director
Midwest Organic and Sustainable Education Service (MOSES)

Preface

For the novice farmer who intends to follow an organic philosophy, this book provides an overview of basic farming practices and organic principles. For those who intend to sell farm products under a "certified organic" label, the book includes discussion of the federal rules for organic agriculture and the certification process.

Experienced farmers interested in transitioning to organic farming will find here the general information they need. But since much of this book is devoted to basic farming knowledge, which they already possess, a more efficient way of getting the technical information they need to make the transition would be to contact one of the education services devoted to organic production, such as the Midwest Organic and Sustainable Education Service (MOSES). General information on transitioning to organic production is also available from various land-grant universities and from ATTRA — National Sustainable Agriculture Information Service, an agency of the U.S. Department of Agriculture. (See the resources section at the end of this book for more information about these and other education services.)

The depth of knowledge you need to be successful with an organic farm enterprise can be obtained only by experience and constant learning. This book provides a starting point, one that I hope will serve to put you on the right road. Your most important resources as you continue the journey will be other members of the organic farming community and your farming neighbors, who know your soils, climate, and local resources.

Acknowledgments

The knowledge anyone would need to intelligently discuss all types of farms in all types of soils and climates can't possibly be contained in the head of a single author. I want to thank all of the farmers who have talked with me or whose farms I've toured, all of the writers, researchers, and conference participants who have contributed to my understanding, and all of the many others who directly contributed to this effort. This book wouldn't exist without their help.

Special thanks to Harriet Behar, who meticulously reviewed the entire manuscript and corrected numerous errors, and to Faye and all the folks at MOSES who connected me with Storey Publishing and then supplied much-appreciated resources and moral support. Thanks also to Deb Burns at Storey for making the path so smooth (someday I hope we play some music together!) and to Nancy Ringer, who clarified and organized my writing so well.

And thanks, Steve, for bearing with me through this long process, and even making dinner once in a while.

Any mistakes, of course, are my own.

What "Organic" Means

Organic farming is family-friendly, economically viable for small and midsized farms, and a boon to our stressed environment — and its products taste great. For the farmers themselves, there's a real joy that comes from working with the natural cycles of soils, plants, and animals, instead of trying to beat them into submission for profit and convenience. Whether you're interested in producing a crop for market or only in growing your own food, organic farming makes sense for your pocketbook, your health, our environment, and our nation.

There is a lot of confusion among both consumers and farmers about what organic farming is and is not. With so many "green" labels now on grocery shelves, it can be hard to figure out what makes organic different from anything else purporting to be, as so many labels brightly say, earth-friendly and natural and all those other good things.

Yet the difference is simple. The "Certified Organic" label means that the grower has:

- Complied with the Final Rule of the National Organic Program of the United States Department of Agriculture (USDA)
- Documented that fact on paper
- Been inspected and approved by a USDA-accredited organic certifier

Other "green" labels may be backed by nongovernment standards and inspection programs, but many are only vague claims invented by marketing teams to dupe consumers into believing they're getting something better than it really is. The only label that comes with a legal guarantee is "USDA Organic."

To have the right to use that label, an organic farmer must follow all the stipulations of the Final Rule and go through a certification process that requires a written farm plan, documentation of all inputs and practices, and an on-farm inspection. In other words, organic farming is different from organic certification. Organic farming is a farming system, while organic certification verifies use of that system. Certification is thus a marketing tool. Certification guarantees the integrity of the product for buyers who aren't in a position to drive out to the farm themselves to make sure the rule is being followed.

Many organic growers follow the rule but don't go through the certification process. Organic certification is a lot of paperwork, it's not cheap, and there is really no need for it if you're selling products to people you know personally, who trust you.

The USDA certified-organic label assures your customers of your organic integrity.

For small-scale farmers the Final Rule contains an exception that allows them to label themselves as "organic," as long as they are following the rule (including having a written farm plan and documenting all farm inputs and practices) and are selling less than $5,000 worth of products each year. There are plenty of these folks around, and probably even more who follow almost but not all the stipulations of the Final Rule.

Many of these farmers who aren't completely organic call themselves sustainable farmers. The term *sustainable* has been tossed loosely around in agricultural circles for many years but has no agreed-upon, clear definition. Generally it is taken to mean farming in such a way that fertility and yields will be sustained or improved from current levels for generations to come. Farms in this country that define themselves as sustainable presumably employ soil-conserving and nutrient-cycling practices such as management-intensive grazing, contour cropping, and composting. These practices are also basic to organic farming, and so organic farming can be defined as a more restrictive subset of sustainable farming, one that is quite a bit more specific about requirements.

Farmers who call themselves sustainable but don't conserve their soil and minimize chemical use are kidding themselves and us — and are not uncommon.

The Organic Approach

Exactly what *organic* means when you're out on the farm is defined in the Final Rule as "a production system that is managed to respond to site-specific conditions by integrating cultural, biological, and mechanical practices that foster cycling of resources, promote ecological balance, and conserve biodiversity." That may sound like so much gobbledygook to the uninitiated, but in fact the concepts behind each of those carefully chosen words are pretty simple to understand, and they all fit together to form the foundation of an organic farming system. The following chapters will walk you through what it all means and how it can work.

To begin, organic farming (as detailed in the Final Rule) generally prohibits the use of manmade pesticides, herbicides, and fertilizers, as well as antibiotics, hormone treatments, genetically modified organisms, sewage sludge, and the feeding of animal by-products to livestock. There are only a few specific exceptions, and they are listed, along with the specific circumstances under which they may be used, in the Final Rule.

Four Principles of the Organic Approach

1. Organic farming is proactive, not reactive. All organic farming activities are directed toward creating and sustaining a living soil and an ecologically healthy, productive farm. Farmers ideally identify, anticipate, and address needs and potential problems before they can compromise the functioning of the farm system.

2. All farm practices and inputs are evaluated for their effect on the entire farm system and ecosystem. The decision to use an input or practice must not be based solely on whether it will solve a particular problem. The effect on every other component of the system must also be evaluated.

3. Good timing and close observation are integral to a successful organic system. Being attuned to seasonal cycles of plants, insects, soil life, and other ecosystem components allows the farmer to time farm practices for maximum effectiveness. Pest controls are applied at the most vulnerable times in pest life cycles; planting and birthing take place when conditions are best.

4. Experimentation and learning to improve the farm system are continual. Working effectively with natural systems requires constant curiosity and experimentation, which results in a continual expansion of knowledge and refining of techniques.

In most people's minds, those prohibitions are the defining characteristic of organic farming. But for the farmer they're just a starting point, the rules of the game. Building a successful organic system really depends on a farmer's ability to put together the mix of allowed farm practices that best enhance the farm's natural ecology, fertility, and long-term sustainability. Choosing and coordinating crop rotations, tillage methods, soil amendments, grazing rotations, and innumerable other details is what makes organic farming so endlessly fascinating and rewarding. The organic farmer may be frequently frustrated, sometimes mistaken, and often short of time, but never bored.

Organic Farms Need Hands-On Management

Enhancing natural processes necessitates, as its foundation, an understanding of how these processes operate, how agriculture can degrade them, and how a farmer can mitigate the unnatural effects of agriculture on the ecosystem. If you're an aspiring farmer who wants to grow organically, you have to accomplish two things: first, learn how to farm, and second, learn how to do it in a way that maintains and improves, instead of degrades, the soil and the ecosystem. Doing both at once is a great adventure, the details worked out in the field to best suit the unique conditions on your farm. In the end, it's your own understanding, creativity, stamina, and resourcefulness that will determine how well you succeed.

If you plan to become certified organic there's one more important step that you should be aware of before you start: you must go through a three-year, chemical-free transition period before you can apply for certification. If no synthetic chemicals have been used on your land for some time and you can prove it, you may be able to shorten this transition period. But "going organic" is not instantaneous. It happens over time, and only with active management.

Because successful organic farming demands such attentive, site-specific, hands-on management, it is especially suitable for small and mid-sized farms. On small farms, even part-time farmers (which is the majority of small farmers) have the time to pay attention to the details that make all the difference. I think every small-farm book repeats some variation of the old saying, "The best fertilizer is the footsteps of the farmer." That's just as true as it ever was.

Diversity — of crops, livestock, and natural habitat — is a cornerstone of organic farming.

Organic Farming Works

Taking a larger perspective, it's interesting to note that for many decades the conventional agricultural community has maintained that organic farming might be very nice but it won't feed the world and therefore it makes no sense for everyone to do it. But interestingly enough, long-term studies at Iowa State University, the University of Wisconsin, and the Rodale Institute comparing organic and conventional cropping systems are clearly demonstrating that organic farming can produce yields equal to or higher than those of conventional farming. Additional studies elsewhere are confirming these findings, as are many organic farmers, and so dispelling that particular myth.

"As it happens, organic farming can feed the world, and in the long run may be the only sustainable method of doing so. Declaring 2015 the "International Year of Soils," the Food and Agriculture Organization of the United Nations states, "The overuse or mis-use of agro-chemicals has resulted in environmental degradation. . . . Agricultural systems and agro-ecological practices that dedicate great care to nurturing soil biodiversity, such as organic farming . . . can sustainably increase farm productivity without degrading the soil and water resources."

The Market Is Strong, and Getting Stronger

Ideally, all farmers would convert to organic production, but transitioning from conventional to organic agriculture on a national scale would take a lot of reorienting of the current food production and processing infrastructure, which is heavily dependent on huge crop monocultures and on moving food long distances from farms to processors to markets. Fortunately, the foundation for an alternative food system based on locally produced food was laid long ago by the organic community, and it continues to expand nationwide with the renaissance of farmers' markets, local food cooperatives, and farmer-owned processing and marketing concerns. And in the past few years, organic food has made the jump to mainstream grocery stores — something almost unimaginable just a decade ago. We've come a very long way in the past quarter century,

and rising transportation costs may accelerate a reorientation of the nation's food distribution system toward more local production. But there is still a very long way to go.

Fortunately, consumer demand for organic products continues to increase at an impressive clip. This increasing demand means that for many years now it's been possible to make a fair wage as an organic farmer. If you are consistent with your quality and well planned in your marketing, you can find a profitable sale for your cash crops. But let me emphasize that last point: Successful sale of farm products depends on more than organic production methods. You must have a good product and a good marketing strategy.

Why Farmers Do (or Don't) Farm Organically

So why isn't every farmer organic? Maybe because it's unfamiliar, because change seems unnecessary, because it takes a lot of time, and because it's scary to bet the farm on a paradigm you haven't seen working with your own eyes. For the past fifty years the U.S. government has actively encouraged chemical farming, so that

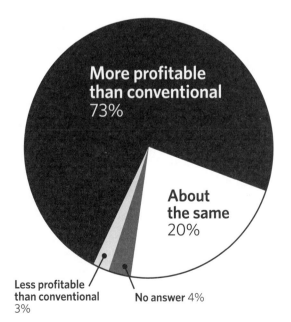

Profitability of Organic Farms Compared to Conventional

More profitable than conventional 73%

About the same 20%

Less profitable than conventional 3%

No answer 4%

Source: "Overview: Experiences and Outlook of Minnesota Organic Farmers — 2007," a survey conducted by the Minnesota Department of Agriculture

most farmers have no experience with or incentive for any other way to do things. Most crops in organic systems require more labor as well, and labor shortages are a chronic problem in American agriculture. And if this year's income depends on the soybeans that are being consumed by aphids today, do you tell your kids there isn't going to be any Christmas this year, or do you call the spray truck? Farmers aren't dumb, and they often, with the caution born of experience, prefer to see their neighbors successfully go out on a limb before they will take a similar risk. But as the evidence of organic farming's economic and environmental success accumulates, more established farmers are transitioning to organic production, and more new farmers are setting up as organic rather than conventional.

For those conventional farmers who have made the leap to organic production, one of the most common motivators is to get agricultural chemicals out of their lives. The hazards of chemical exposure for farmworkers are well documented, and the hazards of chemical residues in your drinking water, laundry, house dust, and kids' play area are evoking increasing concern from the scientific community.

Organic farming's relative insulation from the lows of volatile commodity prices and the better chance of making a profit help convince farmers, too, though it can take five years or more to build soils enough to match the yields of conventional agriculture. Several studies have shown, however, that even with lower yields and no price premiums, organic cropping returns more income to the farmer due to significantly lower input costs.

Be aware, though, that having low inputs does not mean having no inputs. Every time something leaves the farm in the form of crops or livestock, the nutrients and organic matter contained in those products need to be replaced in order for yields to be sustainable. Conventional farmers usually apply synthetic fertilizer every growing season to replace those nutrients. Organic farmers, in contrast, use manure, compost, and other natural amendments to maintain the fertility and structure of their soil.

Organic Farming Is Good for Communities

The economics of organic farming make it more likely that an organic farmer will survive in the farm business for the long term and, significantly, that your community will be a pleasant place to live for the long term. In other words, when it's done well, organic farming is highly sustainable. A study by Dr. Luanne Lohr of the University of Georgia* has found that counties with higher numbers of organic farmers have stronger farm and local economies. That's good news for rural areas, where poverty levels often outstrip those of urban areas.

In contrast to ever-larger conventional farms that tend to buy their inputs in bulk from distant sources and to send their harvests off to huge and distant processing plants, organic farms put more people on the land, who in turn will tend to do more of their buying and selling locally. That's what makes an economy vibrant. There are more kids in the schools and more support for amenities like the volunteer fire department, the library, a good doctor, local retailers, and farmer networking groups.

Beware Snake-Oil Salesmen

Of course, the money you save through organic management can be quickly spent in buying higher-priced certified seed, feed, and soil supplements. And there are plenty of snake-oil salesmen out there who should be avoided. In general, don't buy it unless you need it, and use soil tests and other tools to determine what and how much. If you have livestock, growing all or most of your feed is often essential to a decent profit margin.

---

*L. Lohr, *Benefits of U.S. Organic Agriculture* (Athens, GA: November 2002), a report prepared for the Consumer's Choice Council in Washington, DC.

Farm Business Planning Resources

Most extension services offer free or low-cost assistance with farm business planning or can refer you to organizations that do. The Internet is also replete with farm business planning information.

Building a Sustainable Business: A Guide to Developing a Business Plan for Farms and Rural Businesses, by the Minnesota Institute for Sustainable Agriculture (MISA) and copublished by the USDA's Sustainable Agriculture Network (SAN), is an excellent tutorial in business planning. Especially helpful are the five farm case studies that are used throughout the 280-page guide to illustrate how planning has been successfully applied in real-life situations. The guide is available from MISA and SAN, among other sources (see the resources section).

There's no doubt in my mind that the satisfaction level is higher in organic farming than in conventional farming. In addition, with judicious budget management, the crop is worth more, both nutritionally and in the potential to deliver enough dollars to more than recoup the dollars spent on production.

Support Your Neighbors

A word to the wise regarding your conventional-farming neighbors: Don't think less of them because they are not, like you, set on becoming organic farmers. There are legitimate reasons for not becoming certified organic, and you should respect them. One farmer I know told me that despite following organic practices on his dairy farm, he was not certified for the simple reason that he detested paperwork. Certification takes a lot of paperwork. Another farmer had been under gentle pressure from a mutual friend to "go organic," as he put it. But why, he asked, would he start the steep climb up the organic learning curve when he would be retired from farming before he could become certified? Some farmers are too far in debt to take a gamble on transitioning to organic; they can't afford the possible yield losses during the several seasons it takes to build a fully functional organic system. Most are simply unconvinced that it's a better way, seeing as they are making a living and you probably aren't yet.

Don't be too quick to judge your neighbors. Do respect them for their knowledge, usually acquired through a lifetime of experience, of how to go about making a living at farming. They know a lot about weather, crops, livestock, soils, markets, and the inevitable misfortunes involved in farming, and they can teach you plenty if you are willing to listen. Much of farming is just farming and doesn't have a whole lot to do with whether you're organic. Most neighbors will lend a hand when you need it, as you someday will, and you should be there when they need help. Good neighbors are an important part of a sustainable, organic farm enterprise.

The Balancing Act

As you begin your organic farming enterprise, you'll quickly realize that any type of farming is a constant process of deciding how to allocate available time, money, resources, and stamina among the endless number of things that have to be done, that should be done, and that you'd like to have done. There will always be more to do on your farm than you will ever have the money or time to do. Too many farmers have lost their health, their marriage, the love of their children, or their love of farming from farming too much and enjoying too little.

Though it's not in the organic rules or any guidelines, remember that to be truly sustainable your quality of life has to be sustainable, and that means taking time off from farming. There's a reason why all the world's major religions mandate a day of rest every week. Make sure you get it, too. You aren't getting into organic farming so you can work yourself to death, you're doing it to reconnect with the natural world and have a better, healthier, happier life for you, your family, and your community. Make quality of life part of your plan. The world can't afford to lose any more organic farmers.

Profile: Jim Riddle

Winona, Minnesota

Since 1980, Jim Riddle has been an organic farmer, gardener, farm inspector, educator, policy analyst, author, activist, and avid organic eater. He and his wife, Joyce, live off the grid in southern Minnesota, producing all of their own power from sun, wind, and woods. They raise a huge garden and put up much of their own food.

Jim was founding chair of the Independent Organic Inspectors Association (IOIA) and coauthor of *The IFOAM/IOIA International Organic Inspection Manual.* He has trained hundreds of organic inspectors throughout the world. Jim has served on the Minnesota Department of Agriculture's Organic Advisory Task Force since 1991 and was instrumental in the passage of Minnesota's landmark organic certification cost-share program. In January 2006, Jim was named the University of Minnesota's Organic Outreach Coordinator. Jim is a former chair of the USDA's National Organic Standards Board and is a leading voice for organic agriculture.

Q. Can the organic food sector continue to grow without compromising its original principles?
A. During my five years on the National Organic Standards Board, I saw the influence of corporate lobbyists and lawyers who are in Washington, DC, every day. They know how to play the game, and the USDA is open to those kinds of influences; that's who it normally listens to. They have a deregulatory influence, while members of the organic community have consistently sought strong regulation. For instance, some food corporations have influenced the USDA to delay action to clarify and enforce requirements for pasture, even though it's very clear that pasture is necessary for ruminant animals.

It takes continual engagement and vigilance from the organic community to protect the integrity of the standards and counterbalance "business as usual" influences in Washington, DC. Certainly we've seen the growth of corporate, highly processed organic products, but under people's radar is the tremendous growth of more local organic — farmers' markets, CSAs, and others. The growth of local organic production is at least as significant as the growth of "corporate organic," if not more so, and those markets aren't so dependent on cheap energy and transporting goods across the country or around the world. A lot of those more local producers are not certified organic because their buyers are intimately aware of their farming practices. But that's where the rubber meets the road. Certification of the operation is less important than what's happening on the land itself: ecologically sound and holistic relationships between farmers, land, animals, and consumers.

Q. What would you like to see the government do to better support organic farming?

A. What I would wish for, to begin with, are leaders at the highest levels of the USDA who understand, fully support, and embrace organic agriculture. Strong leadership throughout the USDA, at all levels, will positively impact how the standards are implemented and enforced by the agency. While there have been no changes that have seriously weakened the standards, the USDA's National Organic Program managers have repeatedly issued vague, confusing, and indefensible interpretations and have a poor record of enforcement. The NOP must take the lead by hiring well-qualified staff, being more transparent, and implementing the program as envisioned by Congress. A fully functional NOP will engage with other government agencies, including the Food and Drug Administration, the Environmental Protection Agency, the Agricultural Research Service, the Natural Resources Conservation Service, the Economic Research Service, the Cooperative State Research, Education, and Extention Service, and state departments of agriculture, to maximize the benefits of organic food and farming.

Q. Don't the high prices of organic food make it too expensive for many consumers?

A. If people buy whole and minimally processed organic foods, join CSAs, shop at farmers' markets, and do some canning, freezing, drying, and other food preservation, they can eat organic food year-round for the same money or less. But if they buy out-of-season produce or highly processed organic food products, then it's going to be much more expensive. You must keep in mind that conventional foods do not reflect their costs of production. Conventional producers receive huge government (taxpayer-financed) subsidies, and the true costs of production, such as groundwater contamination, soil erosion, and dependence on foreign oil, are externalized. Plus, numerous studies show that organic foods consistently contain more nutrients than conventional foods, so you are getting more nutrition for your food dollar when you purchase organic foods.

Q. What would you wish for the future of organic farming?

A. I hope that farmers selling into local markets continue to embrace organic practices and convey the values that are represented by organics to their customers. What limited research has been conducted shows that organic has so many benefits: protection of water quality, improvements to soil quality, enhanced biodiversity, increased nutrient levels of both crop and livestock products, less antibiotic-resistant bacteria, better incomes for farmers, and better economic activity in areas with more organic farms. I would like to see the ecological, social, health, and economic benefits of organic food and farming integrated throughout society.

Profile: **Faye Jones**

Midwest Organic and Sustainable Education Service, Spring Valley, Wisconsin

Faye Jones was a market gardener for fifteen years before becoming executive director of the Midwest Organic and Sustainable Education Service (MOSES), a nonprofit education and outreach organization working to support and promote organic agriculture in the Midwest. She oversees education and outreach initiatives, including MOSES's farmer-to-farmer mentoring program, the Help Wanted: Organic Farmers program, and the nation's largest organic farming event, the Organic Farming Conference, held by MOSES late each winter in La Crosse, Wisconsin.

Faye was an appointee to the Wisconsin Governor's Organic Task Force, is a committee member of the National Organic Coalition, and has been involved in numerous other organic organizations and committees. She and her husband, Mark Plunkett, continue to garden and raise livestock at their Morning Glory Farm in western Wisconsin. Here are her thoughts on organic farming:

> *In my experience, there are three reasons why people "go organic":*
>
> *One, because it's the right thing to do. It's a lifestyle choice.*
>
> *Two, because of the higher pay price. It's economic. And that's a fine and valid reason. Farming is a business, not just a way of life. People inevitably become spiritual about it.*
>
> *Three, because of a health crisis of some sort. One farmer had lost half a herd of cattle after they ate from a hay bale that had been stored on top of a bag of pesticide. Other farmers will tell you about their dad dying of cancer, and how he used to mix the chemicals with his bare hands.*
>
> *Being certified organic is a marketing tool. Being just plain organic means following a defined set of practices that seek to build healthy soil, and so healthy crops, healthy livestock, and healthy people.*
>
> *For me organic farming made sense. It was the right thing to do. For me there was never any doubt.*

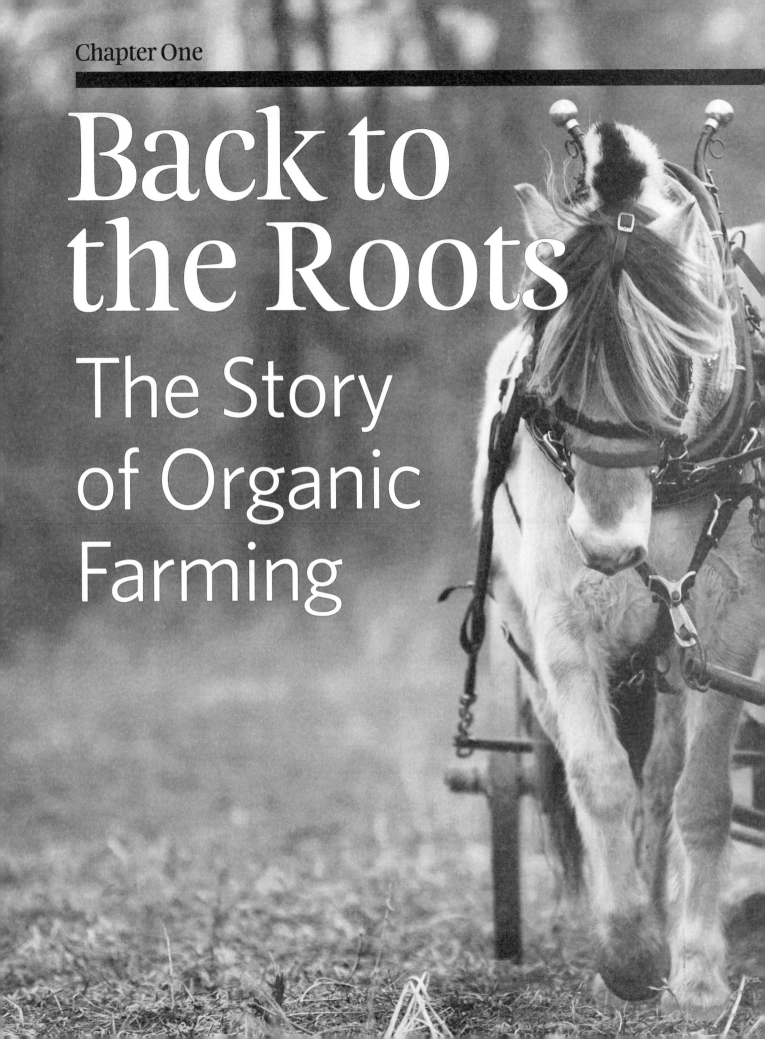

Back to the Roots

The Story of Organic Farming

> *Organic gardening or farming is a system whereby a fertile soil is maintained by applying Nature's own law of replenishing it . . .*
> *a vigorous and growing movement that is destined to alter our concepts of the garden and the farm and to revolutionize them in order to secure for ourselves more abundant and more perfect food.*
>
> **J. I. RODALE AND STAFF**
> *The Encyclopedia of Organic Gardening* (1974)

Organic farming is built on living soil. Deep, nutrient-rich, and above all, alive: teeming with the billions of organisms natural to the ecosystem of the subsurface world. Without healthy soil we would starve, since almost everything we eat either is a plant that grew in the soil or comes from animals nourished by plants. Everything — everything — in organic farming is founded on building and keeping a living soil.

Most people think of organic farming as simply producing food without using man-made chemicals. But organic farming is so much more. Continually evolving and changing, varying in practice from region to region and farmer to farmer, organic farming offers a system for preventing and repairing the detrimental effects that agriculture, by its nature, has on the land, guarding the health of ecosystems so that they maintain their natural richness and can continue to produce food for the human race. Organic farming is essentially an act of self-preservation for the long term, since our health unquestionably depends on the quality of our food, and the quality of our food depends on the quality of our soil.

A Legacy of Damaged Soils

When prehistoric humans first put seed in the ground ten thousand (or more) years ago, they were totally "organic" as most people today understand the term: they used no synthetic pesticides or fertilizers, because they didn't have any. And yet they still managed, in many different civilizations through the course of human existence, to wreck their soils. In many places around the world today, including the United States, many farmers are still wrecking soils in the same old ways.

The simple act of plowing the soil opens it to wind and water erosion. Keeping too many animals on the same ground for too long degrades the

Early farmers realized the need to conserve precious topsoil, and to do so they established erosion-control practices such as terracing, which is still in use in some cultures today.

plant community, compacts the soil, and causes erosion. Careless irrigation causes waterlogging and salination of susceptible soils. Taking crops from the soil year after year without any attempt to replace the nutrients used by the crop eventually makes the soil too poor to grow any crop at all.

The scars of ancient farming practices are still evident in many places around the world, from salinized soils in southern Mesopotamia to the eroded uplands of Greece, Italy, and Israel. Soil degradation has contributed to the decline and disappearance of civilizations from the Easter Islanders of the Pacific to the Mayan civilization of Mexico and Central America. Hundreds and even thousands of years later, these regions still suffer from that legacy of soil degradation.

Innovations in Building Soil

Some early farmers recognized that their land was producing less each year and realized that they needed to figure out ways to stop the decline. Some ancient Greeks, for example, used green manures, crops grown for the purpose of being plowed back into the soil to build soil structure and nutrient levels. Other early farmers developed physical methods such as terracing to stop erosion on steep slopes and discovered that applying organic materials like seaweed, pulverized bones, ashes, and manure put nutrients back in the soil.

Thanks to some innovative early experimenters and writers, much of this knowledge was preserved. These traditional practices are now basic to organic farming, and farmers worldwide still employ them.

Early Organic Stirrings

In 1910 Francis King, a retired University of Wisconsin professor of agricultural physics and chief of the United States Department of Agriculture (USDA) Soil Division, published *Farmers of Forty Centuries*, describing the peasant agriculture of Japan, China, and Korea. Farmers there had maintained soil fertility and crop yields for four thousand years by the use of careful crop rotation, green manures, composting, intercropping, and a host of other sustainable practices.

King's book, still a classic exposition of traditional sustainable farming techniques, greatly influenced Sir Albert Howard, a British botanist who from 1905 to 1931 managed agricultural research stations in India. Howard published several influential books on "nature farming," as he

Combating the Dust Bowl

In the young United States, physical degradation of soils followed European settlers west. As farms were depleted and eroded, their owners simply moved to virgin soils. By the late 1800s students of agriculture were becoming concerned, and the massive wind erosion of prairie soils during the Dust Bowl years of the 1930s finally impelled the federal government to take action. In 1933 the Soil Erosion Service (later the Soil Conservation Service) was established to provide technical and financial assistance to farmers to ameliorate erosion on their land.

The agency's first large-scale project was healing gullies and stopping floods in the washed-out Coon Valley of southwest Wisconsin. In exchange for fertilizer, lime, seed, and technical know-how, half of the valley's farmers agreed to put in contour cropping, use longer crop rotations, turn the steepest fields back into pasture and woods, and implement various other erosion-control measures.

Erosion was reduced by at least 75 percent, and trout returned to the streams. Coon Valley today is beautiful and productive, and its success story has been repeated in many areas across the country. That region is now, and probably not incidentally, also a hotspot for organic production. And technical help and cost sharing for soil conservation are still available for selected projects and farms from the successor of the Soil Conservation Service, the National Resources Conservation Service.

called it, including the 1940 work *An Agricultural Testament*, which advocated returning all organic waste products back to the soil and described how to compost them for maximum nutrient retention and balance. In that same year Lord Walter Northbourne coined the term "organic farming" in his book, *Look to the Land*.

Howard and Northbourne inspired Lady Eve Balfour to launch the Haughley Experiment in Suffolk, England, a project designed to compare organic and nonorganic methods of farming. By 1943 Balfour had sufficient data to write *The Living Soil*, which advocated farming without synthetic chemicals. Balfour also helped establish the Soil Association, which held its first meeting in 1945, with 60 people in attendance. By 1948 the association had members in 32 countries, including J. I. Rodale and Louis Bromfield in the United States.

J. I. Rodale is perhaps the best-known early advocate of organic farming in the United States. He had established an organic farm near Allentown, Pennsylvania, and in 1947 founded *Organic Farming and Gardening Magazine*. This publication later split into *Organic Gardening*, which is still going strong, and *New Farm*, which closed operations in 1995, to the dismay of many organic farmers. (Fortunately, after a hiatus of a few years, it's now available again, as an online publication.) The magazines, Rodale's books, and the Rodale Experimental Farm, all operated by the Rodale Institute, continue to be influential today in organic agriculture.

Rodale's fellow Soil Association member Louis Bromfield was similarly important in those early days when organic agriculture was first becoming clearly different from conventional farming. An internationally known author of fiction, he and his family were living happily in France when the imminent Second World War brought them home to his native Ohio, where he'd spent a good part of his childhood helping his grandfather on the family farm.

Instead of the green, productive valley he remembered so fondly, he found devastation: soils eroded from the hillsides and cropped to death in the valley bottom. Fortunately he had money, farming knowledge, and the inspiration of writers like Howard and Balfour. He bought a number of contiguous farms and applied soil restoration practices with a vengeance, bringing back crop rotations and livestock, plowing down green manures, and letting the grass grow long in the fruit orchard to encourage water retention and beneficial insects. Though Bromfield used synthetic fertilizer when he thought it was needed, his books, *Pleasant Valley* in 1943 and *Malabar Farm* in 1948, were hugely influential in the organic movement. (For more about Bromfield and Malabar Farm, see page 46.)

Retail Chemicals: Farming Made Cheap and Easy

But just as Rodale, Bromfield, and other organic pioneers were building a small following in the United States, a revolution was brewing in mainstream agriculture. New chemicals and bigger, better equipment were drastically changing the farm landscape. And while mainstream agriculture and organic farmers could agree on the benefits of controlling erosion, only organic farmers believed there might be big problems associated with the widespread use of synthetic fertilizers and pesticides. Mainstream farmers saw only the benefits of chemical agriculture.

For thousands of years good farmers have understood the need to return nutrients to the soil and have used a wide variety of natural materials to do so. Good farmers are also always looking for ways to do things better. During the period following World War II, they found them: ammonia plants previously needed for munitions manufacture were diverted to fertilizer production. Synthetic fertilizers became more readily available and affordable, and the results in the field were dramatic and immediate. There were no bad effects that anyone could see, and the return for the dollars spent was terrific.

It was good times on the farm. Fields were producing two and three times as much as before. The increasing availability and affordability of new hybrid seed varieties boosted yields even more, though these demanding crops were much more dependent on synthetic fertilizers. Draft horses had virtually disappeared, and bigger, better tractors meant bigger, better fields. Now you could grow corn year after year in the same huge field; you didn't need to rotate to less profitable crops to maintain soil nutrient levels, and you didn't need to keep livestock so you could haul their manure from the barn to spread on and fertilize the fields. You could do all this more quickly and easily with synthetic fertilizers. Life was simpler and, for a long time, more profitable.

Pesticides had a similar story. Traditional practices for controlling weeds, diseases, and pests were forgotten in the excitement of having easy and effective new chemicals. Imagine years of battling without really winning against Colorado potato beetles, and then one day a spray truck shows up and presto! Problem solved. Who could resist — who would want to resist — results like that?

DDT (dichlorodiphenyltrichloroethane — no wonder everyone uses the abbreviation) arrived on the civilian market in 1943 to great rejoicing. It was cheap and long lasting, didn't wash away in the rain, killed all sorts of bugs, and didn't seem to bother humans. Malaria mosquitoes and crop pests were going to disappear, and the world would be a better place. In 1948 DDT's inventor, the Swiss chemist Paul Muller, was awarded the Nobel Peace Prize for bringing this miracle compound to the world.

Not everyone was convinced that synthetic fertilizers, pesticides, and monocultures were a good idea, however. By the 1950s a small but vocal group of organic farmers, led by Bromfield, Rodale, and other early organic proponents, had sprung up in the United States. They were defined mainly by their refusal to use synthetic chemicals and their interest in natural ways to recycle nutrients to restore depleted soil.

Why Agribusiness Didn't Go Organic

In 1962 an intellectual earthquake hit the world with the publication of Rachel Carson's *Silent Spring*. Carson was already a well-known and respected scientist and author, and her well-documented study of the effects of DDT and other pesticides on the environment became an international best seller and played a major role in the genesis of the environmental movement in the 1960s and 1970s.

From this time on organic agriculture was, in the public mind, defined primarily as pesticide-free farming. There are plenty of stories of folks who jumped on this bandwagon, moved to the country in the 1970s to go "back to the land," but failed because they didn't realize there's a lot more to raising food than not using synthetic chemicals. On the other hand, some back-to-the-landers did their homework and weren't afraid of sweating their way to success. Small nuclei of serious organic farmers began forming in Wisconsin, California, and Oregon and along the East Coast, and an industry began to take shape.

Back-to-the-landers often created difficulties in rural communities. When some unemployed youngster with long hair and weird ideas moves in next door and starts implying that you've been doing it all wrong, there's likely to be a little friction. On the other hand, if your neighbor turns out to be the kind of farmer who thinks that if some herbicide is good then more must be better, and it's okay to raise calves in a windowless shed ankle-deep in their own manure, you might have some objections to his operating methods. Echoes of those days of mutual suspicion and occasionally outright dislike are still around. In some rural communities "organic" is still a dirty word.

The larger agribusiness community was unimpressed with organic farming. Rachel Carson and going back to the land were just blips on the radar compared to the stunning success of the Green Revolution.

The Green Revolution was a response to the desperate and chronic food shortages in many parts of the world following World War II. In a far-sighted response to the problem, the Rockefeller Foundation joined the Mexican government to begin a crop breeding program in Mexico. Over the next several decades a team of researchers led by Norman Borlaug of the University of Minnesota developed new varieties of wheat, rice, and corn with dramatically increased yields. Once the varieties began to be planted in the 1960s, an incredible agricultural revolution occurred. With their higher yields, these new varieties prevented widespread starvation in India, China, and much of Asia. In addition, they reduced pressure to farm marginal land and, in turn, lessened nations' desires to go to war with neighboring countries over food and water. Borlaug was awarded the Nobel Peace Prize in 1970 for his leading role in the Green Revolution and his dedication to making the new hybrids available on a massive scale in the developing world.

In the 1970s, then, there were on the one hand small groups of farmers scattered around the United States working out how to build a modern farming system without modern chemicals and stubbornly continuing to point out that Rachel Carson was right, along with all those others who were raising the alarm about damaged and depleted soils and irreversible harm to the ecosystem. Some consumers believed them, and slowly but surely the tiny demand for organic food began to grow.

On the other hand was mainstream agriculture, enjoying the good times brought on by the new fertilizers and pesticides, high on the success of the Green Revolution, and seeing no reason to do things any differently. Environmental activists could talk all they wanted, but talk wasn't going to feed the world, and besides, the view from the tractor seat looked green and healthy as far as most farmers could see.

IT'S A FACT

Anywhere from 40 to 65 percent of the world's agricultural land is degraded to some degree.

The Chemical Chickens Come Home to Roost

Over the past several decades the environmental and human damage caused by synthetic fertilizers and pesticides, monoculture, sloppy irrigation practices, overgrazing, and concentrated animal feeding operations (CAFOs) has become quite clear. As fossil fuel prices go up and the problems accumulate these practices are also becoming less effective and more expensive for mainstream farmers. Different organizations estimate that anywhere from 40 to 65 percent of the world's agricultural land is degraded to some degree, including the 20 percent of all irrigated land that suffers from salination.

This is a dangerous trend in a world with no good agricultural land left to develop and a population that continues to increase. The bigger harvests produced by the Green Revolution have stagnated, and the dark side has become evident: crops dependent on chemicals, small farmers forced off the land by large-scale mechanized agriculture, and a polluted and degraded soil resource.

Even the most conservative of agricultural authorities are now pointing out that many mainstream agricultural practices, which they'd been promoting for decades, probably aren't sustainable and perhaps farmers should think about doing things a little differently. As an example, the Wisconsin Agricultural Extension Annual "Pesticide Update" experts announced, during their annual state tour in 2002, that, gee, the bugs and weeds were getting resistant to the chemical killers, and the scientists didn't really have any new miracle concoctions coming down the pipeline, so maybe farmers should think about rotating their crops and hauling out Dad's old cultivator to do some mechanical weeding. Coming from mainstream agriculture people, this was rather stunning news.

Pesticides Are Self-Defeating

Resistance develops in plants, insects, bacteria, and fungi because no matter how good the pesticide, it never kills 100 percent of the target. There are always a few that are a little tougher than the rest, and after you've done the spraying, those are the only ones left to breed. Their offspring inherit the parents' resistance, so the farmer has to spray more next time, or maybe a different chemical, and again the only ones left are the very toughest.

After this goes on for a few years or decades, nothing's going to kill those bugs except growing something different and so taking away their homes and food, otherwise known as crop rotation. Nothing's going to kill those weeds except mechanical or flame weeding, or suffocating them with heavy mulch. As of 2015 the Weed Science Society of America has identified 196 species of herbicide-resistant weeds worldwide, and at least 50 of those species are resistant to more than one herbicide (multiply resistant). And an estimated 500-plus species of insects and mites are now resistant to pesticides.

Chemicals Destroy More Than Their Target

Pesticides contaminate water and food worldwide. Exposure to pesticides, depending on their mode of action, can cause blurred vision, diarrhea, vomiting, seizures, coma, breathing problems, and death. Though most states do not require occupational exposures to agricultural pesticides to be reported, the Natural Resources Defense Council estimates that anywhere from 10,000 to 40,000 farmworkers have to see a doctor each year for pesticide poisoning.

Chronic exposure to pesticides, in food, water, and the environment, is now a fact of life for most U.S. residents and has been variously linked to impaired neurological function, cancer, reproductive problems, reduced growth and development, hormonal disruption, and birth defects. Pesticides are known to cause similar effects in other life-forms and are disrupting ecosystems on a large scale.

Synthetic fertilizers are also causing problems. Among the most well known is nitrate contamination of drinking water in heavily cropped areas around the country due to nitrogen fertilizers seeping down to the water table (along with plenty of pesticide residues). High nitrate levels in groundwater from fertilizer and feedlot manure seepage are definitively linked to "blue baby syndrome" (methemoglobinemia, a potentially fatal blood disorder in which the oxygen-carrying capacity of blood is reduced) in humans and animals and to nitrite poisoning in cattle, sheep, and horses. The famous "dead zone" in the Gulf of Mexico, at times covering up to 7,000 square miles (18,000 sq km), is caused by fertilizer runoff from America's heartland into the Mississippi River, which empties into the Gulf.

In addition, fertilizer manufacture is heavily dependent on increasingly expensive fossil fuels. The most common method of fixing nitrogen for fertilizer, the Haber process, requires one and a half tons of natural gas for every ton of nitrogen produced.

Synthetics Derail Common-Sense Agriculture

Synthetic chemicals have created secondary problems, too, for example by making it profitable to give up crop rotations. Row crops like corn, which in conventional agriculture leave much of the dirt bare for much of the season, are especially prone to causing soil erosion and breakdown of soil structure, and continuous corn is a standard practice now in many areas.

The American Milking Devon was once prized for its ability to produce both meat and milk, to pull wagons, and to thrive even on rough forage. Like so many adaptable breeds of domestic livestock, it has all but disappeared from the modern agricultural scene.

Old seeds and breeds that were beautifully adapted to different regions are being lost because most farmers have switched to high-yielding chemical-dependent hybrid crops and to livestock and poultry bred primarily for high production. A global survey completed by the U.N. Food and Agriculture Organization in 2007 found that of 7,600 known breeds of cattle, pigs, sheep, chickens, and other farm species, 11 percent have disappeared, 16 percent are threatened, and the status of 35 percent is unknown. Chemicals in the form of fly sprays, antibiotics, and other fancy stuff have made it reasonable to put up with all sorts of animal problems, as long as the animals produce.

In fact, if you keep your chickens in tiny cages and your cows in the barn at all times, you can control their environment so well that production should go up no matter where you live. Never mind that keeping cows on concrete and feeding them overrich diets has caused so many health problems in so many cows that, though their natural life span is 12 to 15 years, only half of them survive for more than two years in the average milking string. Then it's off to the hamburger plant. The problem is that cows are expensive, and if you have to replace them every year or two it really eats into your profit margin. Not to mention that the cows might not have particularly happy lives.

With both the agricultural world and the larger public becoming progressively more aware of the problems created by mainstream agriculture, the public attitude toward organic farming has continued to change for the better.

IT'S A FACT

Of 7,600 known breeds of cattle, pigs, sheep, chickens, and other farm species, 11 percent have disappeared, 16 percent are threatened, and the status of 35 percent is unknown.

Counterculture Becomes Standardized

Until the mid-1970s or so, organic farmers who were marketing food typically sold to people they knew personally. Though there was no specific definition of the term *organic*, it had a clear connection with fresh, local, and chemical-free. (Though technically everything in existence is a chemical, in the consumer market the term generally refers to synthetic chemicals.) That was good enough for buyers: they knew their farmer and knew how he or she had raised their food.

But as market momentum grew, more middlemen emerged to process, transport, and retail organic products. Fewer buyers met fewer farmers, and questions arose about the meaning and validity of organic claims. Where was the proof that what a farmer claimed was true? And farmers had different notions of what organic meant. One may have thought organic meant no pesticides, but a little nitrogen boost courtesy of the local agricultural chemical dealer didn't hurt anything. Another may have figured that it was fine to use antibiotics to treat mastitis in a dairy cow, because antibiotics aren't a pesticide, right? (Think this through: *anti-* means against, and *bio* means life, so *antibiotic* means against life, specifically, against disease-causing bacteria, which are pests, so that pretty much makes antibiotics an internal form of pesticide. The Final Rule mandates antibiotic use to save an animal, but then that animal is no longer certifiable as organic.)

By the late 1980s, these issues were being dealt with by a third-party certification system that was defining a standard for "certified organic" labeling. As it evolved, this system worked by having independent certification agents, usually experienced organic farmers themselves, visit and inspect the farm that was applying to become certified. The inspector examined all the farm records and then walked the fields and pastures and visited the buildings to make sure there weren't any

Small and Organic Farm Statistics

According to the USDA Economic Research Service and National Agricultural Statistics Service, of the 2,109,363 farms in the United States in 2012 (down 4 percent from 2007), more than three-quarters fell into the hobby or small farm categories. Well over half (56.6 percent) were hobby or part-time farms, with less than $10,000 in annual sales. Another 19 percent fell into the small farm category, with less than $50,000 in sales.

The total number of U.S. farms has been declining for nearly a century. Though there was an uptick of 4 percent between 2002 and 2007, the decline has resumed, with farm numbers dropping by 4.3 percent since 2007. But the number of certified organic farms continues to climb as it has for the past 20 plus years, from 3,536 in 1993 to 19,474 in 2015. And this figure is lower than the actual number of organic farms, since the Final Rule allows farms with sales of less than $5,000 to skip the certification process and still be able to label themselves as organic, as long as they follow the rules.

Sales of organic food have been increasing at a double-digit annual rate since 1994, a trend that is expected to continue now that mainstream grocery stores, including WalMart, are stocking organic products. In 2014 U.S. sales of organic food totaled $39.1 billion, almost 5 percent of total U.S. food sales, and sales of nonfood organic products totaled $3 billion. In fact, there are not enough organic producers out there to supply current demand, especially for organic beef and pork. The non-food categories offer another opportunity for new organic farmers, since organic personal care products, pet food, fibers, household cleaners, and flowers all must come in total or in part from organic farms.

Though the organic market is far from saturated, price premiums for organic food have been falling in recent years now that large food corporations have bought smaller organic food companies and increased supplies and availability. In 2007, five companies supplied 70 percent of all organic produce sold in the United States, a trend some of the founders of the organic movement believe is a travesty of the original philosophy of fresh, local, healthy food. Still, organic food generally commands a 30 to 50 percent price premium over conventionally produced food, though this depends heavily on the food category (dairy, produce, meat, et cetera) and where the product is being sold. Small growers still do best selling locally, either through farmers' markets, deliveries, on-farm sales, food co-ops, or other local retail outlets.

synthetic chemicals or bad farming practices being used. They asked questions to make sure the farmer understood what he or she was doing and had a plan in place for dealing with various issues, asking questions like "How are you controlling prickly ash in this steep hillside pasture?" or "How are you treating mastitis in your dairy cows?"

The inspector then wrote up a report and submitted it to the certifying agency, an entirely separate organization. The agency reviewed the report against its own standards to verify that the farm qualified. Each certifying agency (and there were quite a few at this point) had its own set of criteria for certification. These generally were in agreement on big things like no pesticide use but could vary on such knotty points as whether to allow commercial worm medications (which are used to kill parasitic worms in livestock).

Certifiers also had a tendency to be specialized, so that one would be a lot more specific about vegetable production while another really knew dairy cows. Overall the system did a lot of educating of farmers about what they had to do to be considered organic, and of reassuring consumers about how the food had been produced.

But there were incompatibility issues. For example, a farmer might buy certified organic feed, but the farmer's certifier wouldn't accept it as organic because another agency had done the certifying. There were also some well-publicized incidents of fraud that made the industry's credibility suffer.

Government Steps In

Up to this point, due to lack of interest and frequently active hostility on the part of the USDA and many (not all) land grant universities, the government had not meddled much in organic agriculture. This was just fine with most organic farmers, many of whom tended to have antiestablishment attitudes. Okay, that's an understatement. A lot of them were outright anarchists. In retrospect, the friction seems inevitable. Before the change in administrations in 1981 the USDA did unbend long enough to compile and publish "Report and Recommendations on Organic Farming" for the purpose of "increasing communications between organic farmers and the U.S. Department of Agriculture." That report was quickly shelved and forgotten, and the USDA and agricultural extension services were the last places most organic farmers went looking for help or information. A few states had established their own certification agencies, but aside from these exceptions there was little, if any, research or legislation at any level of government.

But by the late 1980s many farmers, consumers, and government officials were acknowledging the increasing need for standardization, and legislation seemed to be the way to go. In 1990 the federal Organic Foods Production Act was passed. This act required the creation of the National Organic Program and the writing and passage of a uniform rule for organic production and processing.

Organic Supporters Protest

It took seven years for the government to write the proposed rules for organic agriculture, and when they were finally published in 1997 the outcry of protest from the organic community was of historic proportions, deafening and unrelenting.

The proposed federal rules would have allowed antibiotic use, feeding of animal by-products, field spreading of human sewage, planting of genetically modified organisms, and ionized radiation to be used in organic food production and processing. All of these practices were unacceptable to the organic community.

The Final Rule on:
Medications

The Final Rule mandates the use of antibiotics if necessary to save the life of an animal, but that animal, once treated with antibiotics, is no longer certifiable as organic. Organic farmers must manage their animals humanely and prevent suffering or death by using prohibited substances if all else fails. They cannot avoid the use of prohibited health inputs in an effort to preserve the organic status of an animal.

Antibiotics

The objections to these practices begin with antibiotic use. Antibiotics are a common feed additive for livestock for the purpose of increasing growth rates and suppressing disease. This practice has been widely indicted as contributing to the spread of antibiotic-resistant bacteria, and there is concern that antibiotic residues in soil and water affect life in those mediums as well. Antibiotic use, whether in feed or as medicine, is therefore unacceptable to organic farmers.

Animal By-Products

Much of an animal raised for food is not eaten or used by humans. Rather than waste the guts, brains, bones, and blood, slaughterhouses long ago got together with feed manufacturers to recycle a lot of these parts back into nutrient-rich animal feed. Though pigs and poultry don't object to animal-based food, it's not a natural diet for other livestock. But through the miracle of modern processing methods, animal bits have been made palatable for all types of domestic animals. Unfortunately, we now know that animal by-products are a vector for the spread of prion diseases, most notably scrapie in sheep and bovine spongiform encephalopathy, otherwise known as mad cow disease, in cattle. When humans contract the disease from infected cattle, it's called variant Creutzfeldt-Jakob disease and is incurable and invariably fatal.

The organic community objects to feeding animal by-products for two reasons: first, it's not a natural diet for grazing livestock, and second, it can kill our animals and us. In an organic system, animal by-products can be composted (carefully) to return to the soil the nutrients those animals took from it.

Human Sewage

Human sewage has been used, particularly in China, for thousands of years to enrich the soil, and is in keeping with the principle of cycling of nutrients that is fundamental to organic farming. Unfortunately sewage in the United States is contaminated with all the chemicals and heavy metals we use in our households and businesses, and can carry disease-causing organisms as well. Zinc, copper, cadmium, lead, and nickel are present in fairly large amounts in some sewage, and though tiny amounts are necessary for good health, large amounts will damage plant growth and cause human health problems. That's why organic farmers don't use sewage sludge, though many conventional farmers do.

Genetically Modified Organisms

The term *genetically modified organism* (GMO) refers to a plant or other life-form that has had a gene from another organism, whether closely related or not, spliced into its chromosome by humans in the laboratory. This allows plant breeders to select for desired traits that would be difficult or impossible to achieve with natural plant breeding. In the United States the most heavily marketed GMO traits have been for herbicide resistance, such as Roundup Ready corn and soybeans (Roundup is the brand name of a popular herbicide based on the chemical glyphosate, which kills all plants on contact, except those genetically designed to be resistant). This trait allows a farmer to spray weeds in a cornfield without killing the corn, simplifying weed control.

The first GMO seeds were marketed by Monsanto Corporation in 1996, and by 2014, 94 percent of all soybean and 80 percent of all corn acres in the United States were devoted to GMO varieties.

Organic farmers have a long laundry list of potential problems with GMOs:

- GMOs can cross-pollinate with related species. If they cross-pollinate with non-GMO crops, they contaminate that crop and make it unsalable in many organic markets. If they cross with wild species, they contaminate the gene pool, an ominous development for plant research programs aimed at preserving genetic diversity. Such cross-pollination also has the potential to create super-weeds.

- GMO varieties are generally patented, making it illegal for farmers to save their seed for next year's planting. This is not usually a problem for farmers who intentionally plant GMO crops, since, being hybrids, the crops' seeds generally won't reproduce true to the parents, and farmers wouldn't usually save the seed anyway. It is a problem, however, for farmers whose non-GMO crop has been contaminated with GMO pollen. Though they did not plant the patented GMO variety, they still cannot legally save seed from their crop, since it contains traces of the patented biotechnology. There have been several memorable cases in which Monsanto and other GMO patent owners have sued farmers — and won! — for "stealing" their property by saving seed from non-GMO crops that were contaminated with GMO pollen.

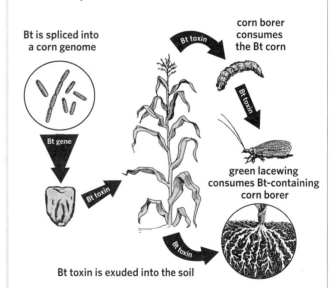

Corn that has been genetically engineered to contain Bacillus thuringiensis *kills the pest corn borer but also affects the beneficial green lacewing, an unintended consequence with as yet unknown ramifications.*

- GMOs have wide effects on ecosystems. For example, Bt corn, engineered to manufacture the natural insecticide *Bacillus thuringiensis* to combat corn borers, also affects the green lacewings that feed on corn borers. The Bt toxin is exuded into the soil as well, and affects soil life. With widespread and indiscriminate use, resistance in pest species to Bt is expected to increase, depriving organic farmers of one of the best natural insecticides available.

- GMOs' long-term effects on human and animal health are unknown. GMO products have not been around long enough for studies to evaluate their effect on human health, but we've all probably eaten plenty of GMO food products by now (food in this country is not required to be labeled when made from GMO sources), so we're the guinea pigs. Stay tuned.

But the essential issue is more basic. In the words of long-term organic activist and now Minnesota Department of Agriculture Organic Outreach Coordinator Jim Riddle, who spoke at the 2008 Organic Farming Conference in La Crosse, Wisconsin, "Genetic engineering is trying to control nature, not work with it, so we reject it at the fundamental level."

Ionizing Radiation

The final item on the organic community's "unwanted" list is ionizing radiation, also called irradiation, which preserves food by bombarding it with electromagnetic waves. If the dose is correct, it kills most disease and spoilage organisms, and extends shelf life by slowing or stopping the normal biological processes of ripening and sprouting. This doesn't make food radioactive, but it does break chemical bonds at the molecular level, spinning off electrically charged fragments, or ions, and fragments with no charge, or free radicals.

The U.S. Food and Drug Administration (FDA) states that the chemical changes caused in food by ionizing radiation are less than those caused by ordinary cooking methods, and that the food is safe to eat. The FDA also notes that vitamin levels may be reduced, and not all pathogens are destroyed, particularly clostridium bacteria, which causes food-related illness. Organic farmers and consumers were, and still are, unconvinced that the FDA knows what it is talking about. Concerns include the damaged enzymes of irradiated food, the lack of research on possible effects on human health, and the temptation for processors to rely on irradiation rather than sanitation to produce pathogen-free food.

Final Rule Definitions

National Organic Program Final Rule (NOP-FR). The USDA rule that legally defines *organic* and all the conditions that must be met in order to be certified organic. The NOP-FR was authorized by the Organic Foods Production Act passed by Congress in 1990.

National Organic Standards Board (NOSB). The advisory board that recommends what goes into the Final Rule. The board is appointed by the U.S. Secretary of Agriculture.

National List of Allowed and Prohibited Substances. That part of the Final Rule (subpart G) that lists all allowed synthetic and prohibited natural materials for certified organic production.

Organic Materials Review Institute (OMRI). A nonprofit organization founded in 1997 that reviews brand-name products for compliance with the Final Rule. OMRI maintains lists of both brand-name (including formulated multi-ingredient) and generic products allowed in organic production, while the Final Rule lists only generic products. The OMRI lists are also easier to access on the Web than the National List. OMRI also maintains a list of organic suppliers of seed.

The Final Rule

The USDA bowed to the wishes of the organic community and the public, and the rewritten National Organic Program Final Rule was officially implemented on October 21, 2002. It prohibited sludge, antibiotics, the feeding of mammalian or poultry slaughter by-products, GMOs, and irradiation.

Though many in the organic community are not completely satisfied, the Final Rule has finally provided a uniform national standard for organic production. What a long way we've come from half a century ago, when a few people were just beginning to figure out that there was a better way to farm, one that would preserve our health and our future.

The Saga Continues

But the story is never over. Amendments to the Final Rule are continuously presented, debated, and fairly often enacted. Ongoing concerns include the use of manures from nonorganic sources, which may contain arsenic and other heavy metals, as well as antibiotic residues, and closing the loophole that has enabled some very large organic farms to circumvent the requirement of access to green pasture for livestock.

In addition, though the establishment of the Final Rule is an important step forward for organic agriculture, farmers must measure proposed Final Rule compromises and changes against the fundamental goals of organic farming. The Final Rule, some say, serves simply as the lowest common denominator. The real test is how responsibly you have acted toward the earth and toward all those who consume the food you produce.

"Our hope was to find a lifestyle consistent with what we believed about the world and what we hoped for the world."

DON ZASADA
Caretaker Farm, Williamstown, Massachusetts

Profile: Sam and Elizabeth Smith

Caretaker Farm, Williamstown, Massachusetts

In the mid-1960s, a job teaching history at the local high school brought Sam and Elizabeth Smith and their three young children to Williamstown, Massachusetts, where they bought a house and immediately planted a vegetable garden in a vacant lot in the neighborhood. The garden turned out to be more interesting than the job, much of the time, and began steadily to expand.

Within a few years the Smith household had also expanded to include a dog, two cats, a rabbit, and baby chicks. They were outgrowing both the house and the garden, so the couple purchased an 1800s farmhouse outside of town, with a barn, sheds, a pig house, a chicken coop, and 35 acres (14.2 ha) of land from what had once been a 250-acre (101.2 ha) diversified farm, and they named it Caretaker Farm. With no background in farming, they had joined a fledgling grassroots movement. No one knew it at the time, but the Smiths, and others like them across the country, were the beginning of the modern organic movement in the United States.

Elizabeth recalls, "The Sixties was a time of great social unrest. We weren't hippies, but we were actively involved in the protests against the Vietnam War and were deeply troubled about the direction our country was moving in, the way our children were being educated, how our food was being grown, and where it was coming from. We realized that our whole food economy was based on oil. A group of us began a local food-buying cooperative, which places monthly orders even to this day."

In 1971 the Smiths and other "back-to-the-landers," led by visionary Samuel Kayman, coalesced to create the Natural Organic Farmers Association (later the Northeast Organic Farmers Association, or NOFA). At the first meeting its membership decided that NOFA would sponsor seminars, seed exchanges, bulk buying, and farm apprenticeships. And NOFA also would publish a quarterly newsletter, *The Natural Farmer*.

In 1973 NOFA held its first summer conference, cosponsored by the Biodynamic Farming Association, "a joyful gathering of like-minded spirits," recalls Elizabeth of the assemblage of influential pioneers of the organic movement. "Scott and Helen Nearing were there, and Eliot Coleman. Wendell Berry was the keynote speaker. We were all passionate about the soil. We left that conference feeling empowered to change the world. It was a spiritual moment."

A year later, Sam resigned as assistant to the president of a local community college and took a job milking cows on a local dairy farm. Quitting a good job to take up organic farming was strange enough to be headline news in the local paper. "When we began to realize that this was what we were called to do, we began to wonder whether Caretaker Farm was too small to support a family," Elizabeth continues. "We began to seriously think we would need a small dairy and more land to get the fertility we needed to grow vegetables."

E. F. Schumacher's book *Small Is Beautiful* and Eliot Coleman's success in Maine with a small-scale market garden changed their thinking. They shouldn't need to get bigger to be successful, they realized. As Elizabeth recalls, "We realized small *is* beautiful. We stuck with the small, more hands-on, very labor-intensive model. We could never have done it without apprentices. For years we had four apprentices who lived with us, and Don and Bridget (who have taken over the farm from the Smiths; see page 27) have continued that tradition." Sam adds:

> Farming was deeply satisfying, and at the same time we were embodying concepts of simple living. We radically cut back on income, but we made that into a virtue and an adventure. There was a great deal of pleasure in rediscovering something that just seemed good: good work, a good thing to do.
>
> There were absolutely no resources. No, that's wrong. There were lots of resources, but they were totally below the radar, hidden. Sir Albert Howard's books were out of print, and the USDA's Yearbooks of Agriculture *from the 1930s — 1937 was a superb year —* were gathering dust. Uncovering these valuable resources was so exciting.
>
> The trajectory of agriculture by the 1960s was toward large-scale farms, very large investments in agricultural technologies, large investments of water and fertilizers and other inputs. There was a complete block about the past.
>
> Those of us who were the early pioneers in the organic movement faced hostility from people. Either you were terribly naive or, worse, you were promoting a system of agriculture that would lead to world hunger. You were subversive.

THE SMITHS PERSEVERED. They began selling NOFA-certified organic vegetables to area restaurants and small markets and seedlings from the greenhouse in the spring. They built a small bakery in a shed off the house, adding fresh bread, granola, and pies to their farm stand, along with vegetables, honey, fresh eggs, jams, and pickles. They hired four apprentices every year. While Elizabeth and Sam shared the work, there were clear lines of responsibility. "I was more the organizer," Elizabeth says, "and had to sell Sam on the idea of becoming a CSA (community-supported agriculture). I was responsible for managing the kitchen and the household, caring for the livestock, maintenance, design, and building. Sam was the grower. We each taught the apprentices in our areas of expertise, and we both had very strong opinions about how the other should do things!"

"Those of us who were the early pioneers in the organic movement faced hostility from people. Either you were terribly naive or, worse, you were promoting a system of agriculture that would lead to world hunger. You were subversive."
Sam Smith

Continued success of the farm was ensured when in 1990 the Smiths, with a core group from their community, adopted the community-supported agriculture (CSA) model, where members annually buy shares in a farm's production and share in the bounty as well as the risks of farming. Elizabeth says:

This idea of a new social structure — of a community being built around food — was at the heart of the CSA movement. It's what we had wanted to do all along. The CSA is the reason the farm has continued to thrive. The more you can create community-based networks of support, the less likely you are to fail, because everyone sees that it's in their best interest to keep the farm going. Certainly our farm members have been converted to local food. They know the taste of it, the feel of it, the health of it.

In 2006, Caretaker Farm was placed in a Community Land Trust to preserve it as an affordable working organic farm in perpetuity. Searching for their successors, Sam and Elizabeth selected Don Zasada and Bridget Spann to be the new farmers (see the profile at right). Sam and Elizabeth retained ownership of a small cottage on the farm, where they are enjoying retirement.

Profile: Don Zasada and Bridget Spann

Caretaker Farm, Williamstown, Massachusetts

Caretaker Farm's 35 acres (14 ha) of vegetables, cover crops, chicken tractors, woods, pigs, pond, and cattle are still in good hands. Don Zasada and Bridget Spann tend the fields and livestock with the help of a willing community of customers, apprentices, and local school groups, in the pattern set by the farm's first owners, Sam and Elizabeth Smith.

Passing a farm on to the next generation is often fraught with stress and complications. But the Smiths found what their farm needed in Zasada and Spann and made their plans carefully. An innovative, graceful transition was engineered that ensures the continuation of the farm, the farm community, and a good life for both the retired and the active farmers.

Yet Zasada and Spann had never wanted to farm when growing up and had no farmers in their families.

"I worked on a farm one summer when I was growing up," Zasada says. "I hated it. I got the impression that it was incredibly physically demanding, just a grind, you made no money, and for what? There was no connection with the living system or the customers. I lasted two weeks, and I never wanted to work on a farm again. I went to college for engineering and religious studies and ended up doing some overseas work."

Likewise, Spann had no interest in farming as a young adult, and after college she also volunteered overseas. She and Zasada met while working as community organizers in a rural village in Chile.

In Chile the two experienced small-scale sustainable farming for the first time, and it changed the course of their lives. "When my wife and I met," Zasada says, "our hope was to find a lifestyle consistent with what we believed about the world and what we hoped for the world. For us there's a high degree of integrity that carries throughout this type of farming system. We're connecting people to the land, and to one another."

AFTER RETURNING TO THE UNITED STATES, Zasada apprenticed on a farm in central Massachusetts for two years before being hired as farm manager for the nonprofit Food Project, a post he held for seven years. Then the couple, like so many other aspiring farmers, began looking for their own farm.

"The crux of the issue is that there are a lot of young farming families who don't have the money to buy land. They want the long-term security. They don't want to jump on that wheel where they rent and then after a while the landowner wants to do something else with the land," Zasada says.

Maybe it was destiny. At that time the Smiths were ready to retire from farming, after decades of building up the organic operation they'd started on part of an old dairy farm in 1969. But while they needed to fund their retirement, they also wanted the farm to continue and wished to make it affordable for a young farm family.

"The reality is that we don't need to own any land at all," Zasada says. "What we need is long-term security. Now we have that. It's a complicated structure involving two land trusts, the previous owners, and ourselves. Equity Trust (a Massachusetts-based nonprofit working on land tenure issues) was incredibly helpful in bringing the partners together."

Under the new ownership structure, one land trust purchased the development rights to the property, which lowered the value enough for another trust to purchase the land. The community owns the land, the Smiths retain ownership of their home and have a 99-year lease on the land it occupies, and Zasada and Spann have a lease on the rest of the land.

The terms of the lease are quite specific. "The idea is that this land will be farmed in perpetuity. We can't live here and not farm," Zasada says. "And we have to live here at least eight months a year. We own our house, the farm buildings, and animals, and we have a 99-year lease on all of the land, which stipulates the practices we can use on the farm. There are two houses on the land and the previous owners, Sam and Elizabeth, still live here. They are a wonderful resource. They're pretty incredible."

THE FARM THAT THE SMITHS BUILT SO CAREFULLY is now being farmed just as carefully by Don Zasada and Bridget Spann. A third generation is under way, as well, with the couple's two children, Gabriela and Micah.

Spann has the primary child-care duties and runs all educational activities on the farm, along with organizing the farm's seasonal ceremonies and teaching winter courses on sustainability issues and food preservation methods.

Zasada handles most of the actual farming, year-round. He says:

One year leads into the next. This being late fall, part of my work right now is signing up CSA members for next season and interviewing apprentices for next year. We're still distributing vegetables, too. We distribute weekly to members from the first of June through the end of October, then we fill the root cellar and distribute every other week till we run out.

Winter is the slow season. It's a wonderful time to spend with my family. We start seeding in mid-February, just onions and scallions. We start most of our seeds in our furnace room in the house, in the dark, until they start to germinate. Then we move them to a wood room attached to the house, and then to a greenhouse, so we don't have to heat our greenhouse until April. Things really start to pick up the first week of April. Our three or four apprentices arrive in April, and at that point we start to do all our field preparations. At about the same time we move all of our animals out of the barn, and they graze our cover crops, cows followed by chickens. We're doing a tremendous amount of watering and transplanting. In the springtime, before the school year ends, students come out and help us. The age range is anywhere from daycare to high school. Every CSA member is required to work, so they're actually out in the fields, involved in the process. They're understanding the system. It helps us labor-wise, but it's even more of a help in involving them in the farm.

Though the heavy tillage and some cultivating are done by the farm's two tractors, most work is done by nonmechanized human labor. "At any given time we're probably hoeing 5 of the 7 acres (2 of the 3 ha) of vegetables and keeping them really clean, and we have black plastic on a quarter acre (0.1 ha). About an acre and a half (0.6 ha) of that is left fallow every year, and an area about the same size is double-cropped," Zasada says.

The cows and chickens are rotated through the 15 to 20 acres (6–8 ha) of pasture on the farm, while the pigs are rotated through a series of pens. The farm also has a half-acre orchard with 15 kinds of apples, a couple of plums, and a couple of pear trees.

Zasada notes, "We milk the Jersey cow every morning, and keep her with her calf in the afternoon. We don't want to spend a lot of time milking; it's just milk for our family and the apprentices, and we make yogurt and cheese."

The farm's soil fertility program is based entirely on compost. Zasada says:

Sam and Elizabeth Smith did a tremendous amount of soil care over the years, and we're doing a tremendous amount of soil testing and monitoring the health of our crops. Organic matter in the soil is between 7 and 12 percent. It grows weeds very well, and our challenge is managing the weed seed bank we have while still maintaining the quality of soil. We don't add any trace minerals, and we're a little on the high side for pH right now.

We don't spread even organically allowed fertilizer; we just use compost. We have a weed-based compost and a vegetable-based compost. We give every farm member a big bucket at the beginning of the season and they're supposed to bring back their food scraps.

Then we have an animal-based compost. Our pigs are turning a wood chip base. We throw cow manure on top of it and bury corn in it, and by the end of the season they've created compost that we use primarily as the base of the soil mix in the greenhouse.

We use winter rye for a cover crop that's not going to winter-kill, or oats for a crop that will winter-kill. I've used vetch and Sudan grass, too. There's no need for legumes right now.

> "Every CSA member is required to work, so they're actually out in the fields, involved in the process. They're understanding the system. It helps us labor-wise, but it's even more of a help in involving them in the farm."
>
> Don Zasada

THE PROPRIETORS OF CARETAKER FARM aim to treat the earth gently, and there are a number of practices on the farm that contribute to that. This past year they put in a 10-kilowatt photovoltaic system that provides 90 percent of the electricity for the farm and house.

"It's walking softly on the land and on our surroundings, and not producing wonderful vegetables at the expense of all this oil and electricity beyond the scenes," Zasada says. "We're living our dream. It's definitely not easy all the time, but it's invigorating, and we're having a wonderful time doing it."

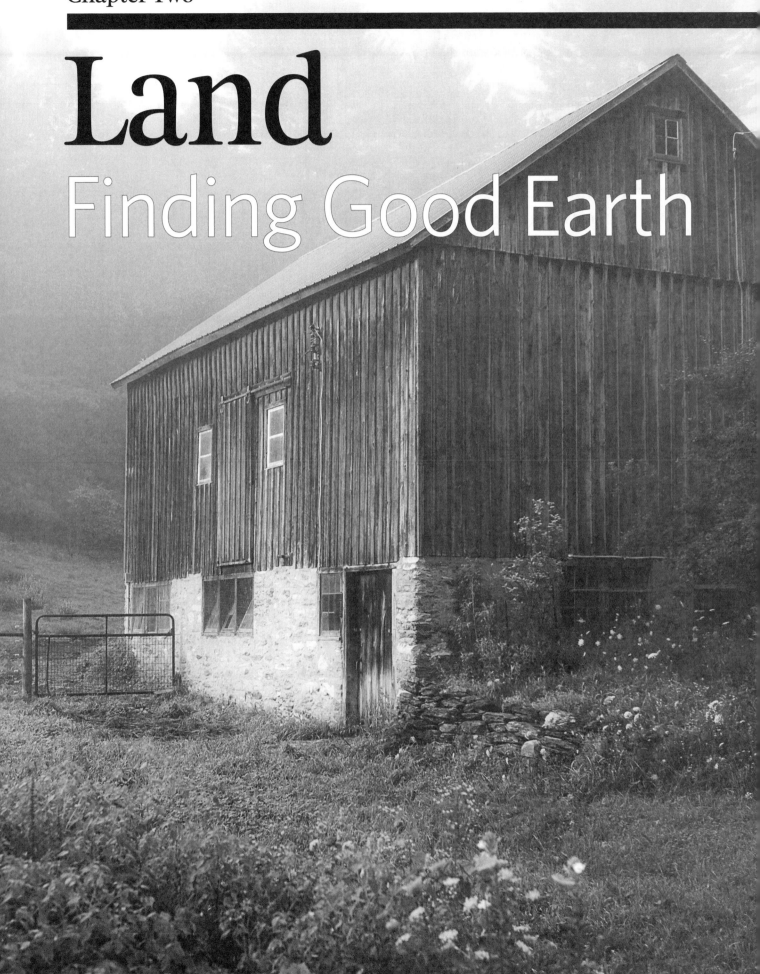

Land

Finding Good Earth

> *Have you become a farmer? Is it not pleasanter*
> *than to be shut up within four walls and*
> *delving eternally with the pen?*
>
> **THOMAS JEFFERSON**
> in a letter to General Henry Knox, June 1, 1795

Finding land can be such a roller coaster ride. You head out on a fine day with high hopes and a realtor only to find that farm after farm was not quite what the ad had led you to believe. That beautiful old farmhouse on the knoll under the oaks? It was just feet from the highway. The old dairy farm with the pond in front? The pond received all the barnyard drainage and smelled like a sewer, plus the outbuildings were falling down.

Do not become frustrated and buy anything just to have something. The first rule for buying a farm is that it's better to keep looking than to buy problems you won't have the time or the money to solve. (The corollary to that rule is that problems always take more time and more money to fix than you anticipated, something you find out pretty quick after you move onto the property.)

So, take your time finding your dream farm and get it right. Only a lucky few find what they're looking for in the first month or two. More often the search can take many months or even a couple of years, and you will spend a lot of weekends looking at farms for sale, and a lot of evenings researching listings and phoning real estate agents.

What to Look For

There are four primary factors to consider when you are looking for a farm:

- **Location.** For the farm, you'll need access to the markets, processing facilities, and other agricultural services necessary to the type of operation you have in mind. For yourself, of course, you'll want to consider community amenities like good schools, medical care, retail stores, a decent restaurant or two, and similar things that will make your life more pleasant. And since it's a rare farm where at least one of the owners doesn't have an off-farm job, a reasonable commute is usually important.

- **Water supply and sanitary system.** Water is the most important limiting factor in farming. The less water that you have available on your farm, the fewer options you will have in deciding what to produce.

- **Farmable land.** In theory, swamps can be drained, deserts irrigated, and an intense program of soil restoration can renew eroded or dead soil. In practice, this might take a lot more time and money than you can afford. If you plan to farm, you have to buy land that is farmable.

- **Condition of the house and outbuildings.** Many old farm homes and outbuildings were constructed before building codes were enacted and enforced, and many have been neglected for a lot of years. All buildings should be thoroughly inspected to determine how much money and labor are needed to make them safe and usable for the long term. Even newer buildings may have problems or not be built to code. Buyer beware!

Let's consider each of these points in more detail.

Location: Home versus Farm Demands

Start your farm search by deciding which is more necessary: having the right location and land for what you want to produce, or being close to jobs and/or family. If the first consideration has priority, then you'll be looking for a region and a neighborhood with suitable soils, climate, support services, and possibly market access for the operation you're envisioning. If the second consideration has priority, then you already know the area where you want to live, and you can figure out later what types of land are available there and what you can raise.

If Type of Farm Comes First

There are six basic types of farm operations:

- Vegetables (also called market gardening or truck farming)
- Fruits
- Field crops
- Poultry and livestock for meat and eggs
- Dairy
- Diversified subsistence production (focused on feeding the farm family rather than selling to outside buyers)

Almost all niche crops and operations, from sunflowers to sheep to shiitake mushrooms, fit into one of these categories. Specialty crops like these are usually easier to get into on a small scale but are harder to market and realize a good return from, no matter what the salesmen say. Since there are only so many hours in a day, most farmers concentrate on just one or two of these five categories to produce a cash crop. What and how much you'll be able to produce and sell depends on how much land and water you have, the soil quality and topography, how close you are to markets, available labor, and on personal inclinations.

What Do You Enjoy Doing?

Think about what you like to do and how you want to live. For example, livestock are a huge asset to an organic farm because they produce manure, the best natural fertilizer there is, and they add value to the forage crops that are so important in crop rotations for building soil. But if you can't bear the thought of vaccinating, castrating, and loading animals onto a truck to go to the butchering plant, or if you can't be there every day to care for them, you won't want to raise them on your farm. There are other options for building soil.

Field crops demand an interest in maintaining and repairing machinery. If you don't have a greasy thumb, paying someone else to make the inevitable repairs can easily eat up much of your profit margin.

Market gardening can involve really long hours working outside in the hottest weather, and getting up in the wee hours to drive the crop to farmers' markets.

Do your research and think it through before jumping into any specific farm enterprise.

Time and Labor

Consider also the time you'll have available when deciding what you're going to produce. Dairy requires the most labor, with market vegetables close behind (though with those you at least get the winter off). Fruit crops and field crops require bursts of intense work through the growing season, interspersed with periods when there isn't a lot that needs doing.

Meat animals require the least amount of yearly labor, although the work needed is in inverse proportion to the size of the animals. Chickens need twice-daily attention and feeding, while with beef cattle on pasture all you need to do most days is make sure they're all there and the water hose is working. But cattle, sheep, goats, and horses all require hay in the winter and in droughts unless your climate allows year-round grazing, and hay is expensive unless you make it yourself. This requires some equipment and a few very long days one to four times a summer (or trading half the crop to a neighbor in exchange for him doing the work).

Organic Learning Opportunities

If you have the time and the freedom, consider a farm internship as a terrific way of learning the type of organic farming you're interested in doing. Start with the national list of sustainable farming internships and apprenticeships maintained by the ATTRA — National Sustainable Agriculture Information Service, a division of the U.S. Department of Agriculture (see the resources at the end of this book for contact information).

Another well-structured internship option is offered by Collaborative Regional Alliances for Farmer Training (CRAFT), groups of sustainable and organic farmers who collectively offer on-farm training combined with field days at other farms. There are CRAFT chapters in New York, Massachusetts, Kentucky, Connecticut, Wisconsin, Ontario, and no doubt other areas, and they all seem to have good Web sites.

If you already own land or are close to buying, you may need a mentor more than an internship. The highly successful Farm Beginnings Project pioneered by Minnesota's Land Stewardship Project has inspired programs to start up in other areas. Farm Beginnings offers a program of classroom training and field trips to beginning farmers, then pairs them with established farmers in their area who will mentor their progress. As of this writing there are Farm Beginnings programs in several states, including central Illinois, Missouri, northern Illinois/southern Wisconsin, and North Dakota.

Other beginning-farmer short courses worth checking out include the Wisconsin School for Beginning Market Growers and the Wisconsin School for Beginning Dairy Farmers, both organized by the Center for Integrated Agriculture Systems at the University of Wisconsin–Madison. Contact your county agricultural extension agent to see if similar opportunities are offered through the land grant university in your state.

An increasing number of both public and private colleges offer sustainable and/or organic college courses and even undergraduate degrees, including Washington State University, Iowa State University, and the University of Minnesota. Inquire at higher education institutions in your region to see what's available.

Last, even if you don't have time for internships, college curricula, or short courses, you should certainly attend field days, conferences, and seminars connected with organic and sustainable farming in your area. These present invaluable learning opportunities and the equally important chance to meet and network with other local organic and sustainable farmers. Finding these folks can take some effort; some groups have Web sites but many don't; some are connected up with agricultural extension services but many aren't. Visit your local farmers' market and inquire of the vendors what they know about upcoming events and local networks, and check the events calendars of national and regional organic and sustainable groups for mentions of activities and networks in your area. Some likely Web sites to check are listed in the resources at the end of this book.

Acreage, Access, and Markets

The amount of land, its physical aspects, and where it is located in relation to potential markets are all important — and too often neglected — factors to consider when looking at farms to buy. Think long term when looking at land. New buyers anxious to get onto an acreage often make light of inconveniences and problems in their rush to buy. But five or ten years later when the honeymoon is over, some of those seemingly small drawbacks can turn into big obstacles to the sustainability of a farm enterprise and a family. Take your time and find the right place for you.

FRUITS AND VEGETABLES

Small-scale organic fruit and vegetable production requires relatively little acreage; 0.5 to 5 acres of garden (0.2–2 ha) is all most growers need to make a profit. But it does demand fertile, well-drained soil, a large amount of hand labor, and plenty of water throughout the season. In many areas irrigation may be necessary during dry spells, while in dry areas (mostly west of the Mississippi) full-time irrigation is necessary. Level land is the best for vegetable production, while fruits from apples to grapes can be produced on more rolling land, as long as it's not too steep to move equipment through. Fruits can require more acreage than vegetables, but not that much more: a 5-acre (2 ha) organic apple orchard is quite a lot for one person to handle.

In small-scale production fruits and vegetables are washed, sorted, and packaged on the farm, so access to processors isn't an issue. But fruits and vegetables are perishable and need to be moved to market and sold promptly after harvest. Thus you need to be close to a population center with a farmers' market, food co-op, or enough people to support a CSA (community-supported agriculture) operation, where you make weekly vegetable deliveries to a pickup point. A processor in a nearby town who will take excess production or less-than-perfect specimens off your hands is also a real bonus.

Think long term when considering market access for these types of operations. A four-hour drive to the farmers' market is not quick and easy, though in the initial excitement of getting your first products to market you may not notice this right away. But after a few seasons it gets harder and harder to get up at 3 a.m. after a long day's harvesting so you can get to the market on time and then be cheerful all day for the customers.

If your plan is to have the customers come to you, with either a "you-pick" operation or a farm store, the same rule of good access applies, only more so. It's essential that you be located close to a sizable population and be easy to find. For every additional mile from town and every additional turn it takes on a back road to get to your place you're going to lose a percentage of potential buyers.

FIELD CROPS

Field crop production is a little different. It's still best to have fairly level, fertile land, but you'll need a lot more of it. In the United States, start thinking in terms of 20 to 120 acres (8–49 ha) of tillable

ground if you're east of the Mississippi, and more if you're west. But because the crops aren't perishable and are usually sold to middlemen or other farmers for livestock feed instead of to retail customers, you can live a lot farther from town, where land is cheaper. That is, as long as you have storage and trucking. On-farm storage is best; off-farm storage for certified organic crops is viable only if the grain elevator has certified organic storage and is within a reasonable distance, and if you can hire a trucker in the area to move the crop who will follow organic rules for cleaning out the truck.

Greater rainfall in the East makes corn and soybeans the main field crops for conventional production through much of the area, while in the drier plains wheat and similar small grains are the primary field crops. The organic rule, however, requires a crop rotation that includes small grains, forages, or other soil-building and fertility-enhancing crops. So don't plan on raising continuous corn or wheat, that is, planting the same crop every year in the same field. Though field crops are less labor-intensive than fruit, vegetable, and dairy production, they do require more and larger equipment, technical expertise, and proactive marketing.

DAIRY

For organic dairy production you need pasture acreage. Pasture doesn't have to be level or even accessible by equipment, so a dairy operation can make use of land too hilly or rocky for crops. The amount of pasture necessary per animal depends

How Continuous Corn Demolishes Dirt

The common conventional practice of planting corn on the same field year after year is really hard on soil quality. The large amount of bare soil between corn rows encourages erosion; the relatively small root structure of corn leaves little organic matter behind; and the same nutrients are used each season, impoverishing the soil and necessitating heavy fertilization. The end results are low levels of organic matter and soil life, poor soil structure, and increased soil loss.

Location: Home versus Farm Demands

on a long list of variables, from what percentage of the diet the pasture is expected to supply to how much grass grows per acre and how the farmer manages the grazing. Here's a place where it's helpful to consult with the neighbors and your county agricultural extension agent, since dairy-cow pasture requirements can range from 1 to 10 acres (0.4–4 ha) or more. (And you can figure roughly that five sheep or goats equal a cow.) Though you can probably get by without tillable land, having good tillable ground allows you to raise the hay and grain you'll need to supplement the pasture.

Until recent decades dairy production in the United States was concentrated in the Northeast and northern Midwestern states, where high rainfall and relatively cool summers supplied the lush forages that made cows happy and dairying profitable. In the last 20 years or so California has overtaken Wisconsin as the country's leading producer of milk, but those massive operations out West generally depend on pumping water faster than it is naturally replenished.

Dairy is the most labor-intensive of all types of farm operations, and the most demanding of infrastructure: you have to be there twice a day, every day, to milk, and organic cow dairying is pretty much impossible if there aren't other organic dairies and a processor in the area to make milk pickup and bottling economically feasible. There are dairies that have installed on-farm processing and bottling plants and are marketing milk, yogurt, ice cream, butter, and cheese to local buyers; these are most often goat dairies, and less often sheep and cow dairies.

MEAT ANIMALS

The final category of farm operation, livestock and poultry for meat, is the least demanding of both labor and land quality. On the other hand, the profit margin is generally lower, too, so it doesn't pay to invest too heavily in expensive land or equipment. That's why beef and sheep ranching in the U.S. has traditionally been concentrated in the far West and other marginal areas on land that is cheap because it's too dry, poor, or inaccessible to produce anything else. The one thing grazing livestock (cattle, sheep, and goats) do need is plenty of acreage in the form of pasture and hayfields, since

the more feed you buy, the slimmer the profit margin.

As for figuring out the amount of land you need, it's the same as for dairy: you don't know until you know what kind and how many animals you'll have, how much your land will produce, how you're going to manage the grazing and soil fertility program, and how much feed you'll be willing to buy instead of raise yourself. But those details can wait. For now just figure that, in general, the farther west you go, and the poorer or drier the land, the more acreage you'll need, up to 20 or even 40 acres (8–16 ha) per cow in the driest areas.

Raising pigs or poultry (usually chickens or turkeys) for meat takes relatively little land, and as with other meat animals the labor requirement isn't so onerous that you can't have a full-time off-farm job.

A livestock operation also needs to have markets and processing plants within a reasonable distance. A day's drive is about as far as most folks like to go. If you're selling to customers you know, they may be okay if you don't have a certified organic processor, but if you plan to label your meat as organic, then you'll need to find an organic meat cooperative or a certified organic plant in the area. This has become an issue in many areas out West, where there are no longer enough small-scale livestock producers around to keep small-scale plants in business.

SUBSISTENCE FARMING

If you are planning only to produce food for your own use, your land options are broader, since farming on such a small scale allows you to make use of odd corners and do all sorts of things that wouldn't be cost-efficient or even possible on a larger farm. Maybe a rock wall around the garden and an intensive soil-building program could make up for a very cold climate and thin, acid soil. Or you could keep a pig and a few laying hens in a pen with a shed at the bottom of the garden. There are all sorts of nifty things to try on a small plot if you have the time and some ingenuity.

How to Start the Search

If you know what you want to raise and if you're willing to consider moving anywhere in the country, you might start with the USDA's Economic

Research Service (see resources), where "resource regions maps" and related materials show what areas produce the most of which crops and livestock. You'll get a feel for the differences in soils, rainfall, and length of the growing season that make different regions suited to different types of production.

State departments of agriculture keep state statistics as well, which can be very helpful, since the primary crops can vary widely in different parts of some states. In Wisconsin, for example, dairy farms are everywhere, but cash grain is concentrated in the southeast and potatoes in the central sands region, and the northern third produces mostly timber. An area where a high proportion of farms are raising what you intend to raise is a good indicator that the soils and weather are pretty good for that type of operation, and also that the necessary truckers, processors, suppliers, markets, and local knowledge are available. I suppose you could start an organic dairy in western North Dakota, but it's doubtful you'd find a trucker to pick up your milk. And given the low population density, it's unlikely that selling organic milk to local customers would pay enough to support the farm.

If Location Comes First

On the other hand, if you're choosing the location first and the type of farm enterprise second, now you can start thinking about what sorts of farm operations might work in the area you've chosen. Be sure to factor in how much time you're going to have to devote to the farm after an off-farm job and family duties; whether you like best working with plants, animals, or machinery; and what kind of marketing effort you're willing to undertake. For example, if you, like so many farmers, don't enjoy constant direct contact with customers, then maybe your spouse will have to do the marketing while you handle vegetable production, or you should consider a different type of farming.

If you're planning on farming full-time, you'll need either a large inheritance to support you while you get up and running, a spouse working off the farm, or an airtight business plan and a lot of stamina. No kidding. Despite all the cheerful stories you might have read, it's a very rare person who can start a farm from scratch with little or no experience and quickly earn enough to support him- or herself without acquiring a mountain of debt.

Basic Farm Requirements

Farm Type	Land Requirements	Infrastructure Requirements
vegetable farm	small acreage fertile soil plenty of water	retail markets close by
fruit farm (orchard, vineyard, berries)	moderate acreage deep soils plenty of water	seasonal labor available retail markets reasonably close by
field crops (corn, soybeans, wheat)	moderate to major acreage, reasonably level moderately fertile soil moderate water	grain-drying and storage facilities equipment dealers and repair shops close by trucking available
livestock for meat (sheep, goats, cattle, poultry)	moderate to major acreage; can be marginal enough water for animals	processing facilities reasonably close by livestock truckers available
dairy farm (cows, goats, sheep)	moderate to major acreage pastureland enough water for animals	milk processing facilities available milk haulers available
subsistence farm	small acreage mixed land use moderate water	variable, depending on farm operations

If You Own Land Already

If you have land and are growing a crop or producing livestock, then you already know the potential and limitations of your land resource. Organic farming practices should make it more diverse, fertile, and productive, although the change won't occur overnight.

Since many small farmers are already close to organic in their methods, you may be much closer than you think to becoming completely, certifiably organic. Certainly transitioning to organic and reaping the resulting price premiums can make the difference between a son or daughter wanting to find a career elsewhere or deciding to stay on the farm and eventually take over from the retiring parents. Organic farming, if the farmer is also a good manager, can and does create greater affluence and a better quality of life as well as a healthier environment, and those are what make it attractive to the next generation.

INVOLVING THE YOUNGER GENERATION

If you're in the fortunate position of needing to expand to make room for more offspring to stay on the farm, you may be either adding to your land base or, if that's not possible, looking for additional cash-generating enterprises that don't require any additional room. For example, any farm with pastures for livestock has room for pastured poultry to follow behind the grazers. Organic pastured chicken can be a nice money-maker and employ at least one person. And with the chickens out there scattering manure, eating bugs, and contributing their own manure, the pastures and soil fertility improve even more.

Farms have traditionally relied on family members to develop value-added enterprises using the farm's cash crop, by making yogurt, say, or, dried culinary herb blends. There seem to be innumerable ideas out there for diversifying the farm, though you should be cautious: no matter how much they're being promoted by various people, not all of them produce enough return for the time invested to make them worthwhile. Seek out those agri-preneurs who have been successful at developing and marketing the category of product you're interested in, and see if they'll give you some pointers on how to make your enterprise viable. Quite often these people are looking for

help in their own business, and you may be exactly the right person for the job!

INVOLVING THE OLDER GENERATION

It's fairly common for aspiring organic farmers to purchase or inherit land that belongs or once belonged to parents or other relatives. If this is your situation, then you already know where you're going to live, and you have the added advantage of knowing your farm's history and potential, and even some of the neighbors. Be sure and sit down with your parents or relatives and get as complete a history as you can of crops and yields, chemical use, the water supply, and the livestock for as far back as anyone can remember, and write it down.

If You're Looking for Land

If you don't already have land or aren't inheriting land, the next step is to get in the car and start driving around your chosen locale. If you don't live in the area already, build extra time into your visits so you can drive around without the realtor and develop a feel for the land on different sides of town, and talk with folks at the feed store, café, or church about what goes on in the community. If you have kids, find out about the health of local schools and layout of the bus routes. Many rural school districts have been consolidated, or will be in the near future. This means very long rides for the kids living farthest out, and it can be difficult for them to get together with their friends or get to school events.

Organic farming makes it easy to involve the entire extended family, and even your local community.

Ideally you'll find a piece of really rich soil surrounded by woods or natural prairie fairly close to a delightful small town where health-minded customers are clamoring for fresh, local food and the schools are first-class. But usually that's not the case. You find the best farm and community you can and then do what you can to make them both better.

Before You Purchase, Do Your Homework

The phone and the Internet are your primary tools for the general fact-finding you'll want to do before you get down to making an offer on a farm. This may take some patience, but it's worth the time and trouble since it'll minimize unpleasant surprises and maximize your ability to find the right place for a fair price.

Property Taxes

Call or stop by the county seat and ask the county clerk a few questions. Property taxes are a major item in any farm's yearly budget, and you'll want to find out what the assessments and rates are in the area you're considering. Most states have some form of use-value taxation, which means the land is taxed according to its value for its present use, not at what it could be used for. This means if you are actively farming, your land will be taxed according to its agricultural value, not at the much higher value it would have as a subdivision. The details of these laws and other tax-relief aid for farmers differ from state to state, and you'll want to know how this will affect your property tax rates.

If the state or local government does not employ use-value taxation, then you may find your taxes rising steeply year after year if development starts creeping in your direction, a common occurrence around towns and cities that are growing. There's no shortage of farmers who have sold the family farm and moved elsewhere rather than struggle to pay ever-increasing tax bills.

Zoning and Development Pressure

Find out how fast the area is developing and what zoning ordinances or other land-use regulations are in effect. Just because an area looks really rural and isn't particularly close to a larger town does not mean it is exempt from development pressure. If it's an easy commute for you to your job, it's going to be an easy commute for a lot of people who might also be planning to move to the country.

A good way to assess the pace of development is to check how many building permits or well-drilling permits have been issued each year in that county during the past decade or so. You should also be able to track down Census Bureau statistics and projections at the county level.

If the area is under development pressure and is not zoned exclusively for agriculture (a rarity), you may buy a farm in the country and find yourself surrounded by a suburb a decade down the road. Friction between farmers and new neighbors who don't want to hear machinery late at night, have manure spread on a field next to their homes, or hear or smell farm animals is a major problem in suburbanizing areas around the country.

Agricultural zoning or "right to farm" laws should protect you from nuisance lawsuits brought by any new neighbors, but if those regulations aren't in place in your municipality, you could someday be forced to change your farming operation or to move elsewhere. That's a hard choice

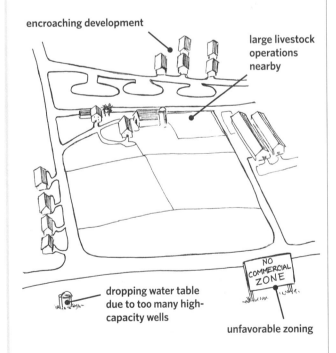

encroaching development

large livestock operations nearby

dropping water table due to too many high-capacity wells

NO COMMERCIAL ZONE

unfavorable zoning

A number of factors are red flags when you're looking at a particular piece of property as a potential site for your organic farm.

to face if you've spent a decade or two loving and building up a farm.

On the other hand, if you're going to grow apples or something else that doesn't have much potential to disturb the peace, getting surrounded by development may present a ready-made customer base for your on-farm market. Unless, that is, local ordinances or zoning regulations limit what you can do. If your area isn't zoned for commercial use, you may not be able to build your farm market or greenhouse business without an exemption or a variance. If farm exemptions aren't already on the books you're out of luck, and variances are not granted automatically, particularly if some of the neighbors think your farm is a public nuisance or don't want any extra traffic in the area.

Either way, if development occurs you'll be sharing the available water supply with a lot more people. If the groundwater is limited in your area for some reason, that could mean eventually having to drill a deeper well to get enough water, particularly if you are irrigating or operating a dairy. There's also the increased likelihood of nitrate contamination in your water from all the new septic systems nearby. These are things to discuss with your county or state land conservation department or its equivalent.

Find out also whether there are zoning ordinances in effect that protect you against having a large concentrated animal feeding operation (CAFO) move in next door, or a commercial operation, like a racetrack. If not, your fresh air and quiet might be replaced with the sweet smell of hog manure lagoons or the roar of the track, and you won't want to stay and will have a hard time selling. It happens.

If you have decided to buy the farm even though the land-use regulations are not the best for your purposes, find out when your local land-use planning commission (or whatever the equivalent is in your municipality) meetings are and plan on attending a few. If there's nothing going on at the local level, find out how the county government is structured and when its meetings occur, or whether a county land-use plan is already on file at the courthouse. Your purpose is to discover what is in place or coming down the pipeline for future land-use regulation in your area. Many rural areas around the country have written

or are in various stages of writing land-use plans, and you will want to know how the plan for your area might affect your farming operation. The sooner you find out what is going on in terms of agricultural, commercial, and residential development and how it's going to be regulated, the better able you'll be to influence the process and protect yourself.

Water Supply

When you finally go to look at a farm, the very first thing to do is to check out the water. There are three things to consider: quantity, access, and quality. A significant problem in any of these areas can be a deal breaker.

Quantity

If there's not enough water for you and the plants and livestock you plan to raise, you can't farm. Though there's never an ironclad guarantee for water supply, if you're in a high rainfall area and there's a reasonably deep well already on the property, there's probably not too much to worry about. If the farm is occupied, sit down with the owners or tenants and ask how careful they have to be about water use, and whether there's much seasonal variation in the supply. Better yet, ask some neighbors who don't have a vested interest in making a sale.

If the farm isn't occupied, then consider the farm's history to get a clue about the water supply. If it's an old market garden or dairy farm that was once successful, it's likely, but not certain, that the water supply is plentiful. If the farm is out on dry prairie and some of the neighbors have quit using their irrigation systems, that's a danger signal that the water table might be a long, long way underground and is dropping.

Turn on the faucets in the house to see how fast the water flows. Five gallons a minute is pretty minimal for a house and garden; a dairy farm should have double this rate or more.

Taste the water. If you're accustomed to city water, well water can taste really different, so don't reject it out of hand. A briny or bitter taste may signal salt in the water, and that's a definite problem for crops. If it tastes of iron, you will probably want to install a water softener to keep your laundry from getting iron stains; you'll also want to pay attention to iron levels in your soil, though this is rarely a problem unless the soil is very acid.

Visit the courthouse again to find out which state government department tracks well-drillers' reports, and get a copy of the report for that well. This will tell you when it was drilled and how many gallons per minute it yielded when it was drilled. The age of the well is important because the submersible pumps now used in most wells have definite life spans, and replacement of a failing pipe and pump can cost $1,000 or more. Anything more than 15 years old is cause for concern. The gallons-per-minute rate is important to know because if it is now much lower than it was at the time of drilling, the well has a problem — either it's becoming plugged up, or the water supply isn't as plentiful as it used to be.

If the land has no well or a new one needs to be drilled, you're looking at investing some real money. Drilling is usually charged on a per-foot basis, so the depth of the well determines the price for drilling it. You can visit the neighbors or ask at the county land conservation department (or its equivalent) to get an idea of how deep you might have to drill to get reliable water. If the land you're looking at is higher than the neighbors', a well may have to go proportionately deeper.

Quality

If you've determined that there's enough water, the next thing to do is put some in a bottle and get it tested for chemical and bacterial contamination. The local county extension agent should be able to recommend a local or regional water testing lab, or you can look one up online. Contact the lab and find out how to label and package the sample and where to mail it. Ask what the sample will be tested

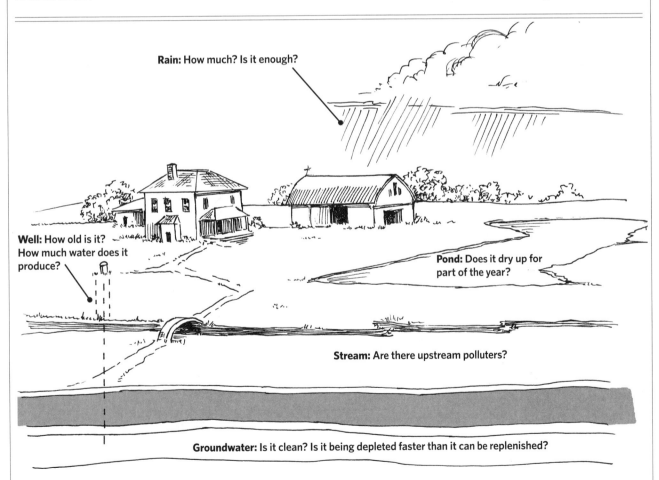

Rain: How much? Is it enough?

Well: How old is it? How much water does it produce?

Pond: Does it dry up for part of the year?

Stream: Are there upstream polluters?

Groundwater: Is it clean? Is it being depleted faster than it can be replenished?

Make sure your water supply will be plentiful and clean year-round, whether it comes from a pond, a stream, a well, or the clouds above.

for: some tests are standard, while others are extra, and not all labs offer all tests. The following contaminants are typically tested for:

- Coliform, which indicates whether water is contaminated with fecal matter from animals or a faulty septic system

- Nitrates, which result from nitrogen fertilizer residues, manure contamination from livestock, or a faulty septic system

- Herbicides and their breakdown products, including atrazine, metolachlor, alachlor, acetochlor, and cyanazine

- Arsenic, in areas where there are large poultry operations, as it is now routinely added to commercial poultry rations

The lab should know whether any other water contaminants might be a concern in your area; if not, contact your state department of health to see what they know.

If any of the lab tests come back with values over the accepted limits, or higher than you're willing to accept, you'll have to research what can be done to ameliorate the problem in your situation, how much it will cost, how long it might take, and what you would do for water in the meantime. Then decide if that farm is worth the effort.

Surface and groundwater contamination is a huge issue in agricultural areas. The National Water Quality Assessment Program of the U.S. Geological Survey found, in testing conducted between 1992 and 2011, that at least one pesticide was present in 97 percent of tested streams in agricultural watersheds, and in more than 50 percent of shallow wells. About one-third of deep wells also had pesticide contamination. Of 83 streams tested in agricultural areas, eight had pesticide contamination greater than the human-health benchmarks set by the Environmental Protection Agency.

Levels below these benchmarks are considered safe for human consumption, though the report notes that "further research is needed" on the effects of combinations of pesticides (a common occurrence) and the breakdown products of pesticides. In the end you'll have to decide for yourself how much risk you are comfortable taking. Given the ubiquitousness of chemical contamination in our environment and the lack of research on its long-term effects, we are all already living with some level of health risk. The USGS Web site is a gold mine of information on what is known about chemical pollutants in water, and many states have excellent information on their health department Web sites.

Fixing the Problem

Fixing groundwater contamination is not easy. Nitrates will break down and dissipate fairly quickly if the sources can be controlled, but other chemicals may linger for years or decades, depending on how slowly the groundwater is flowing, and groundwater is known to move slowly in most cases. Drilling a deeper well to get past the contamination zone might be an option, though there is the risk that if chemical use continues in the area the problem will continue to go deeper.

Water softeners, reverse-osmosis systems, or ion-exchange systems are commonly used to ameliorate drinking water contamination, but might be a little expensive for the quantities needed by livestock or gardens. In addition, using water that's been through your water softener will have a salinizing effect on the garden.

If the problem is bacterial contamination, shock chlorination treatment is usually recommended, though this may only address a symptom rather than the underlying problem. If nitrates are present and there's no obvious source from livestock or crops, the septic system should be checked. Regular cleaning and servicing may eliminate

IT'S A FACT

At least one pesticide was present in 97 percent of tested streams in agricultural watersheds, and in more than 50 percent of shallow wells.

the nitrates, or you may have to put in a new septic system, which will cost anywhere from $5,000 to $20,000 or more, depending on what type of system is allowed by your soils and mandated by county or state regulations.

Access

Access to water is an issue generally peculiar to states west of the Mississippi. Unlike in the eastern United States, you don't necessarily own the surface water that is available on your land.

Because laws and precedents vary so much, it's essential to understand what water rights are attached to a piece of property you are considering for purchase, and to do the figuring to make sure they will support the type of farming you envision. Because these can vary with an individual piece of property, ask the realtor or the owner for written verification of water rights if any surface waters are on or adjacent to the property, or if any irrigation water is being pumped from a source other than a well on the property.

Know Your Septic System

The septic system is often overlooked by buyers moving out from town where they've always had municipal sewage. Don't make this mistake. The septic system doesn't have anything directly to do with the farm operation, except that if it is found to be failing or not up to code, it is expensive to replace and that will put a big dent in your farm supplies budget right out of the gate. A bad septic system can also contaminate the water supply with nitrates. Have a professional inspect the septic system before purchasing a property.

If you've never dealt with a septic system before, it's a good idea to read up on them before you begin living with one. It's far better to anticipate and prevent problems than to have the toilet back up late on some freezing night in the dead of winter.

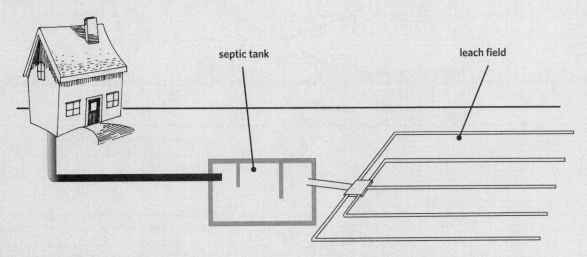

septic tank

leach field

The standard septic system consists of a tank, pipe, and leach field. It is by far the most common liquid waste disposal system in rural areas. Make sure it's in working order before you buy!

Farmable Land

Once you've dealt with the water supply, take a good close look at the land and decide if it suits you and the type of farm operation you have in mind. Farm-for-sale listings typically note the number of tillable (plowable) acres and the number of open acres, which generally means pastures. The difference between tillable plus open and the total listed acreage will be woods, swamp, or ground too rough or awkward for agriculture.

Slope and Steepness

Note in which direction the land generally slopes. If the whole farm is tilted to the north, the growing season will be a little shorter and cooler; it might be a week or more later than the neighbor's land to warm up in the spring. Generally a level or south-facing aspect is considered best, although this will vary depending on the climate and crop.

Also consider the steepness of the land, since this affects what can practically be raised. Very steep slopes are best left alone to grow trees or other natural vegetation. Moderately steep slopes can be used for pasture, if you don't overgraze so that the soil erodes. A thick, well-managed pasture sod will hold the soil better than anything else on a slope. Moderate to mild slopes can be cropped if you use contour strips, grassed waterways, and other methods to make sure the soil won't wash away (see chapter 7). Level land can be used for any crop for which the soil type is suitable. If there's a pond or creek, that's a plus, but if you're planning on keeping livestock you will have to manage the banks and build crossings and watering spots so the edges don't degrade and the water stays clean and clear (see chapter 9).

Soil

Soil type and quality will be discussed in detail in the next chapter, but if you're looking at a farm now, take a trowel or shovel out in the field and dig a hole. What you're hoping to find is deep, dark dirt that just holds together when you squeeze it in your hand, but crumbles into smaller clumps when you tap it with your finger. What you're more likely to find is just a few inches of topsoil,

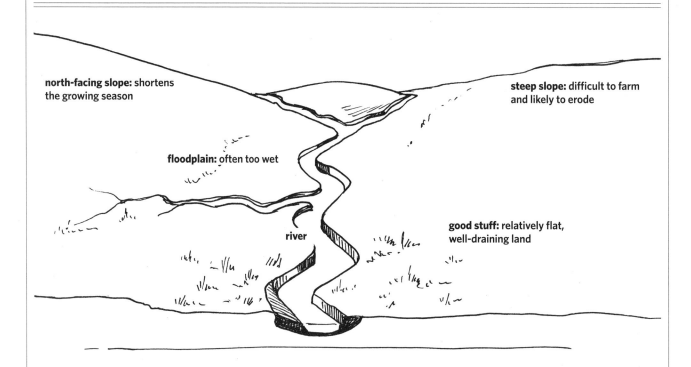

north-facing slope: shortens the growing season

steep slope: difficult to farm and likely to erode

floodplain: often too wet

river

good stuff: relatively flat, well-draining land

The lay of the land is important in farming. All aspects, from the steepness and direction of slope to drainage, affect not only crop growth but also how easy or difficult it will be to farm those crops.

not especially dark, atop lighter-colored subsoil. But unless the topsoil is almost nonexistent or is pure sand or pure sticky clay, chances are that an organic farm program can do much to bring it closer to the ideal.

The determining factor will be the soil test results. Pull soil samples and have them tested at a lab that works with organic farmers and understands their different approach to soil. (Directions for taking good samples are found in chapter 3.) Once you've received the test results (usually one to two weeks), talk them over with a reputable soil consultant or neighboring organic farmers. They can give you an idea of what you can do and what it will cost, over how many seasons, to realize the full potential of your soil. If you can't find any organic soil experts in your area, find one somewhere and give him or her a call.

Midwest Bio-Ag soil consultant Dan Jacobson says that it is fairly common for an aspiring farmer or young couple to purchase a farm with the intention of producing organic food, only to find out that it will take many years and much more money than they can afford to make the soil productive. "They never even considered what 50 years of bad agriculture can do to land. You're not going to fix everything in a year. Phosphorus and potassium are really expensive now. Correcting major nutrient deficiencies on a large piece of land could cost as much as your house," he says.

As with water issues, it's usually a judgment call: First figure out what the soil needs, how much it will cost per acre, and how big an area you need — a 1-acre (0.4 ha) market garden is a much different proposition from 80 acres (32 ha) of field crops or pastures. Then find a pencil and a calculator and figure out if you have the time and the budget to spend on making this soil fully productive. You can make up for a small budget by investing more time and labor, but if your situation demands that right away you need a better yield than this dirt is capable of immediately producing, you will need to look at other farms.

Consider also what sort of soil you like to work. Sandy soils are easier to dig, warm up quicker in the spring, and drain easily after a heavy rain. Clay soils are heavy and slow to warm up, but hold water and nutrients so well that sandy soils can't compare. Some farmers like sand, others prefer heavy soils.

Drainage

Soil drainage is another important factor in the productive capability of a farm. Soils that are wet all or part of the growing season will drown plants rather than grow them, and there is a surprising amount of low, wet land in rural areas. Much of this land has had subsurface drainage installed over the generations, and often a new owner will have to decide between renovating an old drainage

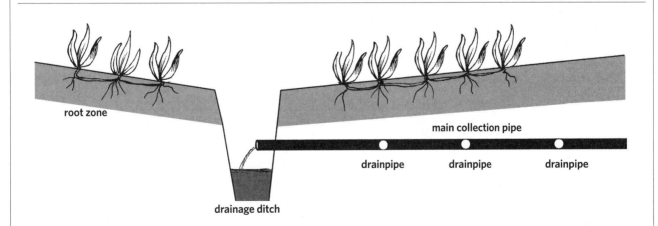

Subsurface drainage is a necessity in some areas for draining land well enough that it can be farmed. In most cases, the system consists of perforated drainpipes that collect water and feed it into a main collection pipe, which ferries the water to a large drainage ditch. If the system was installed some time ago, however, locating all its parts for repair or maintenance may be difficult.

Healing Land: Louis Bromfield and Malabar Farms

Louis Bromfield grew up working on his grandfather's farm in the hills of Richland County in northern Ohio during the first couple of decades of the twentieth century. He had intended to become a farmer, but college followed by an immensely successful career as a novelist intervened, and by 1938 he and his family were happily living in rural France, where he wrote, tended a garden, and often discussed farming with the neighbors.

But by 1938 it was clear that war was imminent, and Bromfield regretfully left France to avoid the conflict, moving his family back to his home valley in Ohio. There, inspired by the works of Sir Albert Howard and others in the nascent organic farming movement, he purchased four run-down farms and began the arduous job of restoring them to full fertility and productivity. Though Bromfield was not organic by today's rules (he was quite willing to use purchased fertilizers that would not now be allowed) his accounts of the successful restoration of the soils on those farms still offer ideas and inspiration for organic farmers today.

What he found when he returned to Ohio, he writes in *Pleasant Valley*, his record of this undertaking, was "a sad history of rich land slipping downhill over a period of more than a century . . . the story of good earth being murdered by carelessness and bad farming and greed and ignorance."

The Ferguson place, he wrote, was a hill farm scarred with gullies whose topsoil had eroded away and whose spring had gone dry. The Fleming place, on once-rich bottomland, had leached soils and low fertility. The previous owners, Bromfield said, "had not farmed it, they had mined it."

"These were the things I wanted to prove," he wrote. "That worn-out farms could be restored again, and that if you only farmed hill country in the proper way, you could grow as much as on any of the flat land."

"The Plan," as he called it, had a dozen major points. The first was to remodel the barns and outbuildings for better efficiency, to save labor and time for more remunerative work. Then livestock, manure, and forages would be returned to the land, integrated into each farm operation for their value in soil building. Only animals that could walk off the farm would be sold off the farm, in order to keep as much fertility as possible on the farm. An ongoing green manure program was instituted, where worn-out fields were planted with grasses and legumes that were then plowed under while still immature to restore nitrogen and humus to the soil. Lime, potassium, phosphorus, and trace minerals were purchased and applied to the soil as well. This made the soil right for growing rich bluegrass and clover pastures for the cattle, and field crop yields rose dramatically. Livestock was fenced out of the woods, so that trees were no longer suffering from soil compaction and overgrazing of the understory, and erosion was stopped.

In those fields where the topsoil was completely gone, "great masses of decaying organic matter" were plowed under, leaving the surface rough, to hold moisture and nitrogen and anchor minerals brought up to the surface with the glacial till subsoil.

The last few points of "The Plan" reflect Bromfield's broad view that "sustainable" has economic and social components as well as ecological. He set a goal to become as self-sufficient in food as possible, including those wild or low-maintenance items that required only having a place to grow: berries, fish, maple sugar, honey, and wild game. He also worked to put more people back on the land, making a good living, by renovating the various farmhouses as homes for the help he hired. He built a large rambling home for his own family. Since he regarded the home as the heart of the farm, he made sure it was of a style and quality where he would happily live his entire life.

Lastly, "The Plan" called for ongoing experimentation, to continue to find better, more effective ways for Malabar Farms to be fertile, productive, economically viable, and happy: "What all of us were trying really to find was the old spacious, comfortable life which farmers and landowners once lived. . . . We all believed that in the soil could we find not only security but the best life in the world."

Bromfield had the leisure and the money for such a massive enterprise on a short timescale, something that is beyond the reach of most beginning farmers. He also had the wisdom to continually seek out advice and information, which anyone can do. Though he spent more than most of us can afford on equipment and help, most of the practices he used are practical over time for any aspiring farmer, and most are still fundamental to good organic practice. These include crop rotations that incorporate forages and especially legumes; the use of cover crops and crop residues to have ground covered in winter; proper drainage; contour strips and terracing on the slopes; spreading manure; buying the lime and nutrients not available on the farm; and leaving trees and brush around the edges as homes for the birds and other life forms who eat insect pests and add to the beauty, biodiversity, and resilience of the entire farm.

In 1976, Malabar Farms became an Ohio state park. It is still operated as a working sustainable farm. The non-profit Malabar Farm Foundation was founded in 1993 to provide resources in support of the vision and programming of Malabar Farm State Park.

system or returning the area to its natural wetland state. Wetlands are terrific for wildlife (see chapter 13), so if you can spare the acreage this can be a delightful option. If you can't, you'll need to learn about drainage.

Wet Soils

Standing water from snowmelt, floods, or spring rains, sitting in the fields for weeks during planting season, means little or no crop that year. If such standing water happens every year, it is a problem you do not want to buy.

A second potential water-related problem is a high water table, which limits the root zone for plants and so will limit plant growth. Soils that are constantly saturated with water don't have enough oxygen for roots or soil life to function effectively. About 25 percent of U.S. farmland is artificially drained to make it useful for agriculture, either by surface ditches or by subsurface tiles and pipes. Some of these drains have been in so long, and the land has changed hands enough, that no one remembers any longer if it's drained or where exactly the drains are. If the land you're considering is low and has been actively farmed, it may have drain tiles. If so, ask the owner or neighbors if they know where the lines and outlet are, since these need to be checked occasionally to make sure they're clear and functional. Otherwise you may find yourself owner of a wetland, which is nice scenery but not good for farming.

FIXING THE PROBLEM

The only way to fix wet soils is to remove the wet, and this is usually expensive. It may be possible to dig deep, open ditches to drain the land, but these require regular maintenance to keep from silting up, are a hazard for children and visitors, and make it inconvenient to move equipment or livestock between fields. The more usual option is to install subsurface drainage. Most agricultural extension services can offer advice on installing drainage in fields. In cases where there is no lower ground to run drain tile out to, or where the expense of tiling or ditching is prohibitive, you're probably better off leaving the land in its natural state and moving your crops elsewhere.

Land that has drains in place but isn't draining is problematic. If you're lucky, the problem is a simple blockage near an accessible part of the drain tile or pipe, and you'll be able to fix it yourself. But if the problem is inaccessible, extensive, or both, you'll have to decide whether to pay to retile the area or restore it to a wetland.

Dry Soils

Dry soils that require irrigation present a slightly different twist on the drainage issue. In the U.S., irrigation is common west of the Mississippi and along the Gulf Coast, and many crop farmers in the rest of the country have backup systems for dry years. When irrigation water evaporates from the soil surface instead of percolating down to the groundwater table, it leaves behind soluble salts of sodium, calcium, and magnesium that are naturally present in the water and soil, and especially so in dry regions with alkaline soils.

Normally the excess salt would be leached away as rainwater percolates down through the soil. In dry climates, however, where there isn't enough rain or irrigation to drive the water down that far, salts can accumulate in the root zone.

Another common situation with dry soils is, paradoxically, a high water table from excessive irrigation, so that the salts that had leached away rise up with the water table and move back into the root zone through capillary action.

Either way you can end up with a saline soil, which most crops won't tolerate. Poor and spotty plant growth may indicate a saline soil, as may a blue-green color in forage crops, or leaf burn and die-off in grain crops. Visible salt on the soil surface is a pretty sure indicator of a problem. If there is any question that the farm you're considering may have a salt problem, have this tested for when you send in your soil samples.

FIXING THE PROBLEM

Of the 16 percent of U.S. farmland that is irrigated throughout the season, an estimated 25 percent, or 14 million acres (57,000 sq km), is currently affected by salinity problems. Reclamation of saline soil usually involves installation of subsurface drainage and overirrigating to wash away the accumulated salts — not a cheap proposition.

The Benefits of Marginal Land

The biggest constraint for nearly everyone interested in buying land is budget. Good farmland is expensive, and more often than not beyond the means of an aspiring farmer. That means you're going to be, more likely than not, looking at marginal land.

There can be some real advantages to *not* buying the very best cropland. In fact, you might not want even to bother looking at land in an intensively cropped area since there's a very good chance the soil is going to be in bad shape and there are going to be pesticides in the drinking water.

Avoid poorly drained land, land showing signs of salination, and land with less than 4 inches (10 cm) of topsoil. On the other hand, even though cheaper farmland may have more slope, more rock, inconvenient topography, screwy field layouts, or less than ideal soil, those difficulties — any or all, in moderation — aren't all that hard for an attentive small-scale farmer to overcome. Marginal land is frequently a bit isolated from neighboring fields or not in heavily cropped areas, and that's a huge advantage to the organic farmer concerned about pesticide drift, groundwater pollution, and pollen contamination from GMO (genetically modified organism) crops. A good natural buffer zone can make up for quite a lot of other drawbacks.

Another advantage of settling in an area with shallow soil, high groundwater tables, lots of steep hills, or other features less than optimal for agriculture is that it may keep out neighbors that you don't want, such as a landfill or a 3,000-cow dairy. You want land that is farmable, but maybe not all that farmable.

House and Buildings

If the location, water, terrain, and soil all check out okay, then look at the buildings. If there's a house, the same considerations apply to it as for any real estate transaction, farm or not. Is it livable, or will you have to budget for upgrades and repairs? How old are the wiring, plumbing, insulation, and roof? Does it have insulation? Are there any signs of dry rot, termites or carpenter ants, flooding in the basement, cracks in the foundation, or other problems that will demand repairs?

Don't even think of buying a house unless you've been in the basement and looked for cracks or bulges in the foundation, dry rot in the beams, wetness, and mold. Often these things won't be a sale-stopper, but they should be taken into account when you're negotiating the price.

Outbuildings: To Keep or Not to Keep?

Far too often the outbuildings are in bad shape if not already falling down, and how you want to deal with them requires some thought. It's often cheaper and quicker if you can repair or renovate to make use of what is already there instead of building new.

If the roof, foundation, and structural support are good, a building is probably worth keeping, even if you can't imagine why you might need it. Most new farmers quickly discover that buildings have a way of quickly filling up with equipment, livestock, supplies, tools, workshops, vegetable-processing or dairy-milking facilities, and grain storage. But if there are rats entrenched under the foundation or the spine of the barn roof has a nice big curve to it, the building should probably go. The expense of repairs, or any necessary demolition and burial, should be taken into account when making your purchase offer.

The Farm Dump

Last, find the farm dump. Before municipal garbage service became available in rural areas, nearly every farm had its own private dump. A hundred years ago there wasn't much to throw away on a farm that would really harm the soil, and if you find a dump that old it might have some nice antique bottles if you're lucky. But many of these dumps were being used until quite recently (and no doubt some are still being used) and you'll want to kick around a little and see if there are any rusted old insecticide containers or other alarming things. Will you be able to bury and forget them, or should the dump be cleaned up, another potential expense?

Many farms will have drifts of old machinery and metal scrap out in the field or in odd corners. These can be a real pain to work around and to clean up, but you may be able to talk a neighborhood scrap dealer into coming by and doing some of the work for you. They won't take everything, but it'll help. Work out the price structure beforehand. Some dealers will pay you; others will insist you pay them. Start with the one that will pay you.

Old farms usually have old dumps, with abandoned derelict machinery. Don't buy someone else's problem. If you're considering a farm with an old dump, check the dump for toxic materials, and assess how difficult it will be to clean up the mess.

A Few More Things to Think About

If you're buying land without any buildings or other improvements, it's essential to make sure of the following:

- **Frontage or right-of-way.** You must either have frontage on a public road or have a legally guaranteed right-of-way (an easement) across other land to get to the road, and the right to construct a driveway on the easement. You may have to obtain a permit to construct a driveway; check with the local municipality and the county zoning office to find out if this is necessary and if so, what restrictions apply and how much it will cost to satisfy any code requirements for construction.

- **Permits.** Check to make sure that you will be able to obtain and afford any required permits for building a home and outbuildings and installing sanitary facilities. In most areas you will be required to install sanitary facilities that are up to code; you may be able to get away with a composting toilet, but in any area where the neighbors are reasonably close you will have to have some sort of acceptable system for handling other household wastewater. If the water table is very high or you are adjacent to a wetland, it may be difficult to get a permit.

- **Mineral and water rights.** The deed to the land should include mineral rights and rights to the groundwater under your land.

- **Property survey.** The property lines should have been surveyed and marked by a professional surveyor, or should be as part of the purchase agreement if that is what is mandated by local or county ordinances. Having rural property actually surveyed used to be unusual in most rural areas and still is in many, but there's a national trend towards land-division ordinances requiring a survey for many, sometimes all, types of land divisions. If it's not required then it's the buyer's judgment call whether to insist on a survey. This can be expensive (from a few hundred to a few thousand dollars), but surveyed property lines will protect you in the event of a property line dispute with a neighbor.

Regardless of whether the property holds a house and other buildings, you should also consider the following:

- **Reliable year-round access.** If you have off-farm jobs or school-age children, it's important to have a driveway that is not routinely blocked by floods or snowstorms.

- **Floodplains.** Determine whether the land and, especially, the buildings lie in a floodplain. If they do, why would you want to live there? And getting good property insurance could be difficult.

- **Livestock laws.** If you plan on keeping livestock, or if the neighbors keep livestock, familiarize yourself with your state's livestock and fencing laws. Most states have summaries available on the Internet; if yours does not, ask your extension agent. The laws will define a legal fence, define liability for damage caused by stray livestock, and define who is responsible for fence construction and maintenance when the fences are on property lines. In Wisconsin, you are legally responsible for the half of the fence on your right when you stand on your side of the line and face your neighbor. But if you are the only ones in your neighborhood who keep livestock, in reality you will probably be the ones who build and maintain the entire fence.

- **Watershed.** In the 12 midwestern states where they're available, buy a county plat book. This will tell you who your neighbors are and what they own, and where any public land is located. You will also be able to trace the local details of the watershed where your farm is located, so you know who is living upstream.

Financing a Farm Purchase

Once you've found a farm and have determined that you can make it work for you, you'll have to find a lender to give you a mortgage unless you have enough cash to buy the place outright.

Most banks have little or no experience with agricultural land and are not willing to make a loan. Or they may offer you a mortgage but at a ridiculously high interest rate. Your best bet is usually a bank or credit union that is locally or regionally owned. Another possibility worth checking is Farm Credit Services, a government-sponsored national network of rural lenders.

There are some federal programs to assist new farmers. Start by contacting the area branch of the Farm Service Agency. The FSA offers direct farm-ownership loans and beginning-farmer loans, though there are some restrictions.

Yet another option is a land contract. This is a formal, legally binding contract between you and the current owner, with no middlemen involved except for the attorney who draws up the papers.

You agree on the price, the term of the mortgage and the interest rate, and make monthly payments until the debt is settled. This is a fairly common arrangement; the downside to it is that if you fail in your payments you not only lose the farm but all the money you've already paid, as well.

A Final Note

If the land you are buying has not had any prohibited substances applied for some time, you may be allowed to certify it as organic more quickly than usual. Normally land must be under organic management for three years before it qualifies for certification, but if you obtain a signed statement from the previous owner and/or operator (the owner isn't always the person farming the land) documenting that no prohibited substances have been applied, you may be able to shorten the transition period. Get this document before you show up for the closing. The date of the last application of a prohibited substance must be stated. The document does not need to be notarized.

"Suit your project to the land. . . . And choose the breeds that are naturally attuned to your area."

NANCY COONRIDGE
Coonridge Organic Goat Cheese, Pie Town, New Mexico

Profile: Nancy Coonridge

Coonridge Organic Goat Cheese, Pie Town, New Mexico

At the age of 19, Nancy Coonridge decided she needed a goat. "The ordinary thing to do would've been to have a baby or get a dog," she says. "I don't think I'd ever even seen a goat before. But I'd realized that a goat wouldn't be a pet. We'd have a relationship — I'd take care of her, and she'd take care of me. I would get food from her."

Though she had an urban childhood in Berkeley, California, Coonridge had two advantages when she chose goat dairying as a career: as a child she'd spent time on her grandparents' farm in southern Illinois and learned to know and love rural life, and she was in California during that state's great dairy goat renaissance, which has since spread across the country.

Coonridge quickly became part of a new community, meeting Esther Oman, who is credited with single-handedly keeping alive the Oberhasli breed of goats in the United States, Jennifer Lynn Bice and Stephan Schack of Redwood Hill Farm and Creamery, and many others who were developing goat dairies and cheesemaking in the area during the 1970s. "I was fortunate. Those people were my peers and good friends," Coonridge says.

Her first goat, she remembers, "gave hardly any milk, and you had to tie all four feet to the stanchion to milk her. But we didn't mind. We thought it was all wonderful and exciting." She soon added more goats, began making cheese, and started selling milk. She then decided to take the further step of learning to free-range her goat herd, and by 1977 she was herding goats part-time in the brush of the California foothills, and beginning to look for land of her own that was suitable, affordable, and remote.

At the time Coonridge was paying her bills by working as a coast-to-coast truck driver. It was a terrific way to see the countryside. "In each place, I'd look around and think, how would my goats do here?" Coonridge says. She considered Tennessee, West Virginia, and many other states, but eventually she decided that land in the high desert with no near neighbors was most suitable for the way she wanted to raise goats and live her life.

In the high desert, Coonridge comments, "you're not going to have the problems with parasites that you have in other places, because it's dry and you have four seasons, with a real winter that kills parasites and other bad things that live in the soil." And, she adds, since goats originated in the dry northern areas of the Middle East and are naturally more inclined to browse on brush than to graze grass, they are inherently well suited to the piñon pine and juniper country of western New Mexico. In 1982 Coonridge bought 40 acres at an elevation of 8,000 feet and two hours from the nearest post office, and she opened her dairy. She says:

> Goats fit this land. It's rich in feed for them. They eat mountain mahogany, wild buckwheat, and different things in different seasons. In the spring they eat the first little grasses that come up, which are really high in protein, and the little oak leaves. They browse a tiny bit of the piñon pine and the juniper, but those aren't favorites. In the fall they eat the acorns and the seed heads of the grasses. I'm so lucky that I get to observe their natural patterns.

"It's not desirable land, so that made it cheap," Coonridge says. The price was low for another reason as well — there was no water source. "The standard wisdom is that if you buy land without a well you are screwed," she says. "But you can save water off your roof.... Even a small roof, say 12 by 12 feet, will harvest around 1,000 gallons of water at 12 inches of rain a year. I don't have a well; instead I have 80,000 gallons of clean water storage. We're getting about 12 inches of rain a year now, and it comes twice a year, in the winter and in the summer. So we can fill our tanks twice a year."

The flip side of little rain is lots of sun, and since Coonridge Dairy is too remote to even consider being on the electric grid, solar panels provide the current for day-to-day living, and propane-powered generators for big jobs. Wood provides heat. Careful conservation of electric power and water enables comfortable living for the family and an intern or two, despite being off the grid.

COONRIDGE DAIRY HAS GROWN over the years to 400 acres, and there are still no fences and no near neighbors. "New Mexico is a free-range state, which means that if people don't want my goats on their land, they have to fence them out," Coonridge says. She keeps the herd, which averages around 70 milking does, in the barn overnight, milks in the morning, and turns them loose to browse during the day, free to wander at will.

The downside of no neighbors and unlimited freedom is plenty of large predators. Coyote, black bear, mountain lion, and — most troublesome to the goat owner — wolves are all fellow residents of the region. Coonridge puts radio collars on her goats to keep track of their movements, and as guard animals she uses Maremma dogs, originally bred in the Italian Alps to protect livestock from wolves.

Just as Coonridge has suited her dairy and her lifestyle to the terrain and the climate, so she has suited her cheesemaking and her marketing. Aged cheeses require high humidity to cure properly, she says. "So we do fresh cheeses. We cover them with olive oil, and the feta with a brine. You can let them sit on the table. They don't need refrigeration." Like the goat's milk, all oils and herbs used in Coonridge Cheese are USDA-certified organic.

Since farmers' markets are too distant to attend regularly, Coonridge focuses her efforts on wine festivals, which are less frequent but bring in higher numbers of attendees. She estimates as much as a twelfth of her total cheese production is used for

> **"You have to follow the seasons. To have all these people around the nation producing exquisite foods in small, organic amounts — you really have something worth the wait."**
> Nancy Coonridge

samples at these types of events, and it is product well spent. "A piece of cheese is not complete until someone eats it. I love to put the cheese in people's mouths, I love to talk to them about where it comes from, and why it's good for them," she says.

For new organic farmers, Coonridge recommends, "Suit your project to the land. If you're in Florida, you might want to have your goats a different way than I do. You could have them on slatted floors and bring their food to them, so you don't get parasites. And choose the breeds that are naturally attuned to your area."

AFTER MORE THAN FOUR DECADES of milking goats, Coonridge remains enthusiastic about sustainable living, organic production, and artisan food. "I love the way my milk tastes, I love the way my cheese tastes. If I didn't, I would do it differently. The taste is different through the year, and that's the beauty of an artisan cheese. You have to follow the seasons. To have all these people around the nation producing exquisite foods in small, organic amounts — you really have something worth the wait."

Profile: **Dan Kelly**

Blue Heron Orchard, Canton, Missouri

Now, more than 20 years later, certified organic orchardist Dan Kelly recalls that in the late 1980s, television actor Eddie Albert did some public service commercials encouraging people to volunteer for the Soil Conservation Service. So Kelly headed down to the SCS office in Lewis County, Missouri, to sign up. The volunteering went so well that after a few weeks he was hired as a part-time employee, and as his interest in soils and conservation work grew, he signed up to audit a soils class at Culver-Stockton College in Canton.

"I couldn't afford to buy the $40 textbook, but they had a copy in the library, so I would go to the library and outline the chapters," Kelly said. At around the same time, he had discovered an abandoned orchard outside of town. "It had about 100 trees. I proposed to the woman who owned the property that I would prune, mow, and maintain it, and her family could have all the apples they wanted. I bought a little sprayer, used some organic insecticide, and actually had a little crop."

That's how Dan Kelly, almost by accident, developed an interest in soils and apples. By this time he was living in a rented farmhouse surrounded by wheat and soybean fields, a mile and a half from the Mississippi River. "I was interested in organics at the time, and I didn't like the idea of the farmer who rented the land being around the house spraying the crops. I proposed to the owners that I buy an abandoned field and bring it back to life. They said they could not sell that field, as it was their access to the back-field acreage, but they were willing to sell the house, the buildings, and about 25 acres that included the house. I stepped off what I wanted — it was 26.4 acres — and they drove up from St. Louis one weekend. For the land, house, barn, garage, deep well, and acreage, they wanted $25,000."

INSTEAD OF A BANK LOAN, Kelly was able to obtain a deed of trust from the owners (similar to the land contract used in some other states), which enabled him to make regular mortgage payments, plus interest, to the owners instead of a bank. Thus was born Blue Heron Orchard.

In 1990 he planted 5 acres to eleven different varieties of apple trees, at the rate of about 100 trees per acre. At the time, "the movement in the industry was toward high-density plantings, anywhere from 500 to 2,000 trees per acre. But you have to pay the same amount per tree, and for me that much was way out of line. And these dwarfing systems all require extensive staking and trellising, and the trees don't live as long, because they have a weak rootstock," Kelly says. "I opted out of that."

Instead, careful pruning to lateral branches keeps Kelly's semistandard and semidwarf trees short enough to make apple picking fairly easy, while the greater strength and size of the trees mean deer browsing isn't the major problem it can be with small trees. (Though when the orchard was young and the trees still small, Kelly sprayed Hinder, an odor-based repellent, to keep the browsers at bay.)

With no real experience in orcharding when he started his enterprise (and no Internet to turn to at that time), Kelly searched in many places for more information. "I'm in an area where apple cultivation was almost nil. I picked up an old book at a Unitarian Church book and bake sale, and I found some extension bulletins," he said. But much of his expertise he's developed simply by close observation and experimentation. Over the years he also has applied for and received a number of grants to research specific aspects of organic apple production.

One of the first was a grant from the USDA's Sustainable Agriculture Research and Education (SARE) program in 1995 to replace brush around the orchard with native prairie grasses to control plum curculio (PC) insects and encourage wildlife.

Plum curculio lay their eggs in small developing apples. The apples abort in June and fall to the ground. The PC pupates in the soil and emerges from the ground as an adult. In the fall, adults fly back into brushy areas to overwinter. I thought if I got rid of the brush — the honey locust, American wild plum — I would eliminate their habitat and replace it with prairie. I could then burn the area in the spring and destroy the PCs before they could come back into the orchard.

Burning prairies is a tried-and-true method for maintaining their vigor and keeping them from being overtaken by brush. Native prairies are an endangered ecosystem and a rich wildlife habitat, so Kelly's proposal worked for a number of purposes, including improving and maintaining biodiversity. "One of the things I planted was little bluestem (a native prairie grass), and quail like to nest between two clumps of this grass. There's a fair amount of quail around now. One time I found one in my office. I came in one afternoon and my window was broken and a quail was trapped between the storm window and frame window. So I picked up the quail and took it outside and let it go," Kelly says.

Kelly mows his orchard just twice a year, allowing grasses and native flowers to grow tall and blossom, furnishing nectar for various beneficial insects. Red clover has time to reseed itself, and the heavy grass mulch from the mowing slows the movement of some diseases, such as scab, from leaf litter into the trees.

I do a lot of insect monitoring, so I have a pretty good handle on what kind of pest problems are out there. I walk around the orchard and I'll find these little larval cases, and I'll bring them inside, put them in a jar, and see what hatches out of them. One time I found a larval case and I was expecting a moth, but then I noticed it hadn't hatched; however, there was a little wasp in the jar. The wasp was exactly what you want in a balanced orchard — you have predator insects doing what pesticides would do. I had released parasitic wasps one year, and I had the sense that maybe I had helped establish the population.

KELLY HARVESTS HIS CROP into bins of native oak that he designed and built. He stores the apples in a straw-bale and timber-frame cold storage structure designed and built with some funds from a sustainable-agriculture grant from the state of Missouri. Help came from friends and participants in a straw-bale building workshop.

Kelly's marketing has been similarly innovative. "Our local area isn't so good for sales," he says. "St. Louis is about 140 miles, and I have family there, and it's a really good area for me to sell into. I've made a really good connection with a CSA in St. Louis — they wanted lots of apples and lots of cider." Kelly also sells to grocery and health food stores, St. Louis University, a buying club, and at farmers' markets. Customers can also come and pick their own apples.

Kelly has built a state-licensed, certified-organic kitchen at the orchard for making apple butter, apple sauce, cider, Pomona's Ambrosia (an apple syrup), cider vinegar, and a habanero cider vinegar.

"Be prepared to do processing," Kelly says, "until consumer perception evolves to understand superficial apple blemishes such as sooty blotch and flyspeck, or why hail hit the apples. We had hail one May that dinged up quite a few apples, and I found myself writing letters to the CSA and the buyer's club to explain why the apples looked like they did. So learn how to make applesauce."

Kelly's latest project, funded by another SARE grant, is a simplified, online organic integrated pest management tool to assist orchardists in site-specific management of the most economically damaging apple pests. He has also been a frequent contributor of technical information to the Organic Fruit Tree Growers Network newsletter since he became a charter member in 2004.

Though Kelly has developed a reputation as a knowledgeable and innovative grower, he recalls that he didn't start that way. "I was so ignorant when I put the orchard in," he says. "I figured by the time I had my first crop I would have a handle on it. It took a lot longer than that."

Soil
The
Cornerstone of
Sustainability

"Soil," says Wisconsin soil scientist Teri Balser, "is the poor man's rain forest. It's more biologically diverse than any large-scale ecosystem on the planet."

It's hard to imagine how a cubic yard of dirt can contain more species and more individuals than a rain forest, but that's the nature of a healthy soil. Every acre of it is inhabited by 2 tons (1.8 t) or more of living organisms, while the number of individual organisms in that space is estimated to be far greater than the world's entire population of humans. Without all that life, the soil's ability to nourish plants falls like a rock off a cliff. "We know that soil organisms are involved in every aspect of soil quality," Dr. Balser says.

Since everything produced on a farm ultimately derives from the soil, good soil management is the cornerstone of a sustainable organic farm. And since you can't manage what you don't understand, good management depends on understanding what soil is, how it functions, and how it stays fertile and healthy in ecosystems that are without human interference. Knowing the ideal conditions for fertile soil and how they happen naturally directs an organic farmer's management plan toward the goal of replicating those conditions. Good crops become by-products of healthy soil, and crop yields are maintained by sustaining the health of the soil.

In fact, maintaining healthy soil is mandated by the National Organic Program Final Rule, which requires organic producers to "improve the physical, chemical, and biological condition of soil and minimize soil erosion."

For decades many conventional farmers have compensated for declining soil quality by increasing their fertilizer use, which hurts soil quality by chemically burning up organic matter in the soil and destroying some soil organisms. The organic farming community, on the other hand, has spent those years fine-tuning tillage methods, natural soil amendments, and crop rotations to develop effective systems for reinvigorating abused soils and maintaining healthy soils. The result is a menu of effective techniques developed by farmers across the country that can be applied as appropriate to the diverse soils and production systems of organic farms.

As you dig into soil science, the lines between physics, chemistry, and biology blur, and whatever you do to influence one will impact all three. Dirt, it seems, is not simple. The rock and mineral particles, organic matter, and living organisms that

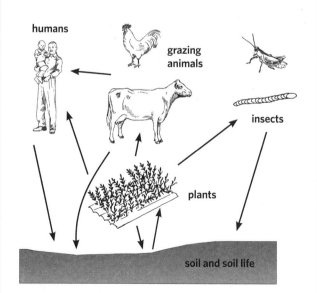

The earth's food web depends on soil. All food ultimately comes from green plants, and all green plants come from the soil.

together make up soil have so many and such complicated interrelationships and interactions that it's possible to take a college major in soil science, and then spend the rest of your life studying it further.

Fortunately, organic farmers don't need a college degree to understand fertile and healthy soils. They do need a working knowledge of what constitutes good dirt, what tools are available to create it and keep it, and when these tools are appropriate. This turns out to be very interesting stuff.

What Is Soil?

Instead of a mysterious rain forest, it might be easier to think of soil as a house construction project. To build a solid, smooth-running, well-supplied home for plants, you must assemble the raw materials, bring in a construction crew to put them together in the correct way, and install systems to keep water, air, food, and power (in the form of solar energy) all cycling through effectively and efficiently, just as in a roomy, comfortable house. Anyone who has raised teenagers will sympathize with the food aspect of this analogy; plants are similarly hungry, and having the necessary nutrients conveniently available is crucial to growing a healthy crop.

Fertile, healthy soil is about 25 percent air and 25 percent water. The remaining 50 percent is made up of three components: tiny rock particles, organic matter, and living organisms.

Rock Particles

Rock particles are the primary component of soil, making up 90 percent or more of the solid components, or 45 percent of the total volume. These are the basic raw materials of a soil: the cement, lumber, and shingles of the soil house.

The tiny rocks come from big rocks, which are made of minerals, sometimes in a pure form but usually in a mix like limestone, basalt, or granite. Frost wedging, root wedging, contraction and expansion from temperature changes, wind erosion, and chemical weathering from rain (which is

naturally mildly acidic and has a dissolving effect on rock) and rock-loving lichens (which exude mild acids) all continuously break off tiny bits of the rock, eon after eon.

These broken-off particles are grouped into three sizes. Sand particles, measuring from about 0.05 to 2 millimeters, are gritty to the touch and don't stick together. The large particles leave large spaces, or pores, between them, so water drains through quickly, at a rate that easily takes loose nutrients with it. Sandy soils warm up quickly in the spring and are pleasant to work, but they tend to be drier and less fertile.

Silt particles, between 0.002 and 0.05 millimeter in size, feel smooth in the hand and also don't stick to each other. Silt soils have smaller pores, though, and so retain water much better than sand: smaller holes make slower drains. Silty soils are generally nice for plants, since they don't drain so quickly as to be overly dry or so slowly that they're always wet, and they retain nutrients fairly well. Roots can work their way through silt easily. The loess soils of the prairies are wind-deposited silt and some of the best soils in the world for crops.

Clay particles, at less than 0.002 millimeter in diameter, are smaller still. Clay is fundamentally different from sand and silt because the

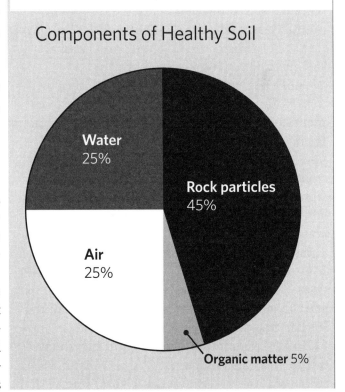

Components of Healthy Soil

Water 25%

Rock particles 45%

Air 25%

Organic matter 5%

Soil

particles are so tiny that they don't have the heft to influence each other much physically. Instead they react with each other chemically to bind together in the soil. They can also chemically bind soil nutrients to their surfaces. This is one of those places where lines between types of forces — chemical, physical, and electrical — blur, since the molecules of different chemicals carry different electrical charges. When a negatively charged molecule meets a positively charged molecule, they bind together (in electricity, if not always in love, opposites attract). The particles actually stick together, and so clay feels sticky. There's so little space between the tightly joined soil particles that water penetrates and drains from clay soils very slowly. Air also has a hard time working its way into clay soils. Since air is necessary for soil life, this can be a real problem. Clay soils tend to be cold, heavy, wet, and fertile.

As rock particles are formed, some stay in place to form a soil, while others are washed or blown away to form soils elsewhere. In the process they are mixed, so that few soils are pure sand, silt, or clay. Soils are technically described by their dominant particle size, and soils that are a mix of 30 to 50 percent sand, 25 to 50 percent silt, and 10 to 25 percent clay will have that lovely word *loam* in their titles, signifying to farmers that they have the best texture for plant growth.

Soil pH

A measure of the alkalinity or acidity of a soil, soil pH has a huge effect on plant growth and nutrient availability. Crops on soils with excessively high or low pH will have significantly lower yields and problems with nutrient deficiencies.

In chemical terms, pH indicates the amount of positively charged hydrogen ions (H^+) versus the number of negatively charged hydroxyl ions (OH). Hydrogen ions make the soil acidic, and hydroxyl ions make it alkaline. In other words, electricity creates chemistry. The pH scale runs from 1, the most acidic, to 14, the most alkaline, with a pH of 7 considered neutral. As you move up the scale, each number represents a tenfold difference, so that soil with a pH of 5 is ten times more acidic than soil with a pH of 6, and one hundred times more acidic than soil with a pH of 7.

Most plants and soil life perform best at a slightly acid or near-neutral pH, between 6.5 and 7.0. The further the pH level is from this ideal range, the more the ion imbalance will chemically tie up some nutrients and make others overabundant. A pH of more than 8 is difficult to fix, and very few crops will do well in such alkaline soil. Very acidic soils with a pH of less than 6 are easier to correct, though it takes time and some purchased soil amendments, such as calcium, magnesium, and sodium.

A soil's pH is at least initially determined by the minerals of its parent rock. Calcium, magnesium, and sodium make soils more basic. The soils of the Great Plains and some areas in the East are derived largely from limestone parent materials, which are high in calcium, and so tend to be neutral or alkaline. Granite, shale, and sandstone parent materials, which are common in New England, the Great Lakes, and the Appalachian states, contain very little calcium and so form acidic soils. Very old soils in high-rainfall climates will be more acid than a farmer would expect from the

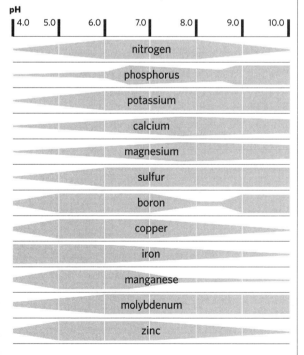

Effect of Soil pH on Nutrient Availability

Adapted from Emil Truog, "The Liming of Soils," USDA *Yearbook of Agriculture,* 1943–1947.

What Is Soil?

parent materials, simply because eons of rain have leached away calcium. (As a side note, the more acid the rain, the more quickly soils are leached and acidified.)

Most organic growers know their soil pH and either grow crops that prefer that pH range or add appropriate soil amendments as needed to maintain pH in a desirable range. In acidic conditions lime is the most common amendment used to increase pH or "sweeten" soil. On alkaline soils sulfur is used to make the soil more acidic, but it's not always effective, and often a better solution is simply to grow alkaline-tolerant crops.

Organic matter will help move pH levels closer to neutral and maintain them there. Humus particles, the most decay-resistant part of organic matter, are negatively charged, so they attract and hold positively charged ions of calcium and other nutrients that increase pH, preventing them from leaching away. Leaching of nutrients is one of the primary reasons soils tend to become more acid over time. Another reason is that each time you remove a crop, you remove nutrients. Adding organic matter in the form of animal manures and green manures replaces nutrients. So organic matter both returns nutrients to the soil directly and helps retain them once they're there.

Mineral Food for Plants

Of the 16 nutrients that plants need, they extract three — carbon, oxygen, and hydrogen — directly from air and water. The other 13 have to come from the soil.

A soil's parent rock is important, since it determines which of the thirteen necessary plant nutrients might be naturally present in the rock particle component of the soil, particularly the subsoil, and which you might have to add on a regular basis to replace what is removed by leaching, crops, or grazing animals. Igneous rocks such as granite and basalt may contain as many as 12 of the 13. Sedimentary rocks vary: sandstone-derived soils tend to be dry and infertile, while limestone-derived soils contain a lot of calcium and magnesium.

Keep in mind, though, that just because some nutrients are present doesn't mean they are available to plants. If you lock the kitchen, the kids can't get to the food. The same goes for soils: if there is a pH or a chemical lock on the nutrients, plants can't take them up through their roots. For example, if pH is above 7 or so, phosphorus will react by chemically binding so tightly to calcium that plants can't pry them apart to get them into their roots. In fact, the primary problem created by pH being too high or too low is that it changes the chemical reactions that occur in the soil. The negative or positive charges get so strong that nutrients become stuck to other molecules and roots don't have enough chemical power to detach what they need. It's as if your breakfast pancake were glued to the countertop. Organic matter and soil organisms, the other soil components, are what unglue nutrients and open the kitchen door.

Organic Matter

The second component that makes up a soil is organic matter: nutrient-rich material deriving from the decay of anything — plant, animal, or microbiotic — that once was alive but is no longer. To continue the house analogy, organic matter is the nails, screws, nuts, bolts, staples, and glues

Acid-Tolerant Crops (pH 4–6)

blueberries

flax

peanut

pecan

most potatoes

radish

raspberry and blackberry

watermelon

white clover

Alkaline-Tolerant Crops (pH 7–8)

alfalfa (will tolerate a pH as high as 8.5)

cherry

most garden vegetables

most grains (including corn, sorghum, wheat, barley, some varieties of oats, and rye)

peach

sweet clover

that hold the structure together. Without organic matter, soil is just a pile of rocks. With it, soil tilth (workability) and water-holding capacity improve, making the soil a hospitable environment for life-forms, including plants. If there's no organic matter, there's no sustainability in agriculture.

Organic matter not only holds the mineral components together but also is the kitchen of our soil house: the oven, fridge, pots, pans, and raw ingredients that provide nutritious food for home dwellers. During their lifetime, all living organisms take up nutrients; soil-borne organisms extract many such nutrients directly from the soil's minerals. After death, organisms decay and release these nutrients in forms that can be easily taken up by plant roots. In this way organic matter is the immediate source of most usable plant nutrients in the soil, particularly nitrogen, phosphorus, and sulfur.

Humus, the fabled ingredient that makes soils friable, fertile, dark, and rich, is the decay-resistant part of organic matter, especially the oils, resins, lignins, cellulose, and waxes left long after everything else has been reduced to more basic compounds and recycled. Though slow to break down and release its component nutrients, humus has a chemical structure that accumulates negative electrical charges at the molecular level, which in turn attract and hold positively charged nutrients, especially calcium and potassium, until plant roots come looking for them. And as humus slowly decays, it releases humic acids, whose electrically charged molecules actually change some soil minerals, especially phosphorus, into a form that can be taken up by plant roots. (This is another one of those places where chemistry, physics, and biology blend.)

Organic matter greatly improves a soil's fertility. It acts as a sponge, holding up to six times its weight in water, which improves a sandy soil's ability to hold water and nutrients. As soil organisms work to break down organic matter, they produce gummy substances that bind soil particles into clumps, or aggregates, helping clay soils become more porous (that is, they will both absorb water and drain more easily) and easier to work. And because it absorbs so much water so readily, organic matter greatly reduces water run-off and thus erosion in all soils. In fact, the

Measuring Organic Matter

A soil with 1 percent or less organic matter is considered biologically dead. Five percent or better organic matter is a good level for farmers to aim for; at that percentage the soil begins to look dark and fertile. Soils that are 20 percent or more organic matter got that way from being water-logged, so that air couldn't penetrate to feed chemical reactions and dead matter accumulated faster than it decayed. These peat and muck soils form in swamps and marshes and have tremendous water-holding capacity. On the other hand, if these soils are drained to make them more suitable for crops, decay speeds up and they shrink, so that the surface sinks.

A laboratory soil test can tell you how much organic matter is in your soil; see page 70.

Universal Soil Loss Equation used by soil scientists calculates that increasing a soil's organic matter from 1 to 3 percent will reduce erosion by 20 to 33 percent.

Clearly, maintaining organic matter levels in the soil is an important part of organic farming. Organic matter can be added in several different ways, from applying compost and animal manures to plowing under crop residue and growing and working into the soil various green manures such as hairy vetch and other plantings. We'll discuss these techniques later in this chapter.

Soil Life

The third component of soil is the organisms that inhabit it. In our house metaphor, these organisms are the subcontractors and hired help: the carpenters, electricians, plumbers, cement layers, and, just as important, cook, maid, and chauffeur. They put it all together and keep everything running.

Soil organisms are so many and so varied that by most estimates the majority of species haven't yet been identified. But soil scientists do have a good idea of the general groups these organisms fall into and the primary functions they have in the soil.

Essential Soil Nutrients

Primary Macronutrients

The three primary macronutrients are those that plants need in the greatest amounts. Most commercial fertilizers are blends of nitrogen, phosphorus, and potassium.

Nitrogen (N). Nitrogen makes plants grow fast and dark green. Growing plants need more of it than any other nutrient. Most nitrogen in the soil is stored in organic matter. But plants can't take it up through their roots until it has been converted to a usable form either by soil microorganisms or by fertilizer manufacturing plants. Too little nitrogen results in pale, slow-growing plants, but too much can result in weak, easily injured plants, delayed maturity, and, for perennials, decreased winter hardiness.
Best organic sources: animal manures, green manures, and legume crops

Phosphorus (P). Phosphorus is important for good root growth, blossoming and fruiting, and resistance to cold and disease. The majority of the average soil's supply of phosphorus is contained in humus. In the soil phosphorus tends to form compounds that don't dissolve in water and can't move in the soil. Humic acid from the decay of humus will act on these compounds to make phosphorus soluble and in a form plants can use. Too little phosphorus will result in stunted plants, often with a purple tinge to their leaves, and delayed maturity. Too much phosphorus ties up other nutrients. Too high of a pH will make phosphorus unavailable to plants.
Best organic sources: manures, especially poultry manure, and other animal products

Potassium (K). Also known as potash, potassium is essential for strong stems, disease resistance, winter hardiness, and efficient use of water. Too little potassium can cause a burnt-looking or yellowness on the edges of lower leaves. Too much may cause a plant to not take up as much calcium as it needs.
Best organic sources: mineral dusts and wood ashes

Secondary Macronutrients

Secondary macronutrients are needed in lesser amounts than the primary macronutrients, but they are no less important to plant health.

Calcium (Ca). Calcium is needed by plants in significant quantities to build strong cell walls. Because so many agricultural soils are constantly short of calcium (due to farmers not paying attention to soil calcium levels), many organic farmers consider it to be a primary macronutrient rather than a secondary one. Calcium storage in the soil depends on the soil's cation exchange capacity, or the clay and humus content of the soil. Sandy soils have the most difficulty retaining calcium. Calcium makes soils more alkaline. Calcium shortages in plants are evident in mushy fruits and shoot tips, hollow stems, and blossom end rot in tomatoes.
Best organic sources: lime, chalk, and gypsum

Magnesium (Mg). Every chlorophyll molecule has a magnesium atom at its center. Magnesium is also important in enabling a plant to take up other elements. Magnesium leaches easily from the soil, so deficiencies are common. Shortages are evident in yellowed leaves.
Best organic source: high-magnesium lime (dolomitic lime, or dolomite), which will also add plenty of calcium

Sulfur (S). Plants need sulfur to make chlorophyll and seeds. This nutrient is held in the soil by organic matter. Many soil experts believe that since the nation's coal-burning plant emissions have been cleaned up, a sulfur shortage is developing in our soils. A plant that is short of sulfur may be stunted and pale.

Best organic source: gypsum

Micronutrients

Micronutrients are also of major importance to plant health, but are needed only in tiny quantities. The Final Rule requires documentation of a micronutrient deficiency through soil testing before a certified grower can add an amendment specific to correcting the deficiency.

Boron (B). Boron is necessary in tiny amounts for new cell development and several other plant functions, but it is toxic to plants in even slightly high amounts. Most of the soil's boron supply is found in humus very near the surface. A shortage of boron results in a variety of diseases and deformations, and often the buds on the tips of shoots die. Most fruits and nuts are very sensitive to high boron, while alfalfa is fairly tolerant. Organic guidelines require that growers use soil testing to document a boron deficiency before applying boron soil amendments.

Best organic sources: borax and some synthetic formulations

Chlorine (Cl). Chlorine plays a role in photosynthesis and is rarely deficient in soils.

Best organic sources: none listed; if you have a documented shortage, talk to a soil consultant in your area who is familiar with organic amendments and the Final Rule

Copper (Cu). Copper is important for the formation of chlorophyll and disease resistance. Shortages are uncommon, and a mild shortage will simply reduce yield. Many copper sources are prohibited by the Final Rule, and certified organic growers must document a deficiency before applying a copper amendment.

Best organic sources: copper sulfate and copper oxide

Iron (Fe). Necessary for a number of plant functions, including chlorophyll formation, iron is common in soil but not always in a form available to plants. The higher the soil pH, the less iron will be available. A shortage is indicated when leaves become pale or yellow between the veins. Too much iron ties up phosphorus.

Best organic sources: animal manure; if a deficiency is documented, ferric oxide and iron sulfate may be used

Manganese (Mn). Manganese is important for seed germination, crop maturity, and the formation of chlorophyll. The higher the pH, the less manganese is available for plants, so deficiencies most often occur in high-pH soils. Symptoms of deficiency include dwarfing, flecks of dead tissue, and pale leaves. Too much manganese is toxic to plants. Certified organic growers must document a deficiency before applying a manganese amendment.

Best organic sources: Manganous oxide and manganese sulfate may be used to correct documented manganese deficiencies.

Molybdenum (Mo). Necessary for nitrogen-fixing bacteria to function and for plants to make proteins, molybdenum is more available at higher pH and is held in the soil by organic matter.

Best organic source: raising the pH by liming the soil may release enough molybdenum for plants

Zinc (Zn). Zinc powers a number of plant functions. Signs of a deficiency are variable, though dead spots on leaves are common. On a certified organic farm, a zinc deficiency must be documented before a zinc amendment is applied.

Best organic sources: zinc sulfate, zinc oxide, zinc silicate, and zinc carbonate.

Microscopic Life-Forms

Sooner or later, everything gets eaten by something else. If we can't see what's doing the eating we call it decomposition, but if we move our observation to a microscope, it becomes clear that decomposition is actually consumption. Like the bigger life-forms, microscopic soil organisms ingest other organisms or parts of organisms, both alive and dead, and in their guts break their meal down into its component parts. They excrete some of these parts and use others to maintain their own metabolism until they, in turn, are eaten.

No matter what we name this process, the end result is that microscopic life-forms recycle nutrients so that they can be used by the next generation of life-forms. A goal of organic soil management is to encourage soil life so the recycling process functions at optimal levels, ensuring that the least possible amount of nutrients are lost to water leaching, off-gassing into the air, or blowing away in the wind.

It's important to note that *optimal* doesn't necessarily mean *maximum*. If microbes decompose organic matter more quickly than it is being replenished, nutrients, particularly nitrogen, will get used up or leached away and shortages will result. This can be an issue in hot, humid climates, so that organic growers in that type of climate may use somewhat different techniques in their soil program to better conserve organic matter in the soil. This could involve using cover crops that don't tend to break down as rapidly in the soil and using more catch crops to hold nutrients in plants rather than the soil. Soil testing to monitor organic matter definitively reveals whether a soil plan is achieving a balance between organic matter inputs and what is lost to cropping or leaching. In the field, if soil color is getting lighter it's a pretty sure indication that organic matter is dropping.

Soil microbes are important not just because they recycle and reconfigure nutrients. They also manufacture and leave behind a variety of vitamins, amino acids, sugars, and other substances beneficial to other soil life-forms and to plants. Additionally, many types of microbes produce substances that help form and hold soil aggregates, contributing to tilth.

Of course, not all soil microbes are beneficial. Oomycetes fungi, for example, bring us potato blight and other plant diseases. But that's the case with just about all types of organisms. Some you encourage, some you discourage, and some you just try to find a way to live with.

BACTERIA

The smallest but most numerous life-forms in the soil are bacteria. They have, among others, three critical functions: fixing nitrogen, decomposing organic matter, and keeping various bacterial populations under control.

The nitrogen-fixing process is absolutely critical to the organic farmer, since it is a primary source of nitrogen, the nutrient needed in greatest quantity by plants. Bacterial nitrogen fixers extract nitrogen from the air and change it into ammonia, a nitrogen compound that plants can take up through their roots. Rhizobium bacteria, the supreme nitrogen fixers, live only on or in the roots of legumes (including beans, clovers, and alfalfa). The many species of actinomycetes bacteria aren't so fussy (or as productive) as rhizobia, and soil scientists continue to find more plants that work with this type of nitrogen-fixing bacteria. Those plants that don't have nitrogen-fixing bacterial partners can still reap benefits if they grow in a field whose previous crop was one that did, leaving a surplus of usable nitrogen in the soil. In practical terms, this means, for example, that it's smart to have alfalfa in the field the year before you plant nitrogen-hungry corn.

Those species of bacteria that don't team up with plant roots to fix nitrogen, which is most of them, are the first cooks in the soil kitchen when it's time to decompose organic matter. It's decomposition, the soil kitchen's equivalent of cooking, that makes most nutrients available to plants. Decomposition uses nitrogen, and decomposing bacteria will compete with plant roots for it if nitrogen is in short supply. This can slow down both plant growth and organic matter decomposition, so one of the challenges organic farmers face is maintaining enough nitrogen in the soil to keep growth and decomposition moving quickly.

While some species of actinomycetes fix nitrogen for some plants and others decompose organic matter, still others have a third function: keeping other bacterial populations from growing out of control. They do this by producing natural antibiotics. In fact, these soil bacteria are the original source of some human antibiotics, such as streptomycin.

FUNGI

Research has shown that many more species of plants, many thousands in fact, team up with fungi rather than with bacteria. Though these fungi don't fix nitrogen, their networks of filaments do entwine with and greatly extend the area of plant roots, allowing them to share otherwise unreachable water and nutrients. Plants in return provide energy derived from the sun to drive the fungi's metabolism. These root-fungi complexes are called mycorrhizae, and many send up mushrooms, which are the flowering parts of the fungi. Mycorrhizal fungi are particularly good at finding phosphorus and making it available to plants. Other species of fungi are important decomposers of organic matter, and perhaps most endearing, fungi are also what makes soil smell good. As they break down organic matter they form a chemical called geomysin, the source of that evocative odor of freshly turned soil.

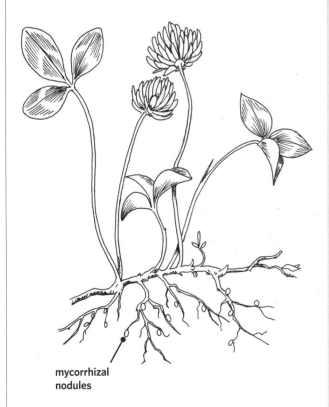

mycorrhizal
nodules

Forming nodules on the roots of legumes like white clover, mycorrhizal bacteria help the plant move nitrogen from air to soil to roots.

ALGAE

Algae, usually associated in the human mind with green coatings on ponds and lakes, are another numerous and important inhabitant of the soil, photosynthesizing the sun's energy into useful compounds, fixing some nitrogen, improving a soil's water-holding capacity, and exuding sticky organic matter that holds soil particles together and improves tilth.

Worms

There's much more to soil life than microbes. Worms come in several different varieties, beginning with the miniscule roundworms and superabundant nematodes, the various species of which consume plant matter as well as fungi and other organisms. (On the other hand, there are many dozens of species of fungi that eat nematodes.) Then there are the familiar earthworms and their tiny partners in plant debris consumption, the pot worms. Pot worms favor acid soils, while earthworms prefer something more neutral and with more calcium. Earthworms are the ones farmers especially want to encourage, since they are big enough and active enough to physically move and mix soil as they push and eat their way through. The plant debris ingested by an earthworm comes out the other end much richer for the experience, concentrated into castings rich in calcium, nitrogen, phosphorus, and potassium. Castings can be seen on the soil surface around the entrance to earthworm burrows; a half dozen of these dirt-colored globs might cover a dime.

Because earthworms are easy to see and study, scientists know more about them than most other life-forms in the soil. Earthworm burrows greatly improve the soil's ability to absorb water, and those that burrow deeply will return leached nutrients to the top few inches of soil, where grateful plant roots can make use of them. Earthworms prefer to be undisturbed by tillage. They obtain most of their food on the soil surface, so tillage will bury their food source and destroy their burrows. Earthworms, as you'd expect, are most numerous in pastures and hayfields, and least numerous in row crops.

Other Soil Creatures

Beyond worms, there are mites, springtails, rotifers, and tardigrades on the small side, and then the somewhat larger snails, slugs, centipedes, millipedes, spiders, pseudoscorpions, endless species of beetles, ants, wood lice, earwigs, and many others that make their livings in and on the soil. Innumerable other insects and bugs spend at least part of their lives in the soil or depend on soil organisms for food, including species of moths, butterflies, flies, bees, wasps (there's nothing like running over a nest of ground wasps with a brush mower for learning how many of them live in the soil), and who knows how many others.

Toads, snakes, mice, voles, gophers, woodchucks, chipmunks, badgers, coyotes, foxes, and other vertebrates make their burrows in the soil. Turtles lay their eggs in sandy areas, and birds work the soil surface for seeds and bugs. Skunks are great diggers, and moles function as subsurface plows. All serve a function in the ecosystem, even if their purpose is not always evident or entirely beneficial to humans. The organic farmer operates on the assumption that each life-form should be left unharmed unless it is demonstrably causing damage to crops or livestock. In other words, a life-form is innocent until proven guilty, and so should be allowed to occupy its natural place in the ecosystem.

The mineral, organic, and living components of the soil combine to create not only their own intricate ecosystem, but also the base of the larger ecosystem that encompasses every other terrestrial life-form. The farmer extracts food from the soil just as every other organism does, but since the human impact on soil can be so huge, the organic farmer carries the responsibility of minimizing this impact in order to preserve all pieces of the system on which we all depend. As the conservationist Aldo Leopold once famously remarked, "To keep every cog and wheel is the first precaution of intelligent tinkering."

Harvests Deplete Soils

Harvesting crops and forages decreases soil fertility. Plants take up nutrients from the soil in order to grow; if the crop is removed, the nutrients are as well. If those nutrients aren't replaced, then soil fertility — which is simply the soil's ability to provide nourishment to plants — steadily declines. Organic farmers replace the nutrients they harvest with organic matter from compost, manures, crop residues, cover crops, and, when necessary, purchased soil amendments.

Nutrient removal from cropping also acidifies soils, first by removing positively charged nutrients, which are replaced with negatively charged, or acidic, hydrogen ions, and second by removing minerals that directly increase the alkalinity of the soil, most importantly calcium and magnesium. Unless those specific nutrients are replaced and the hydrogen ions buffered with sufficient organic matter, the soil grows steadily more acidic as a result of cropping, and yields drop.

When vegetation is not removed annually by humans, soil ecosystems and fertility are more self-sustaining, but they differ according to the plant types. The original vegetation covering U.S. soils is still reflected in the soils themselves. The eastern half of the country was originally forested. Conifer forests make light, acid soils, while hardwoods that drop their leaves each autumn contribute more organic matter to the soil, though it is slow to move deeper than the topmost few inches. Prairie soils are the richest, not only because of their silty parent materials, but also because grasses have more root mass than any other plants. Each fall the grasses die back and so do the roots, and all those dead roots turn into organic matter. This is why prairie topsoil was originally so phenomenally deep, and this is why grasses and other forages are probably the finest soil conditioners of all plants.

Assessing Your Soil

The first step to achieving and maintaining healthy soil is figuring out exactly what's going on beneath your feet, so you can start thinking about how to fine-tune cultivation practices to fit your soil. There are four good tools to use to assess your soil's production capability:

- Soil maps for your farm
- Yield averages for your area
- Simple field tests for soil characteristics
- Laboratory soil testing for pH and nutrient levels

Soil Maps

Soil maps are available for everywhere in the U.S. from agricultural extension offices and the Natural Resources Conservation Service (see the resources). These maps label the soils quite specifically, for both type and slope. The lists that come with a soil map should describe each type and place it in its capability class.

Agricultural soil capability classes are numbered from I to VIII. The higher the number, the less suitable the soil is for agriculture. Class I soils are best suited, while Class III soils will have some moderately severe limitations, such as steep slopes or poor drainage. Class IV soils are not suitable for row crops, and Class V through VII soils should be limited to pasture and woods. Class VIII can't even be used for timber or grazing and should be left alone.

Yield Averages

Your county extension office will also have many years of yield averages for the crops in your area. Even if your neighbors aren't organic, the averages will give you a ballpark notion of what your fields are capable of producing given the natural fertility of the soils in your area. Those crops that are commonly raised in your area should do well on your farm, too, but don't limit your thinking to a small palette of possibilities. Just because all the neighbors are raising corn doesn't mean you shouldn't plant apples.

Field Tests

The third way to assess your soil is to take a shovel and a pint of water out into the field. First, observe. Does the soil have a crust on its surface? Unless your area is experiencing a drought, that's usually a sign of poor tilth, or structure: how easily the soil can be worked, and how easily roots can penetrate and spread to find the water and nutrients they need.

Pour the water onto the soil and see how long it takes to sink in. If it pools on the surface and only sinks in slowly, that is not a good sign. Slow absorption indicates compacted or poorly structured soil.

With the shovel, slice the soil a foot (30 cm) or more deep, and remove enough dirt so you can see the side of the hole clearly. Smell the dirt: a pleasant earthy smell indicates biological activity. Old-timers used to taste the soil. A bitter taste meant alkaline, a sour taste meant acid, and a sweet taste was good.

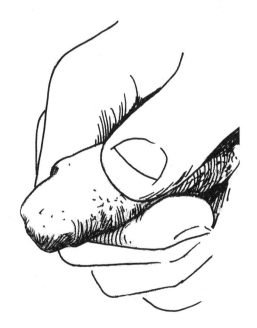

The ribbon test: Take a handful of moist dirt and squeeze it out between your fingers. If you can't make it ribbon, you have sand. If it ribbons and breaks after it's an inch (2.5 cm) or so long, you have loam. If the ribbon grows longer, the soil has a lot of clay in it.

Take up a handful of moist soil and squeeze it into a ball. If it falls apart when you open your hand, you have sand. If it doesn't, squeeze a ribbon of soil out between your thumb and forefinger. If the soil won't ribbon, it's loamy sand. If the ribbon breaks before it's an inch (2.5 cm) long, you have loam. If it gets longer than that before breaking, there's clay in the soil, and the longer the ribbon, the more clay there is.

Measure the depth of the topsoil in your hole. The deeper it is, the better. Shallow topsoils, less than, say, 6 inches (15 cm), call for serious attention to soil building in order to expand the zone of nutrient-rich dirt, which in turn expands the growth potential for crops. Shallow topsoils should be tilled particularly carefully, so as not to turn them over so far that they are buried beneath subsoil, beyond the reach of many plant roots and enough oxygen to maintain soil life.

Take a close look at the plant roots in the hole, if necessary digging down along a plant stem to get a better observation. Healthy roots are heavily branched and loaded with fine root hairs. Roots that grow downward and then turn sideways indicate a hardpan, and a lack of fine root hairs indicates oxygen deprivation. Stick a long nail or other slender metal rod into the soil. If it slides in easily that means the soil is soft and permeable;

if it is difficult to push in, or difficult at certain points, compaction is a problem.

Hardpan is an impermeable layer under the topsoil that restricts the root zone, limiting plant growth. It will also restrict drainage, which could create waterlogging, another drag on production. If the layer is due to bedrock, there isn't much you can do except create more topsoil over the seasons with large additions of organic matter. But sometimes the impermeable layer has formed due to plowing always to the same depth or too much equipment traffic (a plow pan), or it is simply a layer of clay in the subsoil. In these cases, subsoiling with a special plow can break up the pan, though it's an expensive procedure. And if you don't change the conditions that created the pan, you'll be back where you started in a few years.

Laboratory Tests

The gold standard for assessing soil fertility is the soil test. This involves taking soil samples and sending them to a soil laboratory, where they are analyzed for various nutrients, pH, organic matter, and cation exchange capacity (a measure of how well a soil holds nutrients). Soil testing is not mandated by the National Organic Rule, but it's fairly standard among both organic and nonorganic

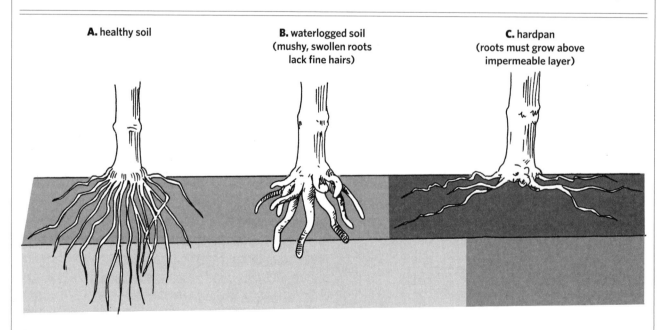

A. healthy soil

B. waterlogged soil (mushy, swollen roots lack fine hairs)

C. hardpan (roots must grow above impermeable layer)

In good soil (A) the roots of a plant reach deep and far. But a hardpan (B) forces roots to grow sideways along the impermeable layer, while a wet soil (C) suffocates roots, so that the fine hairs that it needs to extract nutrients for growth wither and die.

producers to test regularly — every one to three years — in order to monitor soil nutrients and organic matter. Soils, and farm systems in general, are much easier to manage effectively if you're monitoring them regularly.

Agricultural extension services generally offer low-cost soil testing to growers, using state labs that know your region's soils. The results from these labs are fine, but the fertilizer recommendations assume conventional farming methods and generally aren't appropriate for organic farms.

Fortunately there are now many private soil testing laboratories that focus on sustainable and organic growers and can provide results and recommendations based on both soil chemistry and indicators of soil biological activity. ATTRA — National Sustainable Agriculture Information Service is a good resource for locating such laboratories (see the resources).

Taking a Soil Sample

First, find a soil lab that is accustomed to working with organic farmers, and check the website or call the lab to get its preferences about depth of samples, the amount of soil you should send, and any other special recommendations regarding how to take, package, or mail samples.

Second, decide what fields or areas you want to sample. Don't mix samples from different soil types, different field histories, or different topography (for, example, don't mix a hilltop sample with one from the very bottom of the slope). Problem areas should also be sampled separately. The goal is to take a sample that is characteristic of the entire area to be tested.

Third, take the samples. In fields, one sample per acre or two (or a half to a whole hectare) is generally recommended. The simplest method is to borrow a sampling probe or auger from your local extension office. Brush aside any plants or residue on the soil surface, stick the probe in 6 inches deep (15 cm), or whatever your lab recommends, pull it out, and put the sample in a bucket. If you can't get a probe, take a square-ended shovel, mark a 6-inch (15 cm) depth on the blade, then carefully cut holes and pull samples. Once you have all the samples you want for a single test, mix the dirt well and put two cups (or whatever your lab recommends)

in a ziplock plastic baggie or other container that won't burst or open in transit. For example, if you have a 10-acre (4 ha) field that is uniform in soil type, you'll want to pull about ten samples and mix them to create the one sample you send to the lab for that field.

Label the baggie, the lab's record sheet, and your own record sheet with the date and the field or area, and give it an identification number (the lab will use these numbers on the test result sheets).

In smaller areas such as a garden or lawn, it's still a good idea to pull several samples and mix them to get a composite to send to the lab.

Most growers test their soils every two or three years. In the first years, as you get your soil building program established, you may want to sample annually. Growers who have been in the business many years and know their soils well may sample less frequently.

Reading the Results

Soil test results should arrive in a week or two. The soil test results will tell you whether nutrient levels are high or low and supply fertilizer recommendations for the crop to be grown in that field.

soil probe

Most agricultural extension offices will lend you a soil probe for taking soil samples for testing. As you push the probe into the ground, it collects soil in its hollow core.

Fertilizer recommendations from a conventional soil lab will not be in compliance with organic regulations and will vary dramatically from those of an alternative lab.

To build an accurate record of your soils, it's extremely important to sample about the same time of year each time you test, and to use the same laboratory. Spring is a busy time for most conventional labs, while many alternative labs prefer that their organic growers sample in the fall. Since different labs use different methods to get their results, the results will vary significantly between labs on the same sample. In fact, I'd often heard this rumor, so this summer I sent identical sets of six samples to two different soil labs and found that it is true. Numbers for organic matter percentage, pH, cation exchange capacity, and all nutrients except phosphorus were significantly different. This greatly affects recommendations for soil amendments, so it's important to find and stick with a lab that understands organic growers.

Unless the soil conditions and test numbers are very far from optimal, you will be able to grow a crop. The yield will not be as high as it could be, but you will at least be harvesting something (in most cases) while you design and implement your soil program, and figure out how to obtain those nutrients (most often, calcium, sulfur, and the very necessary but very small amounts of trace nutrients) that can't be generated on your farm.

Soil Husbandry

Science and research are wonderful for telling us how things ought to be, but when you're standing in the field looking at a crop, the more immediate issue is figuring out whether that field is as productive as it should be, and if not, what is holding it back. At those moments it can be hard to see how to apply what you know. And even once you've diagnosed any real problem you are having, inevitably the solution will not be as simple as it is in the books, since the real world works with a budget and other constraints. But that's farming. Especially, that's organic farming.

Keep in mind that building soil fertility takes years, so don't think you have to do it all yesterday. Just remember that with soils, there are four general practices you should be doing on an ongoing basis, fine-tuning them over time to fit the unique conditions of your farm system. Those things are:

- Continually add organic matter in the form of manure, compost, crop residues, or green manures.
- Maintain soil cover.
- Purchase and replace those nutrients removed by crops that are not available by recycling nutrients on the farm.
- Rotate crops (unless your fields are in permanent pasture, hay, orchards, or vineyards).

We'll talk about rotating crops in chapter 5; for now, just note that it is an essential component of a soil-building program. We'll discuss the first three components here.

High Salt Levels

If a soil test indicates that a soil has elevated salt levels, the problem may be a water table that is too high. This is most common in irrigated areas with poor drainage. The water is put on faster than it drains off, moving the water table higher. A high water table will not only restrict root growth and limit production but also can bring up salts from the subsoil layer, which kills the soil and poisons plants. The solution is better drainage, either by breaking up an underlying hardpan, improving the texture of an impermeable topsoil, or installing subsurface drainage.

Once the drainage problem is fixed, the soil has to be flooded with water until the salt is leached away. If the soil isn't yet dead from salt, sometimes simply reducing the amount of irrigation water applied will prevent further damage. Trickle irrigation, which slowly drips water into the root zone of the crop plants and nowhere else, is one method for doing this.

Four Rules for Maintaining Soil Health

Rule 1: *Continually add organic matter. Apply compost, plow down green manure, or, as shown here, spread manure.*

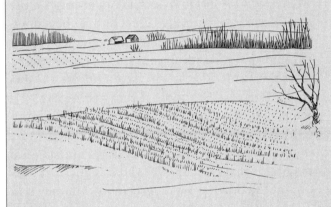

Rule 2: *Maintain soil cover. Except for brief periods when you are planting or doing weed control, the soil should be covered by crops, crop residue, or mulch to prevent erosion.*

Rule 3: *Use purchased amendments to replace nutrients you can't recycle on the farm. Test to find out what's missing, then apply it. On a field scale this is done with a tractor and fertilizer spreader.*

year 1: corn **year 2: soybeans** **year 3: small grain** **years 4 and 5: forage**

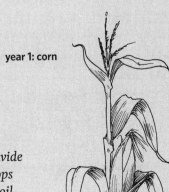

Rule 4: *Rotate crops. Divide the rotation between crops that are demanding of soil nutrients and those that return nutrients to the soil. For example:*

Adding Organic Matter

The first concern of every soil-building program is how to add organic matter on an ongoing basis. Without organic matter it doesn't matter how many amendments you spread or what tillage practices you use; your soil still won't be able to hold nutrients, water, and air to feed soil life and grow crops. With enough organic matter in the soil, nutrients will stay put until plants can use them, and the soil will hold water without excluding air. Plus, more organic matter tends to mean less need for expensive soil amendments to correct nutrient deficiencies.

The organic farmer obtains organic matter from some combination of animal manures, green manures, crop residues, and compost. Roots left in the soil after a harvest also add to organic matter.

Compost

Organic growers wax poetic about composting: how the process turns leftovers, discards and excess, into a sweet-smelling, soil-mellowing magic amendment that creates beautiful tilth and fertile dirt. Pure organic matter loaded with a palette of essential soil nutrients, compost has all the advantages of the things it can be made from — manure, plant residues, kitchen scraps — with none of the drawbacks of smell or difficulty of application.

Compost is the heart of the organic garden, and larger-scale operations have gotten on the compost bandwagon, too. Over the past couple of decades many organic farms have developed and refined methods to produce excellent compost on a big enough scale for field applications and commercial sales. This means the current state of composting in the U.S. ranges from casual heaps next to the garage or in the bushes to closely monitored, carefully built and tended windrows on concrete pads that need big machinery to turn them over.

Compost at its simplest is the process of piling up organic wastes — manure, kitchen scraps, plant matter — in a heap. Soil organisms then do what they do so well: that is, they decompose the pile from a heterogeneous mess into homogenous, dark-colored organic matter. The better job you do of building and tending your pile, the more nutritious the compost you will have, and the quicker you will have it. But though composting can be a science, it doesn't have to be. Carefully built and tended compost piles can produce good compost in two weeks. More casually built and tended piles may require a year or two, and a higher proportion of the nutrients will leach away, but they still

Large-scale composting is done in long windrows, where the compost can be turned and moved with machinery.

produce compost. So suit your composting style to your inclinations and needs.

COMPOST BASICS

Create the right conditions and compost happens all by itself. What are the right conditions? Critical mass (enough material piled up), adequate moisture (neither dry nor saturated), and carbon-rich and nitrogen-rich ingredients.

Compost results from the decomposition of organic matter by bacteria. This is a chemical process that requires carbon, nitrogen, and water. When building a compost pile, the most important thing to remember is to add both carbon and nitrogen sources, more easily remembered as dry brown stuff (carbon) and moist green stuff or animal manure (nitrogen). Carbon, a part of every organic molecule, is necessary to complete the chemical equations that add up to decomposition. Nitrogen is essential energy for bacteria to process the ingredients into the final product.

The bacteria also need enough moisture to power the chemical reactions, but not so much water that air can't get into the pile as well. Aerobic bacteria, which depend on an air supply for existence, are more efficient and smell a lot better than anaerobic bacteria. If there is no air in the compost pile, then anaerobic bacteria will take over. They don't need air, but they work much more slowly and don't do such a complete job of decomposition, and they stink.

As bacteria and other tiny organisms begin busily decomposing the food supply you just gave them, the pile heats up from all the activity. If it feels warm, even hot, when you stick your hand in a week later, then you've built a good pile.

BUILDING THE PILE

You can either build your pile casually, piling on stuff as it becomes available, or build it carefully in a single afternoon, assembling materials and paying attention to the proportions of carbon-rich versus nitrogen-rich components.

Compost piles made solely of nitrogen-poor materials such as sawdust or dry leaves will decompose very, very slowly. Compost piles made solely of nitrogen-rich materials such as kitchen scraps and manure will be too moist; they will not decompose properly and will smell.

As a rule of thumb, compost piles should be three parts dry, high-carbon materials to one part green or fresh high-nitrogen materials. Layers of carbon-rich materials should be alternated with layers of nitrogen-rich materials. So, if you put down 6 inches (15 cm) or so of dry leaves or old hay, cover them with a couple inches of manure, fresh green plant matter, or similar nitrogen-rich material.

The classic Indore method of building a compost pile, developed by Englishman Sir Albert Howard, one of the founding fathers of organic farming, used the following recipe: 6 inches (15 cm) of plant matter (old hay, leaves, coarse sawdust, vegetable matter from the garden) covered with 2 inches (5 cm) of manure and animal bedding, covered with ⅛ inch (3 mm) of soil and a sprinkling of mineral — either lime, wood ash, granite dust, or similar amendment. The soil adds an extra supply of bacteria, so that decomposition starts quickly, and the mineral adds nutrients. These layers are repeated until the pile reaches the desired height. In building your pile, it's a good idea to preserve this general ratio of carbon-rich to nitrogen-rich materials.

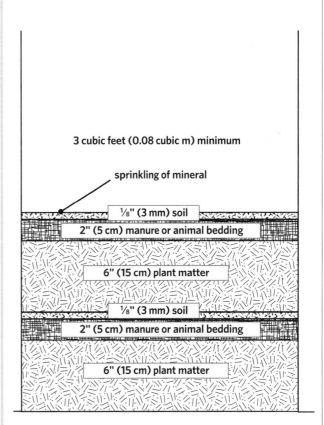

3 cubic feet (0.08 cubic m) minimum

sprinkling of mineral

⅛" (3 mm) soil

2" (5 cm) manure or animal bedding

6" (15 cm) plant matter

⅛" (3 mm) soil

2" (5 cm) manure or animal bedding

6" (15 cm) plant matter

The Indore method of composting is a little easier to start with than the precise ratios mandated by the Final Rule.

You can make the pile whatever size is convenient, but a minimum of 3 cubic feet (0.08 cubic m) ensures that the pile will have enough biological heat and activity to destroy pathogens (disease-causing organisms). A pile of this size or larger will also best conserve nutrients. Since smaller piles have a greater surface area in proportion to total mass, there is more leaching from rainfall, and more nutrients gassing off into the air.

If you are really in a hurry and want to speed the process, you can chop or shred materials before adding them to your compost pile. For example, you could run over a pile of leaves, straw, or similar materials with a lawnmower before adding them to the pile. But it's not necessary: given enough time, all organic matter will decompose.

TENDING THE PILE

The more often a compost pile is turned, the more quickly and completely it will be transformed into usable compost. Turning compost keeps aerobic bacteria well supplied with air, keeping anaerobic bacteria from taking over. It also mixes the ingredients for more even decomposition. But it's not necessary. You can turn the pile never, occasionally, or every few weeks or even days — whatever suits your needs and schedule.

Piles that are too large and not ventilated can become anaerobic, a condition diagnosed by its bad smell. Builders of larger piles may want to insert ventilation pipes every few feet to ensure an adequate air supply for the bacteria.

The Final Rule on:
Compost

If you intend to be organically certified or claim your products as organically produced, you will need to follow the specific composting procedures mandated by the Final Rule. It should go without saying that no synthetic substances may be used in the organic compost pile. Piles in a container or static location must maintain an internal temperature of between 131 and 170° F (50–76° C) for at least three days, while windrowed piles (piles made in long rows) must maintain that temperature for at least 15 days and be turned no less than five times during that period.

The Final Rule also requires that compost piles have an initial carbon-to-nitrogen ratio of between 25:1 and 40:1. This requires a little more calculation than simply alternating green stuff with brown stuff. It's necessary to look up the carbon/nitrogen composition of each individual ingredient (ratios for some common ingredients are listed below). If you're not keeping track of carbon-to-nitrogen ratios when building your compost pile, it's still simple to comply with the Final Rule: just apply the compost according to the guidelines for raw manure (see page 79).

Carbon-to-Nitrogen Ratios of Common Compost Ingredients

Materials from animal sources and green plants have a higher nitrogen content than those from dry or woody vegetation. In general, the greener or fresher the material, the higher its nitrogen content.

Material	Carbon:Nitrogen Ratio
High in Carbon	
hay	15–30:1
dry leaves	30–80:1
straw	40–150:1
corn stalks	60–70:1
wood chips	100–500:1
sawdust	200–700:1
High in Nitrogen	
poultry manure	5–15:1
dairy manure	5–25:1
hog manure	10–20:1
vegetable wastes	10–20:1
grass clippings	15–25:1
coffee grounds	20:1

Source: *The Art and Science of Composting,* by Leslie Cooperband, University of Wisconsin-Madison Center for Integrated Agricultural Systems (March 2002).

The bacteria also need adequate moisture. If the pile is either too wet or too dry it will slow or stop the process. Piles can be watered with a hose as needed. In dry areas piles can be covered to conserve moisture. In wet areas they can be built with a curved top to encourage excess rain to run off.

USING COMPOST

Unfinished compost can be applied in the fall to finish decomposing in the soil. Finished compost is best applied in early spring for maximum conservation of nutrients; it can also be tucked in around a growing crop as a mulch. One to 3 inches (2.5–7.5 cm) a year is good, though there's usually no harm in applying more.

COMPOST TEA

Many gardeners also make compost tea for a nourishing method of watering plants. Simply fill a bucket or watering can half full of compost, then add water till it's full. Stir and pour. You can let it settle for a while first if you like so you're pouring more water and less chunks of compost. The compost can be reused several times, and then dug into the garden.

COMPOST BINS

Compost bins are neater looking than a simple pile and make it easier to control moisture, temperature, and rodents. Bins can be built out of anything handy, from old pallets to hoops of woven wire to fancy plastic contraptions sold in upscale

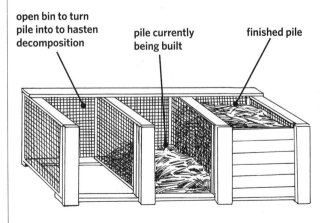

open bin to turn pile into to hasten decomposition

pile currently being built

finished pile

Many gardeners utilize a system of three side-by-side bins: a finished pile in one bin, the pile you're building in the next bin, and the third bin to put a pile into when you turn a pile.

Field Grazing

Cattle, sheep, goats, horses, and chickens are one of the best tools for land restoration when correctly managed. Grazing livestock in the field apply manure and urine directly to the soil, saving the producer the equipment and labor of moving it out there from a barn or feedlot. This also saves the labor of feeding the animals, since they're doing it themselves. However, overgrazing can cause soil compaction and reduce productivity. For this reason, management-intensive grazing (chapter 9) is recommended if you are raising livestock.

Pigs, in addition to having the same benefits as other livestock, are also rooters, so they will dig up and turn over large quantities of soil. They can be useful for weed control and soil aeration and preparation, but they can also make a terrific mess of a field, so equipment should be available to smooth out the pigs' plowing job.

gardening catalogs. Just make sure you don't use any treated wood, which has been shown to leach chemicals and is prohibited by the Final Rule. Generally you want a bin that can be easily removed from the pile, or have a side removed, for ease of turning.

Manure

Manure is one of the best sources of nitrogen on an organic farm, especially when it can be worked into the soil while still fresh. Since the Final Rule restricts when fresh manure may be applied to crops intended for human consumption (see page 79), such an opportunity occurs only a couple times a year in real-life conditions, before spring planting or when doing after-harvest fall tillage. Thus the problem is how to store the manure so all the valuable nitrogen doesn't off-gas or leach away.

Ideally, manure should be composted with a suitable amount of high-carbon material on a cement pad under cover to preserve as much as possible of the nutrients it contains. If this isn't possible, at least pile the manure neatly, mixed with bedding and any other plant materials you have available, so it begins to compost while waiting to be spread. Turning this pile frequently will speed the composting process, further preserve nutrients, and make it smell better. (This is generally too big a job to do with a shovel; most farmers who compost bigger quantities of manure use a skid steer or similar machine to turn the pile.)

Composted manure has the added advantage that it won't burn plants the way fresh manure will. There's a lot of loose nitrogen in most fresh manure that can turn plants brown; composting binds this nitrogen so it's available but not harmful.

Crop Residues and Green Manures

A green manure is a crop that is not harvested but plowed under while still young and tender. Crop residues are what is left above the ground in the field after you have removed the salable or edible portion of the crop, such as corn stubble. Maintaining high levels of organic matter in the soil by tilling in this extra vegetation not only

Cover crops used as green manures are plowed under while they are still young, so they deliver maximum nutrients to the soil.

adds nutrients to the soil but also increases the soil's water-holding capacity, decreasing nutrient leaching. Some growers will set aside a section of garden or field and plant and plow down a green manure two or three times on the same ground during the growing season to accelerate soil building. Different green manures and residues will add different types of nutrients and different proportions of organic matter to the soil, so that organic growers typically fine-tune their green manure use to complement the crops they grow, since different crops remove different amounts and types of nutrients.

Many green manures serve double-duty as cover crops; see the list on page 80.

Maintaining Soil Cover

Keeping soil covered does all sorts of good things: it slows or stops erosion, keeps the soil cooler, moister, and more permeable, and adds to organic matter as well.

Soil loss from erosion is a tremendous problem in agricultural areas, both flat and hilly, and is the primary reason why organic growers insist on keeping their soils covered. Hills make great scenery but complicate farming. Since water runs downhill, and the faster it runs the more soil it can carry with it, the length and degree of slope affect a soil's tendency to erode. The next time you're out for a drive in the early spring, notice the soil color in the fields that are waiting to be planted. Nearly always, the knobs and hilltops have lighter-colored soil than the bottoms. Color is a good indicator of fertility; the variation shows how fertility runs downhill with water.

But flat ground is not immune from erosion either. One little lick of wind strong enough to move a soil particle, which will bounce along and dislodge more particles, is the beginning of wind erosion. The end result can be a dust storm, and less topsoil on the farm.

The best way to deal with soil erosion is to stop it before it begins. Anything that will absorb the force of impact of rain or wind will halt this process. A solid vegetative cover is excellent, while a good mulch of dead plant material will also do the job.

The Final Rule on: Soil Amendments

Certified organic farmers must follow the strict guidelines of the Final Rule regarding the type of soil amendments they can use. In particular, they may *not*:

- Use sewage sludge, anywhere, ever.

- Use any synthetic substances not on the Final Rule's list of approved substances.

- Use any fertilizer or composted material that contains synthetic materials not on the national list of approved synthetic substances.

- Use any amendment in such a way that might contaminate soil, crops, or water with excess nutrients, pathogenic organisms, heavy metals, or residues of prohibited substances.

The USDA's National List of Allowed and Prohibited Substances specifies the synthetic materials that organic farmers are allowed to use under the Final Rule. If a synthetic material is not on the list of approved substances, then it's forbidden.

Soil amendments from natural sources (not made by humans) are allowed in certified organic production *only* if unchanged by heat or chemical processing from their naturally occurring form and *only* if unmixed with synthetic or otherwise prohibited additives, including preservatives, stabilizers, adjuvants, and so on. Any nonsynthetic additives that are not specifically prohibited by the national list are allowed.

Manure

On crops being grown for human consumption, if the edible part of the crop will have direct contact with the soil (which is the case for most vegetables), raw animal manure can be applied directly to the land *only* if it is incorporated into the soil at least 120 days before the harvest. If the edible part will not have direct contact with the soil (for example, in an orchard or vineyard), raw manure may be used if it is incorporated into the soil *at least* 90 days before harvest. However, if there's any chance the manure may splash the plant while being spread, the 120-day rule applies.

Raw manures can be incorporated into the soil for crops not intended for human consumption, such as forages (hay, silage), small grains, and row crops grown for animal feed. Manure is usually applied before tillage or, in the case of hayfields, just after a cutting has been removed, in the fall after the final cutting, or in the spring well ahead of the first cutting.

Manure must be applied in a manner such that it won't contaminate surface or groundwater. Manure should not be applied to frozen ground if there is any danger of runoff into surface waters. Many states have laws regarding manure being spread on frozen ground; check with your agricultural extension agent.

Composted manures can be used at any time, even until the day of harvest, if the composting process meets the standards set in the Final Rule.

Plant Materials

Uncomposted plant materials may be used at any time as soil amendments. Such materials include mulches, green manures, and cover crops.

Compost

Compost made in accordance with Final Rule organic regulations may be applied at any time.

When using amendments or fertilizers from off the farm, certified producers must have documentation of the source, showing that it was not produced using prohibited materials or processes. No matter what you decide to use, check first with your certifier to make doubly sure the formulation is allowed. You can't put the amendment back in the bag if you guess wrong, and even if it was an honest mistake it won't save your organic certification.

Keeping soil covered not only halts erosion, but has many additional benefits:

- It slows water percolation, thus minimizing nutrient leaching in areas with high rainfall.
- It slows the evaporation of soil moisture in dry areas.
- It keeps soils cooler in hot conditions.
- It minimizes the growth of weeds, a boon to any farmer.

In the market garden, where erosion is rarely an issue, these other benefits are more important. In everyday farming a soil cover's reduction of the mud factor in both field and garden is wonderful.

Cover Crops

When green-manure-type crops are used for purposes other than being plowed under while young, they're called by different names. *Cover crops* are planted to protect the soil from wind and water erosion in winter or other fallow times, while *catch crops* are chosen and planted specifically for their ability to retain nutrients, especially nitrogen, that otherwise might leach away during the winter or periods of heavy rains. Obviously catch crops serve as cover crops, too, though the species planted will vary depending on the grower's primary purpose. Both cover crops and catch crops can be planted either after the main crop has been harvested or shortly before the harvest, so that the ground is already covered when you take off the main crop.

Cover Crops

All legumes fix nitrogen in the soil; this is one of their primary benefits as green manures. Grains, grasses, and forbs also make good cover crops. All have different uses and benefits.

Legumes	Characteristics
Alfalfa	Grows everywhere. Deep roots bring up subsoil nutrients. Does best of all clovers in slightly alkaline conditions; tolerates dry weather very well but not wet soils. Perennial.
Clovers	Excellent soil builders; most are perennial in most areas.
• alsike clover	Used in the North; good in wet, acid areas. Short-lived perennial.
• bur clover	Used in the South. Annual.
• crimson clover	One of best winter annuals from New Jersey southward.
• ladino clover	A large variety of white clover, some northern growers use it for forage. Perennial.
• red clover	Used in northern and central states; should grow a full season before plow down. Perennial.
• sweet clover	Grows everywhere; yellow variety tolerates dry conditions better than white; both have deep taproot. Tallest of the clovers. Biennial.
• white clover	Grows everywhere; its low height makes it excellent as a living mulch; will tolerate a more acid soil than red clover. Perennial.
Cowpeas	Fast growing; does well in hot, dry weather. Used in southern and central states. Annual.
Field peas	Used in all areas; can be mixed with a forage crop spring planting if making haylage. Annual.
Hairy indigo	Used in the Deep South; highly resistant to root-knot nematode. Annual.
Lespedeza	40 species; also called bush clovers. Used in the South; good for restoring acid, eroded soil.
Lupine	More than 200 species; be sure to get one appropriate for the South. Used in the South; blue, white, and yellow varieties. Mostly perennials.
Vetch	Annual. Hairy vetch grows in all areas; since it likes cool weather it is a good winter cover crop in the South. Susceptible to root-knot nematodes. In the North can be planted the latest of any legume, up till mid-October. Plant with a nurse crop, since it's a slow starter. Hungarian vetch does better in heavy soils. Purple vetch is not winter-hardy; used on Pacific and Gulf coasts where it grows all winter. Annual.

Low-growing crops like white clover that are planted shortly after the main crop has emerged from the soil for the purpose of smothering weeds are called *living mulches*. These are often used in row crops or vegetable gardens to minimize bare soil during the growing season.

Quick-sprouting species planted at the same time as the main crop to discourage weeds and keep soil covered while the slower-sprouting main crop can gain a foothold are called *nurse crops*. Oats are often planted as a nurse crop with alfalfa, since the oats will suppress weeds without affecting the growth of alfalfa.

Leaving soils bare over winter, a common practice in conventional farming, allows wind and water to erode the soil. It also allows frost to penetrate deeper, increasing the earthworm kill. To protect soil over the winter, leave crop residues in place, or plant a cover crop in the fall that will sprout and then die once cold temperatures hit. In the spring either plow the crop residue under or plant seed directly into this dead mulch.

Mulches

A mulch, simply defined, is anything that covers the soil but is not a living crop. In this section, and most commonly, it refers to using dead organic matter, like straw, leaves, or old hay to cover the soil between crop rows or during the fallow season. During the growing season, many growers use mulch instead of cover crops to cover soil, conserve moisture, and suppress weeds. Mulching is

Small Grains	Characteristics
Barley	Prefers a pH of 7 to 8; use spring varieties in the North. Annual.
Buckwheat	Excellent at suppressing weeds and improving soil structure; can go from seed to seed in 8 weeks. Grows everywhere. Annual.
Oats	Does well almost anywhere except heavy clay. Use spring variety in the North. Often used as a nurse crop for alfalfa. Annual.
Pearl millet	Warm-season crop; good for smothering weeds. Annual.
Wheat	Prefers pH of 7 to 8.5 and fertile soil. Winter wheat is extremely cold hardy; will cover soil through the winter and produce a crop the following summer if not tilled under. Annual.
Winter rye	The only cover crop that can be seeded after a harvest and still have time to establish in cold areas, so very important cover crop for northern growers. Annual.

Grasses/Forbs	
Mustard	Sick-soil cleaner (stimulates soil microorganisms). Annual.
rape	Sick-soil cleaner (stimulates soil microorganisms). Annual.
Ryegrass	Not the same as rye! Two varieties: annual and perennial. Annual ryegrass is an excellent winter cover crop, growing rapidly, dying over the winter, and easy to till under in spring.
Smooth brome grass	Good winter cover crop in the North. Perennial.
Sudan grass	Grows well in hot weather; used as alternative livestock forage by many farmers. Perennial.

Source: *Cover Crop Gardening*, Storey Publishing, 1977.

often easier than getting a cover crop established while the food crop is growing, and an organic mulch can be tilled in or composted at the end of the season to add to soil organic matter. It does not, however, have the high nutrient levels of cover crops.

Plastic mulch, commonly used in vegetable production because it's so quick to put on, is allowed by the Final Rule only if it is completely removed from the field at the end of the growing season.

Other Erosion Control Methods

Mulches and cover crops are not always enough to prevent erosion problems. Sometimes the soil must be bare for a short period of time, since to grow some fine-seeded crops it's necessary to have a fine-textured, bare seedbed, which means the soil is going to be all or mostly bare and open to erosion until the crop has grown high enough to shade it completely. Even after the crop's canopy of leaves has completely shaded the ground, there is still erosion potential in a heavy rain if there is bare dirt between the rows. And in winter, if a cover crop or stubble does not completely cover the soil, wind and spring snowmelt can cause problems.

For these reasons, most farmers employ some additional practices to minimize erosion. Starting at the soil level, having good strong soil aggregates as a result of high organic matter levels and active soil life makes it difficult for wind and water to detach and remove soil particles. A well-aggregated soil absorbs water quickly, so there is less to run off, and such a soil is also not likely to crust in response to a hard rain or a drought.

Windbreaks both slow wind erosion and preserve soil moisture, since wind can be even more drying than sunlight. Windbreaks come in several sizes, beginning at ground level. Small soil ridges can be built with specialized tillage equipment, dramatically dropping wind speeds at ground level. At the field level, trees and brush can be planted along field edges perpendicular to the direction of the prevailing wind, to slow wind erosion. While you're waiting for the trees to grow, rows of short crops can be alternated with rows of tall growing crops to slow wind speed as well.

On sloped land, contour cropping, or running crop rows perpendicular to the direction of slope, will greatly slow soil and water loss from runoff. Some farmers still run their crop rows straight up a hill; the open dirt between those rows offers perfect little canals for accelerating water erosion. On steeper slopes it is now usual to alternate row crops with small grains or forages to slow runoff even more. (See chapter 5 for more discussion of these and other erosion control measures.)

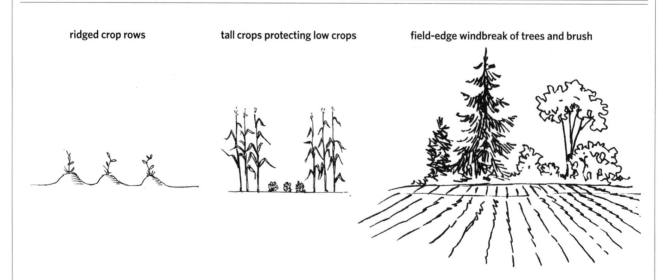

ridged crop rows   tall crops protecting low crops   field-edge windbreak of trees and brush

Windbreaks slow water evaporation and soil erosion. Miniature windbreaks can be formed at soil level by ridging crop rows, larger windbreaks by using tall crops to shelter low crops, and field-scale windbreaks by planting trees and shrubs along field edges. These planted windbreaks are also excellent wildlife habitat.

Using Purchased Soil Amendments

Organic farmers don't necessarily agree on the need for purchased soil amendments. But if your soil test indicates a real deficiency and you can't fix it with on-farm sources such as manures, green manures, or composts, then you'll be in the market for an appropriate purchased source. Finding a supplier of reasonably priced, organic soil amendments who is close enough to your farm to prevent outrageous shipping charges is not always easy. A soil lab that offers advisory services or a soils consultant who is familiar with organic methods and regulations can be extremely useful in sourcing recommended amendments. This is especially useful for field crops, where the large scale demands amendments by the ton rather than by the 50-pound (23 kg) sack that can be purchased at the garden store. If you can't find help in your neighborhood, the Web site of the Organic Materials Review Institute (OMRI) lists suppliers of certified organic amendments.

A word of warning: The USDA's labeling rules are different for fertilizer and potting soil than they are for food. The word "organic" on a bag of fertilizer does *not* mean the product is certifiably organic — it only means that it contains carbon! Unfortunately, some manufacturers have taken advantage of this by labeling their garden products as organic. Don't be fooled. A fertilizer or soil mix is only organic if it is labeled "approved for organic production" or carries the OMRI seal.

The following products may sound like they're "organic" but are specifically prohibited for use as fertilizer or amendments:

- All types of ammonia products including urea
- Gelatin
- Human excrement
- Kiln dust
- Leather by-products
- Lime products that are hydrated, slaked, or burnt (including quicklime)
- Nicotine and tobacco in any form
- Slag
- Sewage sludge

Determining Fertilizer Application Rates

Manure is the primary fertilizer used in organic production. But phosphorus runoff due to manure is a major water pollutant in agricultural areas. Because some manures (especially poultry manure) are very high in phosphorus, the phosphorus level of manure is often used as a limiting factor in determining the manure application rate, so that this nutrient doesn't build up to excessive levels in the soil. In fact, some states and localities require farmers to have a nutrient management plan that uses phosphorus levels in manure to calculate application rates. If the phosphorus level of manure isn't known, a sample can be tested in order to correctly calculate how much to apply over a given area.

To determine how much manure or other amendments you should add to the soil, begin with the results of your soil test, which will tell you how much nitrogen, phosphorus, potassium, and other nutrients you need to grow your planned crop. The results of the manure analysis (or an estimate — most cooperative extension services can offer you a table that gives the average nutrient content for various types of manures) will tell you how much of these nutrients are present in the weight of the manure you plan to spread. Then you calculate how much manure is needed over the area to reach the desired total, usually of nitrogen, since in crop production it's nitrogen shortages that generally show up first and have the greatest effect on yield. You also calculate how much phosphorus that total will yield, for the reasons given above, to see if it is going to keep you from putting on enough manure to reach the desired nitrogen levels. Last, you can calculate what quantities of other nutrients will be applied, and if it's not enough, decide how you're going to make up the difference with other amendments or with green manures.

Organic Soil Amendments

The list below is not exhaustive. Most producers will not use most of these amendments due to price, availability, and cost of transport. Most growers use amendments that are available in their area and reasonably priced. Many of those listed below are too expensive for field use but may be practical in the garden. Those high in nitrogen or price (or both) can be used in compost rather than directly on the soil if you want to stretch their benefits. Certified organic farmers should confirm the status of any amendments with their certification agency before using them.

Animal Products

Product	Final Rule Requirements	Benefits
Blood meal	Allowed; not required to be certified organic	Up to 15% nitrogen; some phosphorus as well
Bone meal	Allowed; not required to be certified organic	Up to 21% phosphorus and 4% nitrogen
Feather meal	Allowed; not required to be certified organic	Up to 15% nitrogen; good slow-release source of nitrogen
Fish products	These are often mixed with additives; see OMRI list for allowed formulations	7% or more nitrogen and 7% or more phosphorus, plus trace elements
Guano	Wild bat or bird droppings *not* from domestic poultry; must be decomposed and dry	Excellent source of nitrogen; can be hard to obtain
Hoof and horn meal	Allowed; not required to be certified organic	10 to 16% nitrogen; some phosphorus as well

Aquatic Plants

Product	Final Rule Requirements	Benefits
Kelp	Allowed	2 to 6% potassium; 1.5 to 2.5% nitrogen; 0.5% phosphorus
Other seaweeds, algae	Allowed	Good source of trace minerals. Has been used as a soil amendment since ancient times

Organic Materials

Product	Final Rule Requirements	Benefits
Ash	Only from plant and animal sources; no ash from manure, garbage, colored paper, or treated or painted wood	Excellent source of potassium; don't use too much. Raises pH; don't use on acid-loving crops
Cannery wastes	Allowed only if certified organic or thoroughly composted prior to use	Good source of organic material and trace minerals
Cocoa bean hulls	Allowed only if certified organic or thoroughly composted prior to use	Good source of organic matter and some nutrients; can be used as a mulch
Compost	Must be made according to the Final Rule; or applied more than 120 days before any crop intended for human consumption is harvested	*The* soil conditioner for organic gardeners; depending on ingredients, rich in nutrients and consisting entirely of organic matter
Cotton gin trash	Allowed only if certified organic or thoroughly composted prior to use	Good source of organic matter and some nutrients

Cottonseed meal	Allowed only if certified organic or thoroughly composted prior to use	Good source of organic matter and some nutrients
Manure	From domestic livestock or poultry; observe Final Rule application rules	Important source of nitrogen, other nutrients, and organic matter
Mushroom compost	Allowed only if certified organic or thoroughly composted prior to use	Good source of organic matter and some nutrients
Paper	No glossy paper or colored inks	Good source of organic matter; often used as a mulch
Wood products (sawdust, chips, shavings, ash)	Allowed only if from untreated and unpainted wood	Good source of organic matter; often used as a mulch or as animal bedding

Minerals*

Product	Final Rule Requirement	Benefit
Basalt (powdered)	Allowed	Excellent source of trace minerals; exact amounts vary according to source of rock
Chalk (fossilized sea shells)	Allowed	Excellent source of calcium
Chelates	Nonsynthetic only	Make soil nutrients available to plants; healthy soils naturally produce chelates
Dolomitic lime	Allowed	Excellent source of calcium and magnesium
Epsom salts (magnesium sulfate)	Allowed	Source of magnesium and sulfur
Feldspar	Allowed	High in potassium, but not in a form readily available to plants (a biologically active soil is capable of converting it to a form usable by plants)
Granite dust	Allowed	Source of potassium and trace elements
Greensand (glauconite rock)	Allowed	1.5% phosphorus, 5% potassium, and trace elements
Gypsum (calcium sulfate)	Cannot be from superphosphate production	Sulfur content lowers pH of soils; high in calcium
Lime (calcium combined with carbon and other minerals, depending on source)	Allowed	The standard material used to raise soil pH; also very important for its high calcium content; improves soil structure
Marl (combination of lime and clay)	Mined only	Same effect as limes, though slower release
Phosphorus	Mined only	Primary macronutrient
Potassium	Mined only	Primary macronutrient

*All the minerals listed here are allowed in their natural forms, with no additives or prohibited processing techniques.

Applying Amendments

With field crops and gardens, compost, manure, and other amendments are usually spread and then worked into the soil with some form of tillage. In pastures, orchards, and other permanent plantings, amendments are broadcast and rain moves them into the soil. With this type of application amendments take much longer to reach plant roots and soil life. A rule of thumb in either situation is to apply smaller amounts often rather than large amounts rarely. This minimizes the leaching that can occur before plant roots and soil life can absorb the amendments and maximizes nutrient availability throughout the season.

When during the season to apply amendments and trace minerals depends on both the crop and what's being added, and some of the options will be discussed in the chapters on garden crops, field crops, and permanent plantings.

Perhaps the most important thing to remember is that a soil program is ongoing, not a one-shot deal. Each year you should return to the soil as much or more as you removed with a crop. Your soil building plan will also be interwoven with your crop or pasture rotations and the seasonal cycle. Once the plan is in motion, you'll have the fun of continually experimenting with different aspects of it. Soil may be dirty, but it's never boring.

"I need the soil life, it needs to be fed, and I can't destroy its home with excessive aggressive tillage."

GARY ZIMMER
Midwest Bio-Ag, Blue Mounds, Wisconsin

Profile: Gary Zimmer

Midwest Bio-Ag, Blue Mounds, Wisconsin

Gary Zimmer has this advice for new farmers with depleted soils who are on a tight budget: "What's the cheapest thing to bring to your farm? Seeds. What hasn't gone up much in price except for trucking? Calcium. And both are major tools for the biological farmer."

Those two items, Zimmer proclaims, are the basis of soil restoration. Why? Because seeds sprout into green manures that can be plowed down to feed soil life, which will turn them into the soluble nutrients needed to grow a crop. And calcium (in the form of lime or gypsum) is the "trucker of all nutrients," as Zimmer says, the element necessary to make all the others get where they need to go and do what they need to do to form healthy plants and flourishing soil life.

Zimmer would know; he is the founder and president of one of the country's premier suppliers of soil amendments and livestock nutritional supplements for sustainable and organic farmers. The company is based on Zimmer's own lifetime of farming experience as well as his extensive research and trialing. He's constantly walking other farms, talking to other farmers, and running experiments on his own place. He and his wife, Rosie, and their son, Nicolas, own and operate the certified 1,200-acre (485 ha) Midwest Bio-Ag in southern Wisconsin, which, despite their unconventional methods, matches or exceeds yields for conventional farms in the area and is a model of biodiversity as well.

Midwest's tight crop rotations feature corn, soybeans, and mixed forages as well as small grains such as oats, barley, and rye. With the help of three full-time employees the farm also milks 200 cows and raises all its own replacement heifers, along with 100 head of beef cattle, 100 head of pastured pigs, and free-range poultry. There's even a vegetable garden. In 2008 the Zimmers were named Organic Farmers of the Year by the Midwest Organic and Sustainable Education Service (MOSES). Judging by the standing ovation the family received from the packed audience, the affable Zimmers are as well liked as they are admired by many organic farmers.

Many of the Zimmers' farm practices are innovative even for the organic farming community. Forages are blends of legumes and grasses rather than straight alfalfa, and in the fall the Zimmer cows graze on peas, oats, and brassicas to give them a more varied diet and more fiber. The Zimmers use subsoiling to open up soils compacted from grazing, at a minimum depth of 12 to 18 inches (30–46 cm). They plant corn at the end of May — in an area where farmers are embarrassed if the corn isn't in the ground by the first week in May. And they wean dairy calves at ten to twelve weeks instead of six to eight weeks, as on conventional farms.

But if yields, quality, and livestock health are any indication, the Zimmers' methods work superbly. "We're intensive managers," Zimmer says. "We're about success and growing healthy."

These days Zimmer is less involved in day-to-day farm management because he is so intensely involved in restoring soils on other farms through Midwest Bio-Ag. A popular speaker at organic and alternative agriculture conferences, he also spends as much time as possible walking farms around the country and around the world — he's been to New Zealand, Australia, South Africa, Canada, Europe, and elsewhere to look at soils and cropping systems. Under his aegis, Midwest Bio-Ag is involved in a number of cropping system trials around the United States, working to find the best combinations of green manures, soil amendments, and crop rotations for healthy, productive crops based on healthy soils. Long ago Zimmer named this approach *biological farming*, which he defines as cultivating minerally balanced, biologically alive, healthy soils.

When you go organic you have to have a mind-set change. The chemical farmer's tool is nitrogen, the biological farmer's tool is calcium. The conventional farmers are dealing with soluble chemistry. The biological farmers are dealing with exchanging nutrients: physics and biology. They need this great soil structure and soil life. The biological farmer has added plant diversity, he understands weeds, he's changed his soil structure. Controlling air and water in the soil and providing lots of minerals are part of the farmer's job. The biological part is the system and management that allow him to farm without chemicals so he can be certified organic.

I have organic down to two questions: First, what are you doing to get your soils healthy and mineralized? I ask farmers to list all the things they are going to do, and to focus on that. Second, what are you doing to get your livestock healthy and comfortable? If soils and livestock are healthy and mineralized, production is not a problem. This is called leading, versus pushing. Pushing crops and livestock leads to stress and problems.

Zimmer says that every year up to a fifth of his farm's 1,000 crop acres (400 ha) are devoted to green manure crops. Often several cover crops are planted and plowed down during the growing season. Planting crops with large, fibrous root systems is especially important for soil structure, and plant diversity stops disease-causing organisms from gaining a foothold, he says. "Those huge root systems are the breathing tubes for the soil, and fertility tends to concentrate near those old roots. The species of plants determine the soil life, since all plants give off compounds to protect themselves, as do the microorganisms in the soil. So if you have diversity, then problems don't exist. If I have 15 different plants (in the green manure plantings), I completely change the soil biology."

One green manure combination used at Midwest Bio-Ag is rye, buckwheat, mustard, and Sudan grass. Zimmer uses oats for nitrogen, along with legumes. "When you put oats in the soil . . . and shallow-incorporate them when they're about 1 foot (30 cm) tall, they're food for the soil. The bacteria are then high in nitrogen when they are consumed by the earthworms and protozoa. The manure coming out of the earthworms and protozoa is high in nitrogen, and that's how we grow it."

After years of observation, Zimmer believes many rotational graziers may be on the wrong track. It's common to graze paddocks when they are short and lush, since this tends to increase milk production. But that younger forage may be detrimental to the cow's and the soil's long-term health, Zimmer says.

You can't expect a cow to be always on that hot, rich food; she also needs effective fiber or dry matter. (Without that fiber) the cow's manure has no structure, so when it goes back into the ground it doesn't build soil structure. Soil carbon levels increase as plants mature. Forages just 8 inches (20 cm) high have too much protein, and the cow doesn't return enough carbon to the soil.

Zimmer says grazing taller, older forages is better for the cow's digestive system. It's also better for soil structure, since the forage plants' root systems are better developed and the cow's manure has more substance to it. It's always about balance, for both the soil and the animal.

He also questions the wisdom of never tilling the ground, whether because it's in permanent pasture or because the farmer is using a no-till system for cropping. In a 2008 article in *Acres USA*, a respected periodical long familiar to organic farmers, Zimmer and his daughter, Leilani Zimmer-Durand, discuss the tillage dilemma and how they think it might be solved. On the one hand, they say, not tilling preserves soil structure, soil life, and organic matter. On the other hand, soils that are never tilled often are compacted by grazing animals or equipment, and any amendments that are added don't reach the root zone, where they are needed. A crust may form on the ground's surface, further inhibiting water and air penetration, and increasing runoff.

Fortunately, Zimmer and his daughter argue, there's a middle ground. Shallow tillage to incorporate amendments combined with subsoiling — which stirs the soil at a deep level without much disturbing the surface — might offer the best of both methods. It "allows for deeper aeration and water infiltration and reduces compaction," they write. "Tillage is thoughtful disturbance of the soil." It should have a purpose.

Zimmer says that to restore soils takes three to five years, though the process can be sped up with intensive planting and plow down of green manures, remineralizing with purchased minerals, and compost. "Time. That's a struggle for the conventional farmer. He wants it, and he wants it now," Zimmer says. The most important tool any farmer has is knowledge, he concludes.

You have to understand where you're going, or it doesn't make any sense to go there. You have to start with knowledge, so you're not jumping all over the place. Do some soil diggings, see how deep the roots go, count your earthworms, get your soil maps, have your soil tested. Get a complete picture of your farm. Then monitor where you're going, including the nutrient value of your crop.

"When you go organic you have to have a mind-set change. . . . The conventional farmers are dealing with soluble chemistry. The biological farmers are dealing with exchanging nutrients: physics and biology. They need this great soil structure and soil life."
Gary Zimmer

The Zimmers practice what they preach. They not only test their soils annually, they have the nutrient levels of all crops and forages tested as well. Zimmer says, "I have to put science in this picture. (For any field) I can do a feed test for $45 at a maximum down to $20 at a minimum. At least once a year on the field I do the expensive test so I know what my trace minerals are doing."

The feed test lets the Zimmers know whether the nutrients in the soil are available to the plants — a foundational concept in biological farming. Nutrients can be unavailable either because they are insoluble, meaning they can't move in the soil, or because they are out of balance. Four minerals are indicators of health for plants, Zimmer says: phosphorus, magnesium, calcium, and boron.

Get plants high in phosphorus, magnesium, calcium, and boron, and those plants will be the highest-quality, healthiest, highest-yielding crop you will ever grow.

Phosphorus is the hardest mineral to get into a plant. It is a major indicator for healthy plants and a large root system, and you need mychorrizhal fungi or you can't buy your way into phosphorus uptake. The next hardest mineral is magnesium, because it's insoluble and in competition with other minerals like potassium. You need to grow plants like buckwheat, which has acidic roots, to get the nutrient into a plant's biological cycle, and you need sulfur to make it soluble. With calcium, just because your pH is fine does not guarantee you're getting calcium. Soluble calcium is what's missing on many farms, and to get calcium uptake you need boron. Remember, calcium is the trucker of nutrients, and boron is the steering wheel.

So let's say the soil is really short of nutrients, and I'm short of money. I want to start with calcium and phosphorus. You can still buy chicken manure from big farms, and maybe other livestock manures. If my pH is low I have to lime, or the soil biology won't work. Even just 1,000 pounds of lime per acre (1,120 kg/ha) is better than nothing; it may outperform the shock of 3 to 4 tons (2.7–3.6 t) in one dose. I could put the lime on top and add a little bit of manure to help make those nutrients soluble, and maybe grow a buckwheat crop afterward.

The details of soil health could fill a book — Zimmer has already written several, and he is working on another. But in the end it boils down to some simple rules: testing soils and crops and providing what is missing in a balanced, available form; maximizing plant diversity; running really tight rotations, tilling thoughtfully and with a purpose in mind; and feeding soil life with organic matter in the form of manures, green manures, and compost.

Profile: **Art Thicke**

La Crescent, Minnesota

On Art and Jean Thicke's dairy farm in the steep hills of southeastern Minnesota, the cows live long and healthy lives, the pastures are lush, and the balance sheet is positive. But Thicke hasn't bought any soil amendments, not even lime, since 1976, he hasn't tilled any ground since 1985, and he's never worried about doing the testing and math to balance the cow ration for protein percentage and total digestible nutrients. By conventional standards, he should not be successful. Even by the unconventional standards of the managed grazing community he shouldn't be successful, since he lets his pastures grow much higher and rest a lot longer than is standard for dairy grazing operations.

But he is successful by any measure: economic, ecologic, and quality of life. This is not only low-input farming, it's low-stress farming. Thicke takes the time for close observation of his farm system and uses that to guide his farm management, instead of worrying about what everyone else thinks is the right way to do things. Long before the National Organic Program defined the term *organic production* as responding to "site-specific conditions by integrating cultural, biological, and mechanical practices that foster cycling of resources, promote ecological balance, and conserve biodiversity," Thicke was doing it on his farm.

And if farm visitors and requests to give presentations are any indication, then both Art and his brother, Francis Thicke, have become a resource for organic and grass-based farmers in their region who are interested in the art of observation and innovative thinking.

"We like what we're doing," Thicke says. "I don't go for really high milk production; the level of milk production doesn't tell your level of profitability. If you want high production you have to feed heavy grain and protein. For me it's better to have lower milk production and cows that breed back easily and have longevity. When the cows are healthy it's a lot more fun working."

Thicke started what he calls "controlled grazing" in 1985. He says:

I don't like the term "management intensive," because I think anything that intensive is too much. I always say "controlled grazing." Everybody was saying you have to go faster, so I started using faster rotations in the early '90s. I found my grasses were dying, because they couldn't take those fast rotations. Then I looked at some of my old photos and saw my cows grazing in grass up to their bellies, and I said, "You know, the cows and the pastures were doing well back then." That's when I started slowing down, and now by May or June I let the pastures rest for 45 to 50 days. I not only want longevity in my cows, I want longevity in my pastures. We have brome, quack grass, orchard grass, timothy, reed canary grass, bluegrass, a little fescue, some red clover, white clover, a little alfalfa, dandelions, plantain. A lot of those forbs are palatable to the cows, and I think beneficial. I like a lot of diversity. With longer rests, all these grasses are able to compete better. There are only certain grasses that can take fast rotations.

I have about 90 acres (36 ha) of pasture for 90 Ayrshire cows, and we graze about seven months of the year, from about April 20 to around Thanksgiving. I'm also supplementing hay at times. A good grazier will never run out of pasture. A good grazier's smart enough to bring in outside feed before it gets too far. Some graziers end up going faster and faster as grass growth slows, till they wreck everything. If you know the pastures are short and you don't supplement, you're just going to abuse the pastures more. It's not rocket science. You just have to be observant. With grass I think rest is the most overlooked thing. I get my cows in and out, and the rest period is really the most important part.

Some people do 15- or 20-day laps all summer, and every three or four years they have to renovate their pastures. I've found that when I have more organic material out there, the cows have a lot of variety, and they need that fiber.

Farm visitors are impressed by the quality of the pastures, especially since they have had no tillage or supplements for more than 20 years. "If you have life in the soil, that releases nutrients," Thicke says. "I do put manure on my pastures, and I purchase all my corn and some hay, so I'm bringing in soil nutrients that way."

Outwintering the cows (feeding them out in the pastures during the winter) also helps the pastures, Thicke says, and when it's done right the cows are healthier, too.

You have to have your cows in good shape and give them all they can eat, because feed is what keeps them warm. All my cattle also have access to buildings any time they want to come in. Cold rains bring them in, and as soon as it's over they go back out. When it gets colder than 20° F (−6° C) we put the milk cows in with a deep bedding of straw.

Outwintering really renovates the pastures. The cows churn everything up, and maybe I'll no-till in some new grass seed in the spring. I rest the area usually until about the end of June or beginning of July. It's amazing how those pastures grow.

Controlled grazing is key to having a profitable dairy, Thicke says. "It's more important how much milk you're producing off pasture than how much milk you're producing per cow. I personally think that grazing is more important than being organic as far as making it farming. That's where your profitability comes from."

Most important of all is finding what works on your farm. "Don't just copy me. Do your own thing," Thicke says. "Every time you do what the neighbor does, you get in trouble. There's no right way or wrong way. We have to suit our farming to our personalities and our beliefs. This is what works on our farm."

"I not only want longevity in my cows, I want longevity in my pastures. . . . With longer rests, all these grasses are able to compete better. . . . With grass I think rest is the most overlooked thing. I get my cows in and out, and the rest period is really the most important part."
Art Thicke

Equipment

Choosing, Using, and Maintaining Machinery

> *For today's small-scale farmer ... used, small-acreage equipment is at the base of his operation. This equipment is often twenty years old and in frequent need of repair. Parts for these machines are sometimes impossible to come by, hence the farmer must turn inventor, designer, and machinist to create the needed parts in his wood or metal shop.*
>
> **KARL SCHWENKE**
> *Successful Small-Scale Farming*

Farm machinery extends and amplifies what can be done with human muscle. Since farmers started using power equipment, they've been able to produce much more crop with much less effort. The economic effect of this has been to lower the unit value of farm products, so it's now necessary to produce more to earn a reasonable return on your labor and capital investment. That in turn has made it necessary for most farmers to use mechanical power to remain economically viable. And for the part-time or hobby farmer with off-farm income, machinery makes it possible to get the farmwork done and still have time for a job and some leisure.

The essential dilemma for an organic farmer is that engines use nonrenewable fossil fuels. There is a lot of tinkering being done by farmers around the country aimed at producing practical, economical, renewable fuels that will run farm equipment.

Some are even working on electric tractors. But until that resource becomes easily available to the rest of us, and even after it does, organic farmers should be actively working on being more efficient with machinery.

If you have a strong back and are farming on a very small scale, say less than half an acre, you could get by with hand tools and little or no mechanized equipment. On a bigger acreage the primary alternative to engines is draft animals, which can power a small farm but demand a high level of expertise and daily attention. Though this is not a practical option for most small farmers, some are successfully using horses, oxen, or mules as their main source of farm power. If you are interested there are a number of written and Web-based resources about draft animals, including *Small Farmer's Journal*, a beautifully produced quarterly devoted to animal power. There might also be hands-on courses offered in your area.

Tractors

On most farms the tractor, by necessity, is the heart of the equipment line. By itself a tractor can't do anything but pull or push things. On the other hand, implements (the machines you attach to tractors) are useless without a power source. If you're going to buy a tractor, the main considerations are finding one appropriately sized to the amount of acreage you plan to work and built for what you are producing and the production system you are using. It's also extremely important to be able to get parts. The guidelines for tractors given below are general and will vary according to your energy level and available time and whether you're working in light or heavy soil. Light soils demand less horsepower for the same job. If possible, also spend some time talking to the many rural people who buy, sell, collect, restore, or simply love tractors. You may learn more practical information in an hour than you could pick up in a week at the library.

Horsepower

Two-wheeled 5- to 20-horsepower rototillers or their close cousins, the "walking tractors," are suitable for perhaps 2 acres (0.8 ha) or less of intensive vegetable, fruit, or flower production or for light work on a little bigger acreage. Different implements are available for some models of these units, though used implements can be hard to find.

A slightly larger "garden tractor," which the operator rides instead of walking behind, will handle up to about 5 acres (2 ha). A tractor in the 30- to 65-horsepower range, the kind most people visualize when you mention "tractor," will handle most of the field work on a small farm of, say, up to 20 acres (8 ha). This is the classic "small farm" size. A 50- to 65-horsepower tractor can pull a three-bottom plow in light soil and move round bales

The rototiller creates a fine seedbed but can be destructive of soil structure if overused.

In the House

One of the most useful things for an organic farmer to own is a computer with Internet access. This is fairly essential for staying current on prices and regulations, finding suppliers and information, and networking with the organic community. There is a tremendous amount of information on organic farming now on the Internet, and it's increasing daily.

The Internet is also great for following weather radar. If you don't want to leave your computer on all day, another handy item is a radio that can receive the round-the-clock weather broadcasts from the National Oceanic and Atmospheric Administration's National Weather Radio Service. These local weather broadcasts, which come from a network of nearly 500 radio stations across the country, can be received on some Citizens Band radios, short-wave radios, and police scanners. Or you can buy a dedicated National Weather Radio from most electronics retailers. Knowing the forecast is important, since all farm activities center on the weather. And since there are no storm sirens in rural areas, it's handy to be able to flip on the radio when the weather looks scary.

Some other items every farmer should have include earmuffs and a supply of sunscreen and hats. Moderate to severe deafness is fairly common in old farmers from too many years of operating equipment without ear protection, and skin cancer is on the increase.

weighing up to maybe half a ton (0.5 t) — more, if the hayfork is front-mounted.

For running a full-sized round baler or bigger heavy-tillage implements, you'd probably need something around 80 to 100 horsepower, depending again on the size of the implements. Tractors larger than 100 horsepower are built for large farms and are a big investment, as is the large equipment they run. (And we won't be discussing them here, as they are generally not used by the typical small to midsized organic farm.)

Fuel Type

Tractors generally run on either gasoline or diesel fuel. Above the 50-horsepower range diesel engines make more sense for fuel efficiency, and they generally require less maintenance than gas engines. On the other hand, older diesel engines won't start on their own in cold weather, so you need to have a plug-in engine heater and park near an outlet if you plan on using one during the winter.

Wheel Configuration

On a basic tractor, the front wheels are set either close together in a "row front" or "narrow front" configuration, or at the same distance apart as the rear tires, called a "wide front." Row fronts are handy for working in row crops and can turn sharply, but they are more likely to tip. For general utility and less chance of tipping, many farmers prefer a wide front.

Body Type

Farm tractors have a hitch at their rear, and tractors built or modified for market gardening and mowing may have front-mounted, side-mounted, or belly-mounted hitches for implements as well, which is a real plus for precision work where you have to keep a close eye on the implement. This is much easier to do looking ahead rather than looking behind, and you won't have a sore neck at the end of the day. Front hitches and belly mounts can be retrofitted onto many tractors. Implements to be used with these types of hitches have to be modified or specially purchased.

Clearance is another important consideration. If you're going to be working with high crops, a tractor with plenty of clearance underneath is handy in both the field and garden.

Hitches

Tractor implements are generally designed for one of two types of hitches: the drawbar (also known as the fast hitch) and the three-point. Any implements you intend to use must have hitches that match the one on your tractor.

Three-point hitches are particularly handy on small farms, since they allow you to lift an

A versatile garden tractor with a selection of implements can do much of the most tedious work in a market garden.

High-clearance, narrow-front tractors are good for working in taller row crops but are tippy on tight turns.

A low-clearance, wide-front tractor is versatile and stable. Tractors

Retrofitting Hitches

Some of the International tractors from the 1950s and '60s have a two-point hitch. If you have one of these it should be fairly easy to jury-rig a three-point hitch with a small hydraulic cylinder. Some tractor models have a drawbar that will fit into the two-point attachment slots.

Any tractor with a hydraulic system can be retrofitted with a three-point hitch. The cost will range from a few hundred dollars to more than a thousand, depending on the make and model of tractor and who is doing the work. If the tractor doesn't have hydraulics but does have a power takeoff, it's possible to add hydraulics, but at that point you should consider whether you shouldn't just buy a different tractor instead of spending a lot of money on retrofits.

Power Takeoffs

Tractors built after the late 1940s generally have a power takeoff (PTO). PTOs power most working parts of any equipment that are not "ground driven," that is, where the movement of the wheels turns gears to operate the other functions, such as the reel on a side delivery rake, rotating discs, and so on.

The short, splined PTO shaft on most tractors is located between the rear wheels just above the bottom part of the hitch. A PTO can be retrofitted onto the front of some tractors, but this is pretty expensive for most budgets. PTOs have killed and maimed a lot of farmers. A PTO should have a safety shield surrounding it on the top and sides, and you should never do any work within several feet of one unless it is completely turned off.

In some cases ground-driven implements can be preferable to those powered by a PTO, since those types adapt better to varying speeds for some operations. But for the greatest utility a tractor should have both a PTO and hydraulics.

Older tractors may have a hookup for a belt pulley, once commonly used to power sawmills, threshers, and other stationary machinery. When manufacturers began adding hydraulics to tractors in the late 1950s, many dropped the pulley. Pulley-

implement off the ground and maneuver it into corners and small areas like gardens. A drawbar, on the other hand, is used with wheeled implements and can be hooked up more easily and quickly. It's suitable for anywhere there's room to turn.

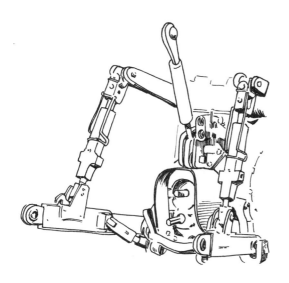

The universal three-point hitch is the standard tractor hitch.

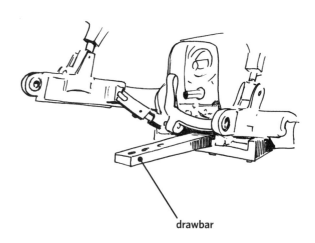

drawbar

The drawbar hitch may be obsolete in the market garden but is still handy for field work and can be combined with a three-point hitch, as shown here.

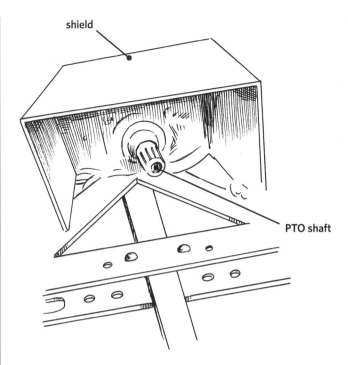

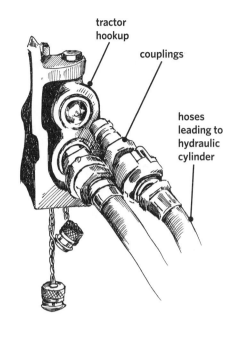

The PTO is the power source for many implements. Make sure yours is shielded, and turn it off whenever you are not on the tractor.

Hydraulic cylinders, hoses, and couplings are built to withstand the immense pressure of a hydraulic system. Check them regularly for leaks and deterioration.

powered machinery is fairly rare and more dangerous to run than most other equipment, due to all the exposed moving parts, while hydraulics are now necessary for a wide range of implements.

Hydraulics

Generally speaking, hydraulics serve to raise or lower an implement, open and close things, adjust how far to the side an implement runs behind the tractor, and similar functions. Dual hydraulics

have greater utility than a single line, since they allow you to control two functions, such as hitch height and an implement function, at the same time.

A tractor's hydraulic system is connected to that of an implement by hoses. These usually originate somewhere behind and below the tractor seat (but not always) and plug into the front of an implement. The hoses and fittings must be manufactured for hydraulic use, as other types cannot handle the pressure level. The hose ends must

The tractor's splined PTO fits into the end of an implement's drive shaft, which transmits power to run the implement.

be compatible, but it's a simple job to buy correct ends, unscrew the old ones, and screw on the new ones. Put some pipe tape on the threads to guard against fluid seepage.

Implements

The range of available implements is mindboggling, but fortunately they organize into just a few categories. These categories are covered briefly below, in the order in which the implements might be used during a growing season.

The one thing all implements have in common is the need for various adjustments and tinkering to make them work correctly with your tractor and your soils. Plows must be set to the correct depth, rake teeth to the proper height, planters to the right openings for the seeds per acre to be planted, and

so on. Owner's manuals or neighboring farmers can help you figure out the quirks of each implement, though your best teacher will be experience.

Primary Tillage

The first job in preparing the soil for planting is to loosen it, bury a cover crop or crop residue when appropriate, and work in any soil amendments that have been applied. This is the job of primary tillage, the heaviest of all soil work, demanding the most horsepower.

Moldboard Plow

The classic tillage implement is the moldboard plow that turns soil completely over to leave a field bare of vegetation. Moldboards come in different sizes and shapes for different soil types and vegetative covers, but an all-purpose type will work in most situations. Smaller moldboard plows with three-point hitches are handy for market gardens.

The classic moldboard plow is built to turn soil completely over and bury any plants on the surface. Different types of points and shares are made for different soils and purposes.

Finding the Right Tractor for the Job

Market gardeners Martin and Attina Diffley were named Organic Farmers of the Year at the 2006 Organic Farming Conference in La Crosse, Wisconsin, for their contributions to the organic community. Mr. Diffley also operates Martin Motor Sales and Service, which deals in used tractors and equipment for market gardens. In general, he says, clearance is less important than the ability to go slow, and front mounts (or tractors that can be retrofitted with them) are ideal.

Here are Mr. Diffley's used tractor recommendations for vegetable growers:

Recommended Models

Allis Chalmers. The rear-engine Allis G is perfect, but parts are extremely difficult to find. The Allis WD45 and D17 gas models are good in the garden, as are the Massey-Ferguson T030, T035, and 65.

Farmall. The Farmall 140 is the best of the offset engine models. Other good choices are the Farmall offset models built between 1939 and 1979, including the Super A, B, Super C, H, and Super M, as well as the Farmall 100, 130, 140, and Cub. The letter-series Farmalls are some of the best tractors ever built, Mr. Diffley says.

Ford. Mr. Diffley recommends the 3000, 4000, and 5000 (post-1965) models.

International Harvester. International Harvester built hydrostatic tractors for 25 years, and Mr. Diffley's favorite of that series is the 656, which can be used for transplanting through harvesting and is still reasonably priced.

John Deere. The HC 900 is a good offset-engine tractor, and Mr. Diffley also recommends the John Deere H, B, A, and G, along with the John Deere 330, 430, 530, 630, and 730 series and the 3010 through 4020.

Others. The Kubota 245 (offset engine) and the Tuffbilt G, the Hefty G, and the Saukville (rear engine) are all excellent for the market garden.

Besides these older tractors, Mr. Diffley notes that a number of small diesel-powered tractors now being built are very efficient. Check out the new models from John Deere, New Holland, and Kubota with hydrostatic transmissions and creeper gears.

Not-Recommended Models

Collectors' items. Avoid the Ford 9N, 2N, and 8N, because they are collectors' items and usually cost more than they're worth as equipment.

Poor performers. The John Deere 2010, 2510, and 2520 are poor performers in the garden, as is the Ford Commander 6000 Selecto Shift.

Poor fuel economy. Tractors with the worst fuel economy are the International Harvester gas 806 and the John Deere gas 4020. The International Harvesters 560 through 860 also have poor fuel efficiency, as do the Farmall diesel MD and Super M and the Allis Chalmers diesel WD45, D17, and D15.

For field crops a wheel-mounted plow with more bottoms (individual plowshares) will do the job much quicker. Moldboards are important for weed control, clean seedbeds, and incorporating organic matter into the soil, but they open the soil to erosion and can be detrimental to soil structure when overused. They should be used appropriately, and the soil should not be allowed to lie bare for long.

Rotary Tiller

A second type of implement used to accomplish the same work as a moldboard is the rotary tiller, also called a rototiller (the walk-behind type) and rotovater (field-sized, tractor-powered). These rapidly spin L-shaped shanks through the soil, burying trash and leaving a fine seedbed. The smaller garden units are self-propelled, while the tractor-mounted units demand a lot of horsepower, are fairly expensive, require quite a bit of maintenance, and do a beautiful job. Rotary tillers can eliminate the need for any secondary tillage.

Chisel Plow

Unlike rotary tillers and moldboard plows, a chisel plow is designed not to bury surface trash. This type of plow uses shovel-like points on long shanks to penetrate and stir the soil without turning it over. It's used in "conservation tillage" where the goal is to leave enough crop residue on the surface to prevent most erosion. Chisel plows are big, but some small-scale farmers have cut them down to suit their farms and tractors.

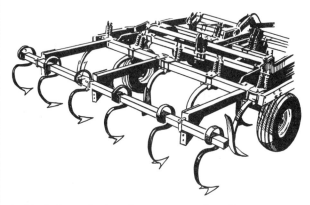

A chisel plow stirs but does not turn the soil.

Discs

Discs are built either heavy for plowing or lighter for harrowing, an operation done after plowing to smooth the field. A disc plow will have a single line of large metal concave discs that are offset to the line of travel, so they cut and push the soil. They can be used in conditions where moldboard plows are less effective, including very dry, very sticky, or stony soils.

Secondary Tillage

Secondary tillage implements are used to break up clods and smooth the soil surface after the heavy work of primary tillage. A smooth soil surface provides a more uniform planting bed, increasing the chance that each seed will make good seed-to-soil contact at the proper depth and so have optimal conditions for sprouting and early growth.

Disc Harrow

The disc or disc harrow is most effective when the discs are "ganged," or have one set of discs running behind and at an opposing angle to the front set, so the trailing gang will cut the ridges made by the first gang. These are also commonly called tandem discs and are a good choice for most small farms, since they're much more effective than single discs and can accomplish a multitude of tasks. Different sizes and configurations of discs are also used to build irrigation borders, create raised beds, chop heavy surface trash, and other jobs. Discs are just plain handy for all sorts of stuff. A disc with wheels and hydraulics is much easier to move around the farm than one without, unless it's a little one mounted on a three-point hitch.

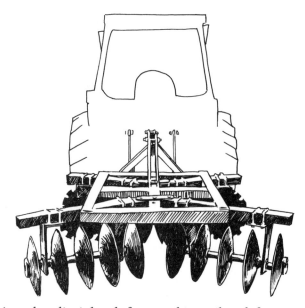

A tandem disc is handy for smoothing soil, and if there is no crop residue to be plowed under, it is all you'll need to prepare a seedbed.

Field Cultivator

An alternative to the disc is a field cultivator, with either rigid or spring teeth. (Spring teeth will bounce off an obstacle such as a rock instead of breaking.) Besides seedbed preparation, field cultivators are used for destroying weeds like quack grass and thistle. Like chisel plows, these usually big implements can be cut down to a size suitable for a small farm if you can weld or know someone who does.

Harrow

For creating even finer, smoother seedbeds, discing can be followed with a spike-tooth harrow, spring-tooth harrow, or even brush tied in a bundle. If possible hook the harrow behind the disc (use heavy chains) to save another pass over the field. Old harrows with no wheels are called "drags," are usually cheap, and come in all shapes and sizes.

Packer

Cultipackers, packer wheels, and similar devices with different names are variations on a rolling weighted cylinder used to firm up the seedbed for good seed-to-soil contact to improve germination. These can be pulled behind discs or after a planting implement.

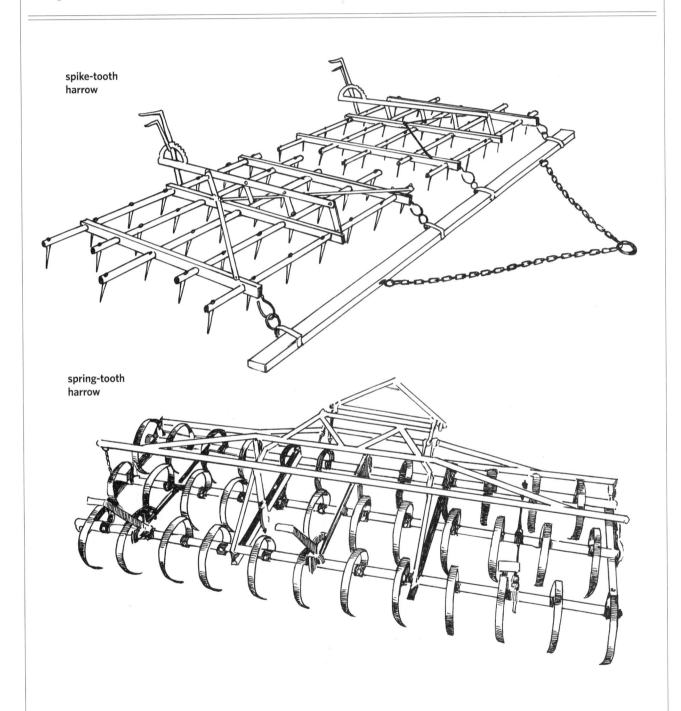

spike-tooth
harrow

spring-tooth
harrow

Harrows are used for final seed bed preparation. If set up for working in rows, they can also be used for cultivation.   Implements

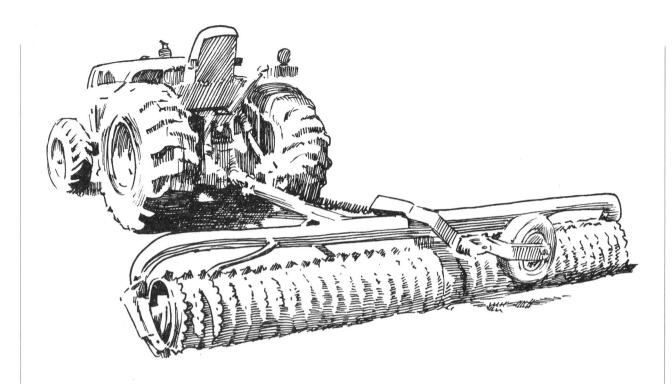

A cultipacker can be used after planting to firm the seedbed and ensure good seed-to-soil contact.

Planting

Seeds are either large and planted in wide-set rows, or small and either broadcast or planted in very close rows.

Grain Drill

Small seeds include the small grains, clover and alfalfa, and grasses. The implement that plants these is called a grain drill; it consists of a long rectangular hopper for the seed with holes spaced along the bottom, some sort of feeder mechanism to keep the seed flowing, and, at ground level, one or two discs for each hole to open a shallow furrow,

Grain drills are used to plant small-seeded crops in close-set rows. Be sure the drill is correctly calibrated, or you may run out of seed before you run out of field.

with a chain dragging behind to close it lightly over the seed. Some grain drills come with a grass seeder attachment, which is an additional hopper for mixed seedings. Others may have a fertilizer box so you can fertilize as you seed.

Grassland Drill

Grassland drills are used for planting seed into sod or other cover. For the especially light, fluffy seeds of native prairie grasses, the Truax drill is a favorite of many prairie restoration experts.

Planter

Implements for planting large seeds like corn and soybeans are called planters. Instead of adjusting the size of the openings at the bottom of the hopper as you do in a grain drill, old planters will have different plates for different-sized seeds. Plateless types use pneumatic pressure, belts, or other methods to plant the seed evenly at the correct depth.

Hand-pushed or small tractor-mounted planters for market gardens may use plates to meter out seeds or belts for planting a solid row. The plate type is more expensive. Other specialty planters are available for such diverse crops as cotton, potatoes, pine trees, and so on.

Push-type planters are ideal for a market garden, with plates for different sizes of seeds. They're quick and accurate and save a lot of bending over.

Cultivation

Since they don't use herbicides, organic farmers do a lot of cultivation, or mechanical weeding. So having the right types of cultivators is critical to organic field and market garden operations. The types of cultivators available are as varied as the crops they're used on and the people who tinker with new methods. For cultivating by hand, there are hoes of all shapes and sizes. For mechanical cultivation implements, the classic reference work is *Steel in the Field: A Farmer's Guide to Weed Management Tools*, edited by Greg Bowman. (In fact, the details of the illustrations on pages 106 and 107 come from *Steel in the Field*.) This succinct, detailed discussion covers rotary hoes, tine weeders, harrows, and the various types of field cultivators used in row crops, as well as the more specialized Spyder, Torsion, Spring Hoe, finger, basket, and brush weeders used in vegetable and horticultural operations. The book also includes interviews with farmers from around the country; they clarify what tools work best in which types of farm operations.

Broadcasting

The simplest method of all for planting small seed is to broadcast it by hand. It's hard to do this evenly, so if you are overseeding a pasture or planting grain in a small area, a hand-cranked spinner seeder will usually do a better job with less work. This is essentially a widemouthed canvas bag with a loop that goes around your neck and shoulder. At the bottom of the bag is an opening that you adjust for the desired seeding rate, and under that a spinner device that throws the seed when you turn the handle.

Using a hand-cranked seed spinner is a pleasant way to plant small grains or a cover crop. Be sure to take into account the side-to-side throw of the seeder and the pace of your walk so you cover the ground evenly and at (more or less) the correct seeding rate. And choose a calm, not windy, day.

Types of Cultivators

(Many of these illustrations are based on ones found in *Steel in the Field*, edited by Greg Bowman; see the resources.)

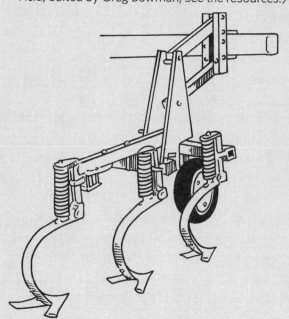

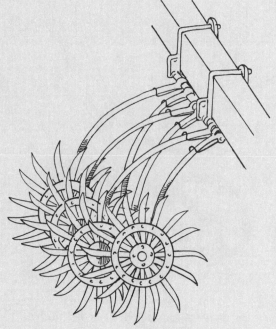

Mechanical cultivators are used to weed row crops. The standard cultivator has C-shaped shanks with triangular tines, called sweeps, for uprooting weeds.

The rotary hoe is aggressive and should be used carefully so as not to damage crops.

The flex-tine weeder should be used when weeds are still tiny. It works by vibrating the tender seedlings out of the soil.

The Spyder Weeder can work close to crop rows, tilling up weeds and pushing dirt either into or away from the crop plants, depending on which way you set the steel wheels.

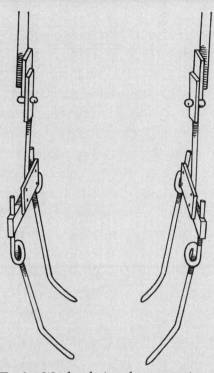

The Torsion Weeder, designed to uproot in-row weeds, should be used when crop plants are well established and the weeds have just or not quite sprouted.

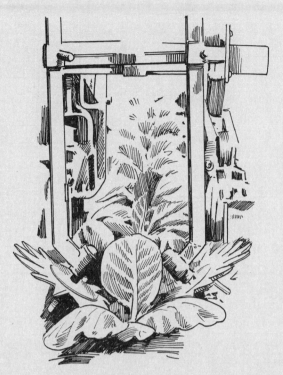

The finger weeder works just under the soil surface to dislodge weeds in and next to the crop row — not a tool for fragile crops!

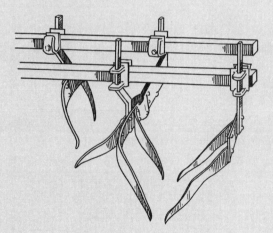

The Spring Hoe Weeder stirs the soil between crop plants to kill small weeds.

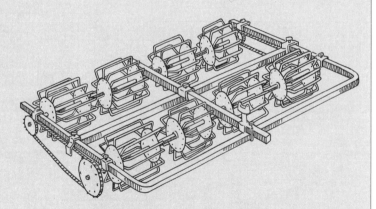

The basket weeder works between the rows, effectively killing small weeds.

Field Harvest

Very small plantings of corn or small grains can be cut, shocked, threshed, and winnowed by hand with hand tools. In Europe and some other areas where small-scale grain production is more common, small-scale machinery for mechanized production is commonly available. In the United States, tractor-powered or large self-propelled units are the norm for grain production.

Combine

Field crops are harvested with combines built, adapted, or adjusted to handle corn, soybeans, small grains, or any of many dozens of other crops. Combines are so called because they combine several operations that a few generations ago were performed separately: cutting, threshing (separating the grain from the stalk), winnowing (separating the grain from the hulls and chaff), either windrowing or scattering the straw (the stalks), and putting the grain in bags or a bin. Corn combines will have picker heads to funnel the rows into the cutter, while grain combines have a rotating reel in front to hold the stalks while they're cut.

Modern self-propelled combines are generally too large and expensive for most small farms. Pull-behind corn pickers and small grain combines common in the 1950s should work if you can obtain or fabricate parts. Combines with canvas

Small-scale combines are useful for small acreages and often available cheap. Just make sure you can get parts for the make and model you're looking to buy.

Specialized Stuff for Serious Market Gardeners

Farmers have a well-deserved reputation for inventing or building what they need but can't find or can't afford, and it's amazing what some of them will come up with to make their farming easier and more efficient. Some of the staff at the University of Wisconsin-Madison Center for Integrated Agricultural Systems and the Department of Meat and Animal Science have gotten on the do-it-yourself bandwagon, too, developing a number of do-it-yourself plans for putting together practical small equipment especially targeted for market gardeners. These include plans for building a hands-free washer, a pull-along seated harvest cart, a motorized lay-down work cart, a strap-on stool for field work, and a rolling dibble marker for easy transplant spacing.

The Healthy Farmers, Healthy Profits project (see the resources) offers sources for many hard-to-find but extremely handy tools and small equipment, such as mesh produce bags, standard containers for garden crops, and a nifty little rolling device that picks up nuts.

aprons are probably best avoided unless you have a source for canvas and know how to fit it on a combine.

Corn Picker

Instead of a corn combine, many small farmers still use corn pickers, which remove the stalk and husk from the ear of corn but don't shell the corn from the ear as a combine does. Ear corn is traditionally stored in a crib, and it may then either be fed on the cob or run through a sheller.

Windrower

When small grains fail to ripen evenly, they may be cut and windrowed, which means being laid in neat rows on top of the stubble, until dry enough for combining. This operation is done with a windrower.

A one-row corn picker may not be easy to find in many areas, but using one is a lot easier than picking by hand.

A swather or windrower is used to put cut small grains in neat rows for drying in the field.

Hay and Forage Harvest

The difference between hay and straw is that straw is the dry stalks of mature small grains, with the grain removed. Hay is grass or legumes cut while green and growing, before fully mature. Straw is golden or brown and used for bedding. Hay is used for animal feed, and good hay has a green color to it, though very old or rained-on hay will be pretty brown. Except for poultry and pigs, farm livestock need hay for feed whenever pasture isn't growing.

Making hay generally involves less equipment and better weather than field crops, but you still need implements to cut, rake, and pick up the hay. If you want to try ancestral methods, you can cut with a sickle, rake it by hand into windrows, then fork it onto a wagon and stack it loose in the hay-mow or in an outdoor pile carefully assembled to shed rain. This is possible if you have a big work crew and just a few animals to feed.

Mower

Options for tractor-powered hay mowing begin with the side- or rear-mounted cutter bars, found in 5- to 7-foot (1.5–2.1 m) lengths, that are still in use here and there around the country. Though the lack of safety shields demands that the operator be careful, these are mechanically simple and handy for small acreages. Some farmers mount a low-slung steel bar ahead of the cutter bar to scare out fawns and other small wildlife so they don't get accidentally maimed. More common now are pull-behind mowers or haybines, which have a reel in front of the cutter bar and rollers behind it to crush the stems for quicker drying. Though these can still be found in 7-foot (2.1 m) widths, it may be hard to find parts. A 9-foot (2.7 m) haybine is small enough to be handy and is just as easy to fix, and parts are still easy to locate. Larger units are common, though the bigger they get, the more awkward and expensive the repairs.

Flail Chopper and Forage Harvester

Flail choppers and forage harvesters are designed to harvest silage, though they are also used to knock down tall green manure crops before they are worked into the soil. Choppers and harvesters differ from mowers in that they make several cuts on a plant instead of just cutting it off at the base, so you end up with a pile of little pieces instead of long stalks. Since silage depends on anaerobic fermentation, it's necessary to have short pieces of plant that will pack tightly and keep air out. Short piece are also much easier to work into soil than unwieldy big stalks.

A flail chopper is mechanically more simple than a forage harvester, but the end result is the same.

Hay Rake

Hay rakes are used to rake the cut hay into windrows for pickup by the baler. The most common type on small farms is still the side-delivery rake, which is either ground-driven or PTO-driven

A haybine has three basic parts: a reel in front to sweep up the crop, a cutter bar to cut the stems, and rollers to crush the stems for quicker drying.

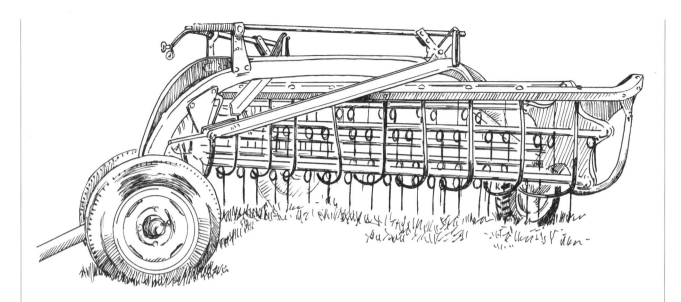

Though side-delivery hay rakes have been largely superseded by faster, more modern rakes on large farms, they are still an easy-maintenance, low-cost choice for smaller acreages.

and runs at an angle behind the tractor, flipping the hay forward and to the side. Much wider, and so much faster, are the wheel rakes, also called V-rakes, in which rake teeth are mounted on spinning wheels set in rows along a steel bar. They cost twice as much or more than a decent side-delivery rake, but they do two or three times the work in the same amount of time.

Baler

Balers make either small squares, large squares, small rounds, or large rounds. The large square bales demand more equipment for moving, storing, and feeding than is usually found on a small farm. Small square balers are traditional, but demand a crew to get the bales made, loaded, and stored before the next rainstorm. Round bales can be stored outside and tolerate rain. In a wet year the outside few inches may go bad, but if the bales are tight the inside will still be good.

If you decide on a small square baler, you'll also need a hay wagon, preferably two or more, to move the bales from the field to the barn or shed. A second tractor is really handy, too, so one can be baling while the other is moving bales. If your hay storage is in a loft that you can't drive into, you'll need a hay elevator to get the bales up there.

On the other hand, if you have a round baler you can leave the bales on the field until you've put

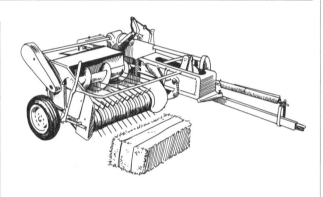

Traditional balers should never be left out in the weather, since rust and dust are enemies of the sensitive knotting mechanism. A round baler does in hours what will take a square baler days to accomplish. But round bales are too heavy to move by hand.

Equipment for Market Gardens, Orchards, and Vineyards

These crops all demand the same operations as standard field crops: preparation, planting, cultivating, and harvesting. But because they are so variable in how these operations are best performed, there's a corresponding plethora of equipment available. There are also numerous operations unique to fruits and vegetables, so there is a corresponding number of implements that do things like lay plastic mulch, transplant seedlings, vacuum bugs, and mechanize crop harvests. In general, an implement that can be utilized for more than one operation is going to be more economical than one that is only hauled out once a year. Do your research thoroughly before buying to make sure a particular implement fits your soil and production system, and that it will pay for itself in labor savings.

the baler away, then attach a hay fork or front-end loader with a fork to the tractor and pick up the bales to move them out of the way. This eliminates the need for wagons and elevator.

Spreaders

Manure spreaders save a lot of shoveling on any farm where animals are routinely inside for shelter or milking. If you have an old dairy barn with a working barn cleaner, it's wonderfully quick to flip on the cleaner and let it move all the manure and soiled bedding out the end of the barn and dump it into the waiting spreader. You can also load spreaders by hand with a shovel. Many farmers use skid steers for loading as well, but you can't get a skid steer inside many old barns. Manure spreaders come in various capacities and generally have a wood bottom with chains or bars that scrape over it, moving the manure to the back end, where the beaters fling it out. When spreaders get too decrepit for manure, they can be used to haul firewood and other odd loads.

The other type of spreader is used for lime or fertilizer. This type of spreader is either an

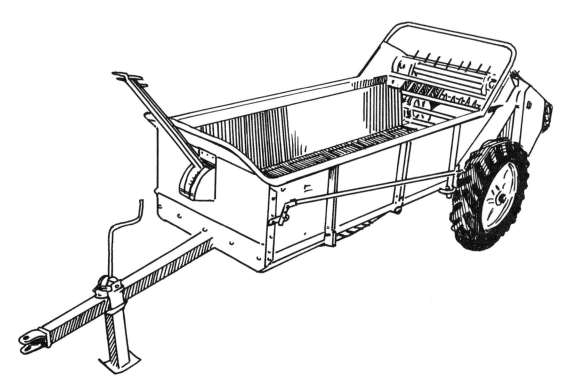

A manure spreader is necessary if you have enough livestock in a confined area often enough to accumulate piles of manure. A spreader is also handy for spreading compost and even hauling firewood.

open metal box on wheels that feeds the amendment onto a spinner at the rear or a grain-drill-type hopper with spaced holes along the bottom. These spreaders come in various sizes and are very handy to own, though in many field and pasture situations it's easier to rent or borrow one, or, quickest of all, to have the lime or fertilizer spread by custom operators or the dealer you're buying from. (As is the case with all off-farm machinery, if you're certified organic, the spreader operator must deal only in certified organic materials or the spreader must be thoroughly cleaned before it is used on your farm.)

Miscellaneous Equipment

Some very handy pieces of equipment don't fall into a specific category because they're used for so many purporses. In fact, these pieces often are used more than any other equipment on a small farm.

All-Terrain Vehicle

At the top of the list for being all-around useful on small farms is the all-terrain vehicle, or ATV.

The ATV with room to haul vegetables, fencing materials, calves, or all the many other types of things that need to be moved around on a farm is an extremely handy piece of equipment.

The farm and utility types, which have a miniature truck bed in back and golf-cart-style seats, can carry two people and a lot of stuff. Some models also come equipped with forklifts and power takeoffs, with a corresponding range of implements. (It's easy to spend more money than you really need to on an ATV.) Livestock farmers use them for checking their animals, carrying fencing materials, and moving newborns up to the barn. Market gardeners transport seeds, tools, and produce. If you're out working in the field, an ATV makes it easy for someone to bring you a picnic lunch. I used to think ours was a luxury item, but over the years I've become convinced that we do a lot more, a lot more efficiently, than we would accomplish if we had to rely solely on our feet or our tractors.

Skid Steer Loader

Another handy piece of large equipment is a skid steer loader. Most larger farms have one, and a fair number of small farms as well. You can buy all sorts of implements for them, and they do all sorts of jobs quickly and well, from plowing snow to augering postholes to smoothing the dirt driveway to moving round bales. They are not a necessity in most situations, but they can be a big help if you have enough ongoing work for them to justify a purchase. They can also be rented for occasional jobs. The 40-horsepower unit is popular, big enough to do some serious work, and easy to find used. If you decide to purchase a skid steer, stay away from older units with variable drives

A skid steer with a selection of attachments is pricey, but it can accomplish all sorts of heavy labor on a farm.

Ergonomics

The most important piece of equipment on the farm is you, so if you're serious about being sustainable, start with yourself. The U.S. Department of Health and Human Services' National Institute for Occupational Safety and Health publishes a handy little booklet titled *Simple Solutions: Ergonomics for Farm Workers*. The booklet states that "backaches and pain in the shoulders, arms, and hands are the most common symptoms that farmworkers report," a fact that any farmer will sympathize with. To reduce the pains and injuries associated with repetitive gripping, lifting, bending, twisting, kneeling, squatting, and using vibrating equipment, the key is to either redesign the tools or redesign how the job is done. The publication offers detailed suggestions, targeted at market garden operations, for 15 different work situations that commonly cause problems.

The publication's tips for a healthy back apply to all farmers — indeed, everyone:

- When lifting, keep your back straight and lift with your legs.

- Work close to the job whenever possible instead of reaching.

- Carry more light loads instead of a few heavy ones.

- Change position or task often.

and those with more than 4,000 hours. If you're looking at a used unit, warm it up thoroughly and drive it to make sure it drives evenly. If not, the hydrostatic drive motor may need replacing — an expensive repair.

Brush Mower and Chain Saw

Brush mowers, either walk-behind or tractor-mounted, are great for mowing fence lines, trails, and ditches. If you have any number of trees on the property, a chain saw is handy for making firewood, clearing branches that come down across the driveway or fences, and trimming and thinning trees. If you're buying a chain saw, you should also be buying chaps, ear protection, and a hard hat, and getting some basic instruction from the dealer or local woodland owners' organization on how to start and operate it safely.

Buying Equipment

Given the price of new equipment, most farmers will end up buying most, if not all, of their equipment used. This requires a little more caution when making a purchase, but on the whole, farm tractors and implements are built to last, so plenty of used stuff has a lot of years left in it. If you can weld, or know someone who does, you'll have even more buying options since you may be able to modify a piece to fit your operation.

Where to Find Equipment

There are three places to buy used equipment: dealers, private sellers, and auctions. Ads and notices for equipment can be found in local and farm papers, as well as on the Internet. If you're looking for a particular piece, you could let a couple of dealers know since they might have a better chance of finding it sooner.

Dealers

Dealers are a good place to start. You may pay more, but since they have a reputation to maintain they're generally pretty good about taking care of any problems that arise after the purchase. Dealers know plenty about what's used and what's available in your neighborhood, and if you can catch them on a slow day it'll be a worthwhile chat.

Private Sellers

Private sellers are more risky. Many are guys that just like equipment, and buying and selling is simply a hobby. Some are farmers who are cutting back, but others are trying to make a quick buck on a bad implement. Consider what you know about the seller, and go over the piece extra carefully before buying.

Auctions

Auctions are either consignment, where a lot of sellers pool their items, or farm auctions, where all the items come from a single farm (except maybe for a few brought in by neighbors). Don't start by buying at a consignment auction. Consignment items vary considerably in quality, and it's difficult or impossible to find the previous owner and get a history of the piece you're looking at. Consignment auctions are not for the inexperienced buyer.

Farm auctions, on the other hand, can be a real source of bargains. Or not. It all depends on the crowd and the day. Even if you're not planning on buying, if you have the time it's fun to stop by any auctions in the neighborhood. You'll meet neighbors, find out what's happening in the area, and get a feel for the bidding process. Bidding can be a real adrenaline rush. Protect yourself by knowing your spending limit before you start and by sticking to it. Auctioneers may use various ploys to move the bidding higher; that's why it's nice to see a few auctions beforehand to get a feel for them.

Parts and Prices First

Before you begin looking at equipment, make a list of dealers within, say, 30 to 45 minutes of your farm, and what brands they stock parts for (the easiest place to start is the phone book). This should be a deciding factor in most equipment purchases. If you can't get parts fast, you are setting yourself up for a lot of frustration and wasted time. You can order parts for most equipment online, but even so, if the dealer you're ordering from is far away, the parts will take at best a day or two to reach you.

Buying the Farm Plus

If you are buying a farm and its equipment is in good condition, consider negotiating to have some or all added in as part of the sale. It's much nicer to have a line of equipment on hand that is compatible and suited to the particular farm than to have to assemble it all from scratch.

Remember that an older but popular model may be easier to find parts for than something newer that had a limited production run. It pays to ask around.

Before you go to look at or bid on a particular piece of equipment, check prices in dealer ads and online. That lets you know approximately what you should be paying.

Giving It the Once-over

Before buying, look over the equipment bit by bit. In particular check the belts, chains, and other parts that get a lot of wear and need to be replaced periodically. All those things should be easily accessible and simple to take off and replace. If it's going to take several hours just to disassemble enough parts to access the problem, you might be better off looking at something else that will be quicker to fix.

Look for unusual wear patterns, bends where things should be straight, and metal cracks. These could identify defects in the equipment. Identify any part that is badly worn or rusted, and decide whether it will be worth fixing when it breaks.

In addition:

- Look at the depth of tread on the tires, and check for weather cracking on the sides. Check the rims for looseness, rust, cracks, and rough edges.

- Check chains and sprockets for excessive wear; ditto for the splines on the PTO shaft. Chains should be oily and roll easily.

- Wiggle any moving parts to see how tight they are, including the wheels. If they have too much play side to side, their bearings may need replacing.

- Look for signs of oil, transmission fluid, radiator, or hydraulic leaks.

- Check the belts for tightness and condition. If they are loose or cracked, they will need to be replaced.

Many, if not most, of these problems are usually fixable, but if there are a lot of things wrong you should either reduce your offer or look elsewhere. Some implements are just plain badly engineered or are not suitable for your soils and purposes. Avoid these!

Cleaning Equipment for Organic Use

Organic farmers have to be particular about using any equipment not used exclusively for organic crops. Any crop residue from nonorganic operations will contaminate the organic harvest and make it unusable for certified organic sale or feed. The standard method for cleaning harvest implements that were previously used on nonorganic crops is to run them down the first row, then dump that row as nonorganic before harvesting the remainder of the field. You must document this cleaning if you are certified organic. See chapter 7 for more details.

In fact, if you can bribe, kidnap, or otherwise persuade a mechanically minded friend or an experienced farmer to come along with you, it's worth at least a good steak dinner for what you'll learn and the problems you'll avoid.

Getting It Home

Once you've purchased a tractor or implement, the next problem is getting it home. Dealers will deliver for a price, and some private buyers may do so as well. At an auction and with most private buyers, you're on your own. At auctions you typically will be expected to remove your purchases within 24 hours.

Anything with wheels can be towed (you'll usually need a pin hitch instead of the standard ball hitch found on most vehicles), and you'll probably have to drive slowly to keep it from fishtailing all over the highway. Have a triangular orange "slow-moving vehicle" sign and some wire to attach it to the rear of the implement. Anything without wheels will have to be put in a truck bed or on a trailer. For all these reasons I like to buy as close to home as possible.

Alternatives to Buying Equipment

If you aren't quite ready to start buying machinery or have too few big jobs to justify the expense, consider bartering with or hiring a neighbor, hiring a custom operator, renting what you need by the hour or day, or leasing for a season. If you are certified organic, any bartered, rented, or hired equipment must be cleaned thoroughly before it is used on your farm, and this cleaning (its date and method) must be documented (see the box at left).

Custom Operators

A drawback of hiring a custom operator is that you are now on his schedule, not yours, and may have rained-on hay or a poorer-quality crop due to a late harvest than you would have otherwise. On the other hand, sooner or later your hay will get rained on no matter who is making it, and if your machinery breaks down you could wind up just as far behind in fall work. The upside of custom work is not having to buy, operate, maintain, repair, and store a lot of machinery. One common glitch, however, is that in areas with bigger farms there may be plenty of custom operators, but they won't want to handle small acreages. If you can time your field operations so they don't coincide with everyone else's, you'll have better luck finding a willing operator. Many state extension services publish custom rate guides that list the going rate for various field operations. If you can't find one of these, ask your extension agent or neighbors what they have heard are typical charges.

Rental and Leased Equipment

Rental equipment is available from many dealers, and is an especially good idea for when the job is something you'll only need to do rarely, like mixing cement, moving big rocks, or augering fence postholes.

Leasing farm equipment for a season is not always an option, but if you can talk an equipment dealer into it, it's a good way to get to know how to maintain, operate, and do minor repairs on equipment before deciding if you want to buy.

Equipment Operation and Maintenance

For anything more mechanically complicated than a drag or an old disc, the first thing to obtain after purchase is an owner's manual. If you're lucky you'll get this from the previous owner, but most often you'll have to order it through either a local dealer for that brand or from someone who buys and sells manuals (through the Internet or ads in farm publications). The owner's manual will tell you where the lubrication points are and how often they should be lubed, as well as capacities, belt sizes, and all sorts of other useful specifications. Go over the equipment with manual in hand, checking that everything is in good condition and working properly. And before you operate any newly purchased engine, make sure you know how to stop it and how to turn it off.

The "Preflight" Check

Before using any equipment, do a "preflight" check. On a tractor, this means checking oil and radiator levels, topping off the fuel tank, checking tires, and greasing any zerks advised by the manual. (Zerks are pea-sized spring and ball valves with a threaded base that screw into the moving nonengine parts on a machine; see the illustration on page 121.) Make sure the precleaner or air cleaner and the radiator grill are clean. On implements, check all necessary adjustments, and make sure the hitch is at the proper height and distance and is secure.

After you start up your equipment, begin slowly, watching to make sure that everything is running correctly. Listen to the rhythm of the implement

The Farm Shop

If you have equipment you'll need some mechanic's tools and a place to keep them. Once you've assembled the basic ratchet set, wrenches, screwdrivers, pliers, and so on, other essential items will naturally accumulate over the years.

One item that is often missing but is extremely useful, especially to the organic farmer, is a record book. Kept in the farm shop, an entry should be made each time machinery is maintained, repaired, or cleaned. This is useful for knowing when routine maintenance like oil changes are due, and it impresses buyers if you ever sell any equipment. If implements are being used on both organic and nonorganic crops, having a record of when and how cleaning was done between nonorganic and organic use will be essential to organic certification.

and the pitch of the engine, as any sudden changes are likely to mean a problem. At that point, stop, disengage or turn off any moving parts (especially the PTO!), and only then look to see what the problem is. Attending to a problem sooner rather than later will save time and repair bills.

Storage

When at all possible, store your equipment under cover. Sun, rain, and wind accelerate deterioration of paint, electrical connections, and belts, and increase rust. In particular, a square baler with an exposed knotter should never be left in the elements. Strap a tarp over it if nothing else.

Maintenance

Equipment maintenance should be a regular part of your farm routine. Follow the owner's manual recommendations for maintenance intervals and procedures.

At the manufacturer's recommended intervals during the season, and again before storing implements for the winter, grease all zerks and oil all chains. A retired farmer once told me that he used

Safety First

Safety is the primary consideration when operating any farm machinery. Farming is consistently ranked as one of the most hazardous occupations in the United States, and the majority of farm accidents occur in and around machinery, with tractor overturns alone killing an annual average of 100 people. Older equipment has fewer safety features, so it's even more important to be diligent about safety procedures. The basic rules are as follows:

- Turn everything off before working on or examining machinery.

- Do not drive tractors up or along steep slopes. Back up slopes and ramps.

- Do not drive tractors and equipment near sharp banks or drop-offs.

- Do not allow passengers on farm equipment.

There is much more to know about the safe operation of equipment. Tractor safety courses are offered in many counties by 4-H programs to students 12 and older. Send your kids, and if you don't want to attend yourself, at least obtain and read the course manual.

to work with a grease gun strapped to his thigh. Probably a slight exaggeration, but not by much — keeping things greased and oiled saves a lot of trouble. A grease gun with a flexible rather than rigid nozzle makes it easier to reach awkwardly placed zerks. I like to put a film of oil on all bolts every season, since sooner or later I will probably have to take them off, and it's so much easier if they aren't rusty.

When something breaks, as it inevitably will, you have to diagnose the problem before you can figure out how to fix it. The problem may or may not be obvious, but eventually you or the neighbor will figure it out. Usually the repair involves removing the broken part and heading to town to replace it or have it fixed. (If the problem is with a belt or chain, before you remove it, take a photo of it to use as reference so you can route the replacement belt or chain correctly.) Over the years most farms build up an inventory of commonly needed parts, from zerks and chain links to a collection of belts, bolts, rivets, and cotter pins. Speed up repairs by labeling which parts fit which machines.

For those times when you can't fix it yourself and can't get it to town, like if a back tractor tire blows, you can usually find an on-farm repair service. Such a service is expensive, but sometimes you don't have much choice.

Fluid Nutrition

Because equipment is so important to a farm operation, keeping it running is a priority. The easiest and simplest preventive maintenance is frequent oil changes and ample grease. However, there are a lot of types of oil and grease to choose from out there. Just as organic farmers care greatly about what they eat, so they should care as much about what they feed their engines if they want them to run well and last long. This will sound just like your basic agricultural extension publication, but it's a critical point: knowing and using the correct types of oil, fuel, and grease will greatly extend the life of your equipment.*

Oil

The most important fluid in an engine is oil. It lubricates the rapidly moving parts and keeps them from wearing each other into oblivion. Oil also carries away the soot and carbon left over from combustion and the grit that finds its way into the engine from outside.

But oil isn't just oil; additives make up anywhere from 1 to 33 percent of the mix. Additives can do a superb job of preventing rust, removing heat, stopping corrosion, preventing foaming,

*This section draws in large part from an article the author wrote for *Backwoods Home* magazine in the mid-1990s. *Backwoods Home,* based in Gold Beach, Oregon, publishes articles offering practical, hands-on information on a wide range of self-reliance topics.

sealing compression, and absorbing vibration. In addition, the mix has to perform at temperatures ranging from below zero (-18°C) in a cold engine to over 250°F (121°C) in a hot one. That's a lot to ask of a single fluid.

With oil you generally get what you pay for, and paying for premium is usually worth it, since the superior additive package and the care that has gone into refining the oil pays off in oil performance and long and trouble-free engine life. Once you have found a brand that works well with your engine, stick with it. A couple of mechanic friends told me that because of differences between additive packages and oil quality, switching brands may result in an engine that leaks or burns oil. (Okay, a lot of old engines burn oil, but you don't want to exacerbate the problem.) It seems a little odd that a nonsentient entity can have a brand preference, but it does. If the maker of your engine also makes an oil specifically formulated for that engine's configuration and alloy types, that's the one to use, but that isn't common.

Synthetic oils are excellent but a little expensive for use when you're changing oil as often as you should.

Reading the Bottle

The numbers and letters on oil bottles indicate viscosity and quality. Viscosity is the measure of the thickness, or flow rate, of the oil. Lower numbers signify thinner oil, higher numbers mean thicker. Thin oils work well in cold temperatures or under light loads; thicker oils are necessary for high temperatures and heavy loads. Engines are generally engineered for specific oil weights, so it's important to check your owner's manual for the correct oil viscosity. Small gas engines tend to work at high temperatures and full loads and generally need oil in the SAE 30 range (SAE stands for Society of Automotive Engineers, which sets viscosity standards).

Oils that have two numbers on the bottle, such as 10W-30, are multiple viscosity oils and are suitable for a range of engine temperatures. The W means this oil is appropriate for winter use. Some chemical magic in the form of additives causes these oils to be thin at lower temperatures and thicken as engine temperature rises. This is a boon for cars and tractors operated through a wide range of loads and temperatures, but these types of oil are not appropriate for a small gas engine. Unless an owner's manual specifically recommends a multi-viscosity oil, don't use one.

The other letters on the bottle are the service rating, or the engine-protecting quality of the oil. The American Petroleum Institute (API) publishes the guidelines, which are listed in alphabetical order from the least protective, SA (service classification A), to SJ. All categories but SH and SJ are now considered obsolete; either of those two are fine for small engines. The letter C, instead of S, means the oil is formulated for diesel engines, not gas ones.

Changing the Oil

Check the oil level every time you use the engine, and take note of how dirty (dark and opaque) it is. Oil can hold only so much soot and grit before the excess begins to be deposited on engine parts, abrading and clogging the cylinders, pistons, and other moving pieces. Oil additives break down over time, leaving the engine unprotected. My mechanics have said a good rule is to check the manual for recommended oil change intervals, and then cut that time interval by a third. Most small engines don't have oil filters, but on the larger engines it's recommended you change the filter each time you change the oil. Filters can get a little pricey on some tractors, so consider the grit factor: if you've been working under clean conditions, maybe you can get away with only changing the filter every other time. But remember that a filter clogged with grit can't move oil through fast enough, and that's asking for trouble.

Changing oil takes about ten minutes. Do it while the engine is still warm after use. It will drain faster and so carry off more grit and glop. Park on level ground, disconnect the spark plug wire if it's a small engine, slide a bucket or pan under the engine, and unscrew the plug. Let all the oil drain out; it's those last few drops that carry off the most sludge. Put the plug back in, being careful not to overtighten or it may not come off next time. If there is one, remove the oil filter with a filter wrench and replace, just snugging down the new one and then giving it an extra quarter turn. Now add the amount and type of oil called for in the manual and no more. Too much oil tends to burn off around the piston rings, making them sticky.

If the engine is brand-new, the manual usually recommends changing the oil after the first five hours of use. This is because tiny metal shavings come off the engine parts as they break in, and that first oil change removes those little abrasives that can cause engine wear.

Giving Oil to Two-Cycle Engines

Two-cycle engines don't need oil changes. Chain saws, handheld weed whackers, and similar types of tools are usually two-cycle; if you're in doubt, read the manual or ask a mechanic. In these engines oil is mixed with the gas. Two-cycle oil is specially formulated to work at higher temperatures, thicken under pressure in the crankcase to lubricate the bearing surfaces, and then burn off with the gas. Two-cycle and four-cycle oils are very different and should not be substituted for each other. The gas-to-oil ratio varies from engine to engine, so you'll need to check the manual when mixing.

When you mix gas and oil in a can, it works best to add half the gas and all the oil, shake, then add the rest of the gas and shake again. Write on the gas can which engine this mixture is for, so there are no mistakes. (And with chain saws, don't mistake the bar oil for the engine oil.)

Fuel

Gas and diesel are expensive and getting more so. So get the most for your buck by using the correct type and using additives where appropriate. Making sure air filters are clean, keeping tires at the correct pressure, and not using implements too big for your tractor also maximize fuel efficiency.

Gasoline

In cars, a high-octane fuel allows a higher compression before combustion, delivering more power and less of the preignition that causes engine knocking. As a result the engine runs more smoothly, more efficiently, and more cleanly. In a small engine, though, a high-octane fuel will burn the valves and pistons and ruin the spark plugs. If you are getting some pinging and power loss in a small engine it's probably because the octane rating of the fuel is too high. Fuels containing ethanol or alcohol also burn hotter and may have the same effect. Stick with 87-octane regular gasoline.

Leaded Gasoline

In old tractors with engines built to use leaded gasoline, add lead replacer each time you gas up. This costs just a few dollars a bottle at any farm store and protects the valve seats from rapid wear. Adding more than the recommended amount per gallons won't hurt, but don't skimp. If your tractor was built after 1975 or has had an engine rebuild, it probably has hardened valve parts, so lead replacer isn't necessary.

Diesel

Diesel is fairly simple compared to gas; you generally don't need any additives and there's only one choice for tractors. It keeps well. The only real problem is that it gels at low temperatures, so you need a way to keep it warm if you want the tractor to start, either with a plug-in block heater on the engine or in a heated shed.

Fuel Storage

Gas and diesel can be stored in tanks or plastic cans. If you have enough machinery running to make a large metal tank worthwhile, make sure it is placed at least 50 feet away from any building or your property insurance may be voided. (Check with your insurance agency for its specific fuel storage specifications.) You will need to have it mounted in a tank containment area to protect from spills and leaks. A cement slab with a curb around the edge works; be sure to seal any cracks that appear over the years. Underground fuel

The Quick Reference List

If you have a number of engines in the machine shed, instead of searching through manuals all the time, it's easier to make a quick-reference list of which engines get how much of which type of oil or mix. Jot this information down in the front of the notebook where you record all the maintenance and repairs.

tanks, considering the potential for leaks, are not an option for the organic farmer.

Your fuel delivery service will help you to set up the right size and type of tank and give you tips on how to keep it functional. Make sure there's a filter on the hose to keep dust and dirt out of your fuel lines. Before fueling, tip the nozzle downwards to let any rain or dew run out. Whether you're filling the engine tank from a large metal tank or a small plastic can, it's a good idea to keep the nozzle touching the rim of the tank to keep a static charge from building up. And never fuel a running engine or smoke while refueling.

Winterizing Fuel

When storing gas engines for the season, add a gas stabilizer to the tank to keep the gas from getting gummy. After adding the stabilizer, run the engine for five to ten minutes so the additive has time to work its way into the carburetor. The needle valve on the carburetor float bowl is what sticks and keeps the engine from starting in the spring. If the additive doesn't reach the valve, it hasn't done you any good. If you do use the machine during the winter, top off the gas tank before you put it away again. A full gas tank keeps water from condensing in the tank and fuel line as the engine cools, another frequent cause of engine problems.

Air

Gas and diesel can't combust without air, and though you don't personally have to add air to the engine, you need to make sure it can get there. Up to 9,000 gallons of air are necessary to burn 1 gallon of gas. If the air is full of dust and grit, this will scratch and abrade the combustion chamber and work its way into the rest of the engine to cause more wear. All engines also depend either partly or wholly on air to keep them cool.

After checking the manual to determine what parts of the engine are involved in air intake and exhaust, check them daily and clean them on a regular basis. This may include prescreeners, radiator screens, cooling fins, exhaust ports, and the air filter. Old tractors may use an oil-bath filter, which can be changed and cleaned when you change the oil. Newer tractors and small engines use paper, foam, or metal mesh filters. Paper filters can sometimes be reused if serviced often but should still be replaced on a regular basis. Foam filters are usually washed in soap and water. Some are replaced dry and others oiled; only your manual knows for sure. Your manual will also have specific directions for servicing a metal filter.

Grease

Grease is the fourth major food group (after oil, fuel, and air) for engines. It's also critical to nearly every farm implement, from plows to balers. Grease is a combination of oil, soap, and additives that lubricates and rust-resists moving parts through a wide range of temperatures. There are hundreds of types of grease available, but the only kind needed on most farms is any brand of MP, or multipurpose, grease, either lithium- or petroleum-based. Farm stores sell grease in tubes that fit into grease guns. For greatest utility, get a gun with a flexible nozzle, not a rigid one.

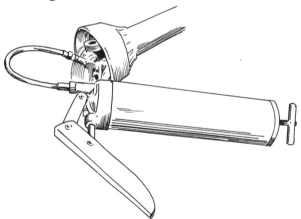

To grease a zerk, fit the nozzle of a grease gun over the zerk and hold it firmly in place while pumping the grease gun handle.

Zerks turn up in the oddest places. Check your owner's manual and look over equipment closely to make sure you find them all.

Grease is applied through zerks. It's not easy to find all the zerks by just looking; so check the manual if you have one. In a new machine, check the recommendations for amount and frequency of grease. In an old machine, it doesn't hurt to grease more often. Those old loose joints have more room, and it's better to fill them with grease than with dirt and water.

Hydraulics

If your hydraulics are working slowly you should check the fluid level and top it off as needed. It's generally a good idea to use the brand of hydraulic fluid formulated for your brand of tractor. Otherwise, hydraulic fluid needs little attention unless it becomes contaminated. Then you have to drain the fluid, find the problem, and put in fresh fluid.

This is not a job for anyone short of time, and new fluid costs roughly what you would pay for liquid gold.

Radiator

If the engine is big enough to have a radiator, check the level daily before use and top it off if necessary. If you are topping off the radiator more than occasionally, you probably have a leak somewhere and need to get it fixed. (I just had to have the radiator on my Case 930 recored, for about $1,000. It's these sorts of surprises that put a real crimp in the farm budget.) Most engine manuals recommend a half-and-half mix of antifreeze and distilled water in the radiator; follow the specifications in your owner's manual.

"Really good equipment makes life easier."

RALPH MOORE
Market Farm Implement, Friedens, Pennsylvania

Profile: **Eugene Canales**

Ferrari Tractor CIE, Gridley, California

When Eugene Canales bought a small farm in the 1970s and then went looking for the right-sized equipment to run it, he found that most U.S. manufacturers were abandoning their small-scale machinery lines. The trend toward ever-larger farms was accelerating, and the equipment industry was focusing on building bigger tractors and implements. Finding what he wanted in the United States proved difficult.

But in Europe, where small-scale farms continue to hold their own, manufacturers have kept investing in the research and development of small-scale equipment. Canales eventually solved his equipment difficulties by going to work for an importer and then, in 1987, taking over the Ferrari agricultural equipment dealership near Gridley, California. These days he still raises beef cattle, but his primary business is selling small-scale farm equipment.

Much of the time, Canales says, new farmers don't know exactly what they're looking for. "There are all kinds of decisions to be made, depending on the climate and on how much ground you intend to work at one time. The farther north you are, the more you're limited by the growing season, and there's less time to do things. Most farmers plan to market as well as grow, and that limits your time, too. Contrary to popular belief, it is the smaller grower who most needs labor-saving equipment."

Equipment must be sized to get the job done on the acreage you till during the time you have, he says. "A lot of farmers think they're going to do everything with a walk-behind tractor. I don't know anybody who can do 5 acres (2 ha) with a walk-behind. In my view, 2 acres (0.8 ha) is the upper limit. Most of the implements cover only a 30-inch (76 cm) strip, and 1 acre (0.4 ha) is a lot of ground to walk over in 30-inch strips."

Larger operations should look at a mini-tractor or something even bigger, Canales advises.

It depends on the crop and the acreage. If you're in the vineyard business, you're going to need a tractor with 35 horsepower or better, because you're going to be doing spraying and things that require more power. In an orchard, you might want something bigger for mowing, spraying, and chewing up wood prunings. If you're doing vegetables and over-the-row cultivation on a large scale, then you have to start looking for more specialized tractors.

Terrain is an issue, too. If you're in hilly or mountainous conditions — orchards and hayfields tend to be on less ideal croplands — then you need the right kind of tractor to be safe. One with a low center of gravity and equal-sized wheels, such as a Ferrari, Antonio Carraro, or Goldoni tractor, is best. Most of the people I deal with tend to have a mixture of crops, so they have to have something that is adaptable and versatile.

Once a tractor is decided upon, implements follow. "Most people think of a walk-behind as just a rototiller, but . . . there are literally a hundred attachments. For the mini-tractors, there are miniature balers that run with as little as 20 horsepower. People who raise small flocks of sheep or goats like the small bales," he says. There's a full range of market-garden implements for riding tractors, and even small-scale implements for grain production. And obtaining parts for specialty implements isn't as much a problem as it once was, Canales says, and usually takes no longer than for common U.S. makes and models.

There is no good way for aspiring farmers to learn how to run equipment, Canales notes, except by apprenticing on someone else's farm. "It's a good way to find out if you really have the tolerance for the amount of work involved." The rest just have to jump in and try to avoid expensive mistakes.

"Be realistic about the ground you want to cover," Canales advises. "A walking tractor and implements aren't a cheap investment, and then you may discover you really can't do all the work you wanted to do and you will have to reinvest in implements for a four-wheel tractor."

Canales believes in the old adage that you get what you pay for. "If I don't think it's high quality, I don't handle it. I don't sell cheap stuff," he says. "A spading machine may cost twice what a rototiller costs, but over time a rototiller destroys soil structure, while a spader will continue to improve soil structure."

He advises new farmers to go to organic conferences and talk to successful farmers about not only equipment but farming in general. "Conferences often have a panel of growers, and you should listen to them and see what might apply to your scale and circumstances. Look for some niches that may suit your area," Canales advises.

Profile: Ralph Moore

Market Farm Implement, Friedens, Pennsylvania

"We made the mistake of buying a tractor. That's where it all starts, because you can't do anything with just a tractor. You need implements," says Ralph Moore, owner of Market Farm Implement in Pennsylvania. After Moore and his brother started their organic farm in the early 1980s, they began purchasing and renovating equipment to use, first for themselves, and then for others.

"We had a mechanical background, and we liked working on cars and so on," Moore says. "After a while the equipment became more full-time, until one year it came around to spring and we didn't have time to farm, because we were so busy trying to take care of our customers. We haven't farmed as a business since 1989, though we still grow some crops and test out new machines so we learn what works."

The Moores specialize in equipment for vegetable crops, and over the past two decades they have built an excellent reputation and a national customer base. Besides field-testing the equipment they sell, they also take time to discuss with customers how to find the right combination of equipment for their operation. One of the most common mistakes, Moore says, is starting with a tractor that is too small.

The tractor is the last thing you should buy. Select your implements first. How many rows do you want to cultivate at a time? You're investing in a production system — don't invest in a slow one. We started with a 35-horsepower tractor. We were going 1 mile per hour and it was lugging the engine. We ended up with a 65-horsepower tractor.

On the other hand, buying all of your equipment at once not only is hard on the wallet but can make for an exhaustingly steep learning curve that first season. "Look at the best way to allocate your money to save you the most amount of work," Moore advises. "Maybe you can buy one machine a year. It's easier to hand-plant than it is to hand-weed, but planting comes first, so people tend to buy a planter. But really they should tackle the hardest job first and buy a cultivator."

Of the field operations required for a vegetable crop — tillage, planting, cultivation for weed control, and harvesting — tillage requires the most horsepower. But it doesn't require specialized equipment like so many other jobs in vegetable production. For that reason it's usually fairly easy to hire another farmer to do your tillage, Moore says, and save your equipment budget for other items.

Of the other three operations, weed control is the biggest challenge for organic growers, and harvesting is the most expensive to do by hand, Moore says.

Machine harvesting a crop, that's fun. You don't have to deal with a crew of harvesters, which is usually the most expensive part — having them harvesting the crop and put it in a box. Planting, tilling, cultivating — each takes about an hour per acre with even a slow tractor. But harvesting . . .

In our work the most profitable machine we've found is a green bean harvester. You can till an acre in an hour, plant it in an hour, cultivate it in an hour, and then harvest it with the harvester. In 21 man-hours we could grow an acre of green beans and put it in the box, and it was worth $50 to $100 per man-hour. That's how you should equate one crop and one piece of equipment to another: figure out the income for each crop per man-hour, based on using a particular piece of equipment, so you can compare them.

On the other hand, a good cultivator might be a better initial investment for an organic grower, since it's often easier to find willing hands to help with harvesting than with weeding. It's common for vegetable growers to own a few different types of cultivators to accommodate different crops and conditions, but, Moore says, "now there's a good cultivator that will do anything you want it to do. A lot of people use plastic for weed control, but it's overkill. There are cultivators that can weed an acre in half an hour, and they will get all the weeds and you don't have to deal with all that plastic. Every farmer has a horror story about weeds, but nobody complains too much about planting, even if they have to do it by hand."

Moore recommends starting an equipment search by looking at domestic brands. Of course, some types of equipment aren't made in the United States. "A lot of the European stuff is really good quality," Moore notes, "but for small growers that's often not as important, since they aren't going to wear things out."

Used equipment is fine if you can find it, Moore says. "Vegetable equipment is pretty specialized, and there's not much of it available used. It depends, too, on how handy you are at fixing up tractors and stuff. If you're not a good mechanic, secondhand equipment is going to get pretty expensive," he notes.

"Really good equipment makes life easier," Moore concludes. And, being a farmer, equipment tester, master mechanic, and nationally respected dealer, he should know.

Plants 101
Basic Organic Growing

> *The land gives all that man can wish, if he*
> *cultivates it as he should.*
> **PHILIP OYLER**
> *The Generous Earth*

Though there are many differences between vegetable, fruit, and field crop production, they all deal with plants, and there are some general principles that apply when growing any type of plant. There are also specific organic practices to follow no matter what type of crop you're producing. We'll cover those topics first, then move on to the differences between gardens, fields, and orchards.

Choosing the Right Crops for the Environment

There's a reason no one grows rice in North Dakota or ginseng along the Texas coast — you'd never get a crop. Different plants have definite preferences about growing season length, soils, heat, and precipitation. And different varieties of the same plant species will grow better or worse in different locations. For example, a tomato variety adapted to the short, cool growing season of Maine will produce happily there but would probably sulk under the hot sun and high disease pressure down in Georgia.

In general, most species of fruits, vegetables, and field crops are similar in liking loamy, well-drained soil, a pH between 6.4 and 7, and no shade.

(So do most weeds, for that matter.) To be at their best, vegetables need even moisture throughout the growing season, but some types of small grains, like wheat, do very well in drier climates. Most fruits and nuts are particular about winter temperatures, while many annual crops will shut down production if the summer days are too hot or too cool to suit them. Numerous varieties of plants have been bred for specific climates or to be resistant to specific diseases. So, the first decision facing the new organic grower of any type of crop is choosing the species and varieties of species that are best suited to growing in your soil, climate, and rainfall zone, and that are resistant to the pests and diseases most common in your area.

Climate

Start with the climate zone map developed by the U.S. Department of Agriculture. Zones are designated according to how cold it gets in the winter. Zone 1, the coldest, denotes areas where winter temperatures routinely reach 50 below zero (-45° C). Zone 11, the warmest, is where temperatures don't go below 40 (4° C). In any reputable seed catalog, each listing of a tree, shrub, or perennial plant (a plant that lives for many years, as opposed to an annual, which lives and dies in a single season) should indicate the appropriate climate zones for that variety. For annual plants (field crops and most vegetables), the catalog should list the number of frost-free growing days needed by that plant to bear a crop. How hot your summers get is also important, though figuring out which species and varieties like hot or cool weather takes more research in gardening books or talking with extension agents and seed dealers.

Daily temperatures are important because they affect a plant's ability to photosynthesize and thus to grow. Plants adapted to warm climates grow more actively in higher temperatures, while plants adapted to cooler growing conditions will slow or

USDA Climate Zones

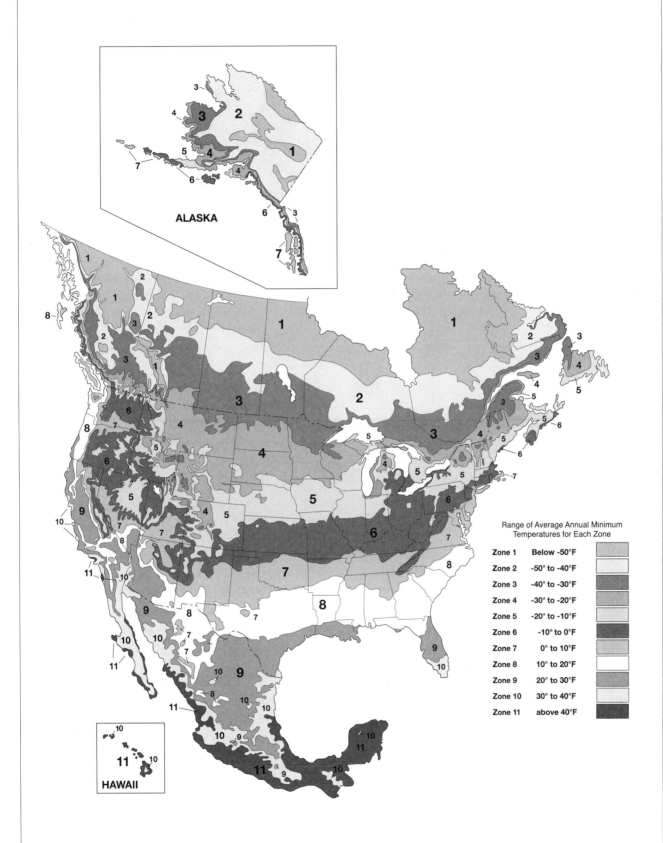

ALASKA

HAWAII

Range of Average Annual Minimum
Temperatures for Each Zone

Zone 1	Below -50°F
Zone 2	-50° to -40°F
Zone 3	-40° to -30°F
Zone 4	-30° to -20°F
Zone 5	-20° to -10°F
Zone 6	-10° to 0°F
Zone 7	0° to 10°F
Zone 8	10° to 20°F
Zone 9	20° to 30°F
Zone 10	30° to 40°F
Zone 11	above 40°F

Frost-Free Days

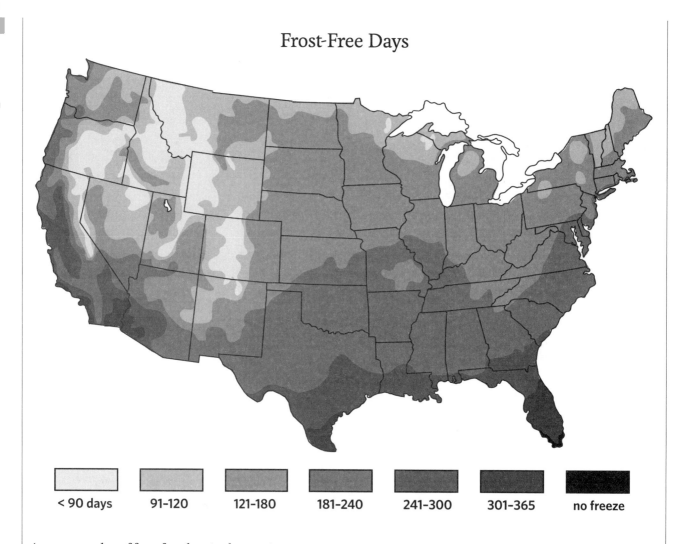

| < 90 days | 91–120 | 121–180 | 181–240 | 241–300 | 301–365 | no freeze |

Average number of frost-free days in the growing season.

stop growth when temperatures exceed their tolerance. This is quite noticeable in grasses, so that some livestock producers in the South will even plant different pastures to either warm-season grasses or cool-season grasses. The cool-season pastures are grazed spring and fall, and the warm-season pastures carry the herd through the hot summer weather.

In a garden, where most growers are producing at least some vegetables that originated in more tropical climates, several techniques can make the growing season longer and warmer than it really is, such as starting plants indoors. (See chapter 6 for more information.) This is not feasible for field crops.

Soil

Soil type is a little less critical, since crops will do better or worse on different soil types — rather than producing nothing or simply giving up completely, as they will when planted where it's too hot or cold for them. Some exceptions to this guideline are fruit trees, which won't tolerate a poorly drained site, and blueberries, which won't grow except in a very acid soil. Still, it's helpful to develop a working knowledge of which species and varieties of the crops you intend to grow will do best in your soils. Root crops like carrots love loose, sandy soils, but will need extra attention in the form of organic matter and maybe raised beds in a heavy soil. Alfalfa is happiest on a lighter soil, while red clover prefers something heavier and slightly more acid.

Average Annual Precipitation

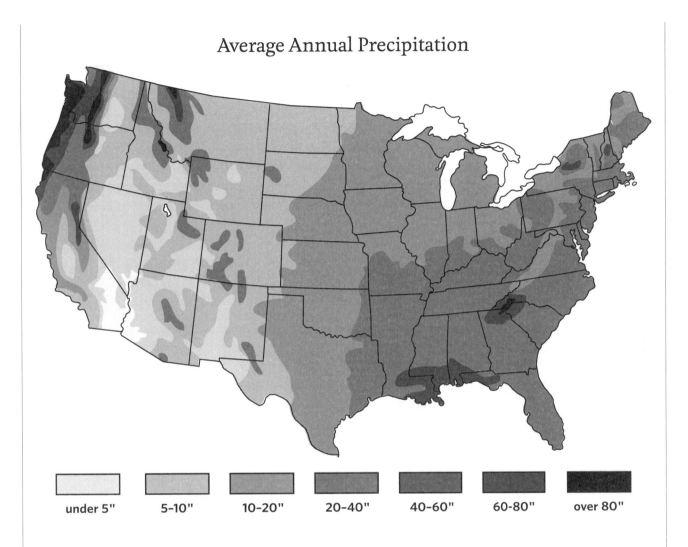

| under 5" | 5–10" | 10–20" | 20–40" | 40–60" | 60–80" | over 80" |

Average annual precipitation (in inches), 1961–1990.

Rainfall

The third consideration in plant selection, the amount of precipitation during the growing season, is not an issue if you are irrigating and have a reliable well or other water source. Irrigation is used by most market gardeners, but field irrigation is expensive, and most small growers of field crops find it more feasible to adapt the crop to the expected annual rainfall. Precipitation differences are why most corn is grown in the high-rainfall area east of the Mississippi, and most wheat in the dry plains. Tree fruits and other perennial crops should be planted in areas with ample rainfall, as well as mulched or cover-cropped. If you're growing just a couple trees or a few brambles, these can be watered by hand in a dry climate.

To figure out how much precipitation you can expect in your area, consult an annual precipitation map (see above, or look one up on the Internet) or check with local conservation resource organizations.

Disease and Pest Resistance

Selecting disease-resistant varieties is critical in an organic system. Since your treatment options for disease control are limited once the disease takes hold, your focus has to be on preventing disease. The first step, obviously, is picking plants that aren't likely to catch it. You don't hire a kid with a hay allergy to stack bales, and you don't plant tomatoes that are prone to fusarium wilt when that fungus is present in your soil. Which leads to another important point: it's a good idea

Irrigation

When enough water doesn't fall from the sky, irrigation can save the crop. There are three basic ways to irrigate: subsurface, surface, and overhead.

Subsurface irrigation is expensive to install and and not worth the trouble in all but very specialized circumstances. It's not practical for most farmers.

Surface irrigation done with drip hoses is economical and efficient, and it is the method of choice for many market gardeners. Drip hoses are simply water hoses with a lot of little holes. They are laid along a plant row and deliver water directly to the root zone, with almost no loss from evaporation. Laying the hose under a mulch conserves water even better. Drip hoses are widely available in garden supply catalogs and stores, and with appropriate couplings they can be easily configured to suit any garden.

Overhead irrigation units are simply high-powered variations on the standard lawn sprinkler and can be feasible for gardens, field crops, and even pastures. Systems vary widely in price and complexity, beginning with standard lawn sprinklers, which work fine in a garden, and "pod" sprinklers that are simply more powerful lawn sprinklers and can be placed and moved by hand for garden, pasture, or field irrigation. You can spend more for industrial-strength sprinklers that really shoot a lot of water a long way, but you'll need the hoses and water volume to match. For larger areas, or where irrigation is necessary for longer periods at greater volumes, central pivot and side roll (lateral moving) irrigation systems are used. These large-scale systems are a major investment and are not found on the typical small-scale farm.

to find out what diseases and insect pests are the biggest problem in your area. Plant varieties are never resistant to every possible problem, so you have to pick those that are resistant to your particular problems. Talk to other growers, call your local extension agent, or get on the Internet and do a little research. Growers on the West Coast will have a quite different set of problems in most cases than those on the East Coast, while northern growers will have more problems related to weather and fewer insect and disease issues than growers in southern areas.

Maximizing the Plant Environment

Selecting plants to suit local environmental conditions is key to success in organic agriculture. Improving the even more local plant environment to maximize plant growth and productivity is equally important. The big factors that a grower can affect are:

- Soil health and fertility
- Light availability
- Airflow and humidity immediately around the plants
- Competition from other plants (weeds)
- Diseases and pests

Soil Health and Fertility

We discussed soil fertility in chapter 3, and the importance of building a good soil home becomes evident as soon as you move in its new occupants: your crops. Healthy soil harbors few soil-borne plant diseases and grows vigorous plants capable of shrugging off pests that would quickly decimate a weaker crop. The importance of healthy soil can't be overemphasized.

Light Availability

Light availability is an essential factor for plant health and for disease and pest control. All our food ultimately comes from sunlight, since we eat either plants or meat from animals that ate plants.

Only plants have the ability to convert sunlight into usable energy for the rest of us. This process, called photosynthesis, depends first on sunlight hitting green leaves, and second on the leaves having certain chemical ingredients, trace minerals that come from the soil. The minerals are necessary to run the chemical reaction that converts light and heat energy from the sun into the carbohydrates, proteins, and sugars that the rest of creation uses for food: grains, fruits, vegetables, and greens. Though many species of plants are adapted to thrive in shade or partial shade, as a rule nearly all food plants and livestock forage crops prefer full sun.

Photosynthesis also requires water and oxygen; if either becomes less available, the process slows and may even come to a complete halt. That's why soil that is too dry (not enough water) or too wet (not enough oxygen) will stunt growth and finally kill the plant if not corrected in time.

Light availability for an individual plant is most dependent on plant spacing. Plants that are too crowded will shade each other, and none will get enough light to reach full growth potential.

Airflow and Humidity

Airflow and humidity are also dependent on plant spacing. Most diseases and insects thrive in conditions of high moisture and little air movement. Plants crowded close together impede airflow and don't dry off quickly after rain, watering, or heavy dew. Plants that are crowded together also compete with each other for soil moisture and nutrients.

Photosynthesis is the chemical process by which plants convert sunlight to mass. For photosynthesis to work, plants must have adequate water and nutrients.

When planting, follow the spacing recommendations for the variety you're growing. Ask your supplier about any variation from the recommendations that might need to be taken into account for your area. For example, standard-sized apple trees grow quite a bit bigger in southern states than they do in northern states, and apple growers must take this disparity into account when laying out an orchard.

Weeds, Diseases, and Pests

Though you don't want to crowd plants so closely together that they are competing for light, water, and nutrients, you also don't want to space them so far apart that there's plenty of space for weeds to move in. Ideally, when the plants are mature their leaves will just touch their neighbors, shading the ground completely. Until then, a grower needs to be diligent about not letting weeds get big enough to compete with the crop. Weeding, called cultivation when you use a machine to do it, is critical. The chore can be lightened by the use of mulches or cover crops that won't hamper the main crop's growth.

Pests and diseases in the well-nourished, well-spaced, well-watered, and well-weeded garden of the organic grower are only rarely a major problem, since everything is so healthy that it's hard for a problem to find a weak spot to get started. But there are a few pests and diseases (Colorado potato beetles and fusarium wilt are two of the most familiar) that seem to pop up no matter how perfect the growing conditions are. Crop rotations, eliminating winter habitat for pest insects, and using green manures with disease-fighting properties are important weapons against the more intransigent pests. Crop rotations and cover crops are also integral to organic growing.

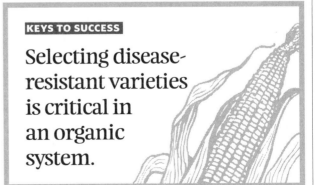

KEYS TO SUCCESS

Selecting disease-resistant varieties is critical in an organic system.

Best Management Practices for Organic Agriculture

While organic growers are thinking about choosing the right varieties and optimizing conditions for them, they can't lose sight of maintaining soil biodiversity and fertility. Fortunately, crop rotations and soil covers, the gold standards of organic practices, address these concerns simultaneously.

Crop Rotations

The most important (and also required) management practice of organic growers is crop rotation. In organic production, you never grow the same crop on the same ground two years in a row. Most certifiers won't even allow an every-other-year rotation, such as corn/soybeans/corn, and most growers prefer four years or longer before you plant that ground to that crop again.

There are two major reasons for rotating crops. The first, mentioned earlier, is that rotation is a primary tool for pest and disease control. The second reason, as discussed in chapter 3, is to build and maintain soil fertility. Each crop differs in the amounts of nutrients it needs from the soil and in what it gives back to the soil. Growing the same crop on the same spot year after year ensures that the nutrients most needed by that crop are in smallest supply, while the population of pests and diseases that prefer that crop will be growing explosively. Following a heavy nitrogen feeder like corn with a nitrogen builder like clover or alfalfa cleans and restores the soil. Adding a year of small grain to the rotation adds more organic matter and confuses pests even more.

Crop rotations should include one or more cover crops used as green manures that are plowed down to increase soil fertility and organic matter. Having a forage crop in the rotation will increase earthworm populations, add organic matter, and improve tilth. A legume as all or part of the forage planting will fix nitrogen in the soil for any row crops the following year.

Different species of plants grow roots to different depths, and those that grow deepest can be very valuable for bringing up nutrients that have leached beyond the reach of other plants. Alfalfa, for example, is valued for this reason: its roots can reach 30 feet (9 m) or more, which also makes it much more drought-tolerant than more shallow-rooted plants, a significant consideration in dry sandy soils. In contrast, other plants like grasses have spreading, bushy root systems that hold soil tightly. Having both deep and spreading types of root systems in a crop rotation maximizes soil benefits.

Garden and field crop rotations differ significantly because the plants grown in each situation are different. The following chapters will address rotations in each situation, but the principle remains the same: build a rotation that will build fertility and break up disease and pest cycles.

Soil Covers

Bare soil is an opening for erosion, nutrient leaching, weeds, and moisture loss. Bare soil also represents a lost opportunity to build fertility. As we discussed in chapter 3, soil covers, equating

Low-growing ground covers can be seeded between tall-growing row crops as a living mulch. Legumes such as clover, which fix nitrogen in the soil, are a natural cover-crop partner for corn, which is a heavy feeder.

generally to cover crops and mulches, are integral to success in organic farming. Recall that in the garden, mulches and cover crops are used to cover soil, while in the field only cover crops may be practical, especially in winter or during a fallow season.

Cover Crops

Cover crops not only protect soil, they build it. Any crop planted but not harvested is a cover crop, whether its primary purpose is to be plowed under as a green manure, hold nutrients and soil through the winter, or stifle weeds during the growing season. Chapter 3 addressed all the lovely ways in which cover crops can help a farm.

There are additional methods for maximizing soil coverage and plant productivity. Interplanting, also known as companion planting, is less common than cover-cropping but also useful for making full use of ground. This practice involves planting two types of crop plants in the same field or close together in the garden. These companion plantings are selected to be not especially competitive with each other and have the additional benefit of producing two crops in the space of one — either two crops for you, or one for you and one for the soil. The classic American Indian combination of corn, beans, and squash is one example. These species have different root systems, somewhat different nutritional needs, and different aboveground structures that minimize competition. This allows plenty of room for each of the crop plants, and there's no bare soil.

Mulches and Water Conservation

Managing soil moisture is not required by the Final Rule but is essential in a market garden and important elsewhere. Maintaining high organic matter levels in the soil is the first step, since among its many benefits organic matter acts as a sponge and holds water. Where water is naturally in short supply, as it is almost everywhere during at least part of the season, growers working in gardens or other small spaces use mulches to conserve moisture. A winter-killed cover crop serves the same function in fields, though crops should still be selected to suit the average annual rainfall. And in both garden and field, a mulch is as effective as a cover crop in preventing soil loss from erosion.

Required Management Practices for Organic Certification

In addition to crop rotation and maintaining soil cover, the Final Rule has further requirements for certified organic plant production:

- Land used for organic crops must not have had any prohibited substances applied for at least three years before the first organic harvest.

- An adequate buffer zone must be maintained around all organic crops to protect them from chemical or pollen drift from nearby conventional crops.

- Tillage and other practices must not cause either erosion or water pollution.

- All seed, including seed for cover crops, must be certified organic, unless you can document that certified organic seed for the crop you want is not available. You are not required to use open-pollinated seed; hybrids are fine. If legume seed

Mulch under squash keeps the vegetables clean, suppresses weeds, conserves moisture, and prevents soil erosion.

has been treated with nitrogen-fixing rhizobium bacteria, the bacteria must be documented as a non-GMO strain. All planting stock that will be treated as an annual (purchased, planted, and harvested in one season) must be certified organic; there's no exception for unavailability, as there is with seed. For perennial planting stock (trees, shrubs, and so on), if you can't find an organic source, you may plant nonorganic (but non-GMO) stock, but the harvest can't be sold as certified organic until the plants have been under organic management for a full year.

Additionally, to be able to market your crops as certified organic, you must have a written, ongoing, long-term farm plan for building and sustaining soil fertility and plant health. The plan must include crop rotations and the specifics of how you will use compost, manures, green manures, purchased amendments, or a combination of these to achieve that goal. Your methods and techniques for dealing with pests and disease and maintaining biodiversity (see chapter 13) must also be included. You must also maintain records of all activities in the field and keep receipts for all purchases and other transactions in connection with the crop.

If you have a system, this paperwork is less onerous than it sounds. Chapter 14 will cover the pencil-pushing details. For now, let's discuss how these rules are applied in the field.

Transitioning to Organic

The Final Rule mandates that growers follow organic protocols for three years before they can market their products as certified organic. The three-year transition period from conventional to organic production allows time for prohibited chemicals in the soil to either leach away or break down into inoffensive chemicals. Growers should use this period to assess and build soil fertility by use of soil testing, grazing, crop rotation, composting, allowed purchased amendments, green manures, or any combination thereof.

Buffer Zones

The Final Rule requires buffer zones between your crops and those of your neighbor to prevent

The Final Rule on:
Seeds Used during the Transition

Use of certified organic seed is not required during the transition period. But you still can't use genetically modified seed or seed that has been treated with a nonapproved synthetic (which is surprisingly common). Legume seeds (clovers, beans, peas, and alfalfa) are usually treated with a nitrogen-fixing rhizobial bacteria, and if these bacteria have been genetically modified, you cannot use these seeds either.

If your land has been left fallow or has been in the Conservation Reserve Program or a similar government program, *and* you can document its status, you may be able to shorten the wait. But check first with your certifier. It's no fun to go through all the paperwork, pay the fees, and get the organic inspector out to your farm only to find out that you'll still have to wait the full three years.

chemicals or GMO pollen (from genetically modified organisms) from drifting onto your crops. Buffer zones will be assessed by your organic inspector and certifier according to whether they are sufficient to prevent the Final Rule's stipulation against "unintended application of prohibited substances."

The first order of business is to check with your certifying agency to see what it will require for a buffer zone. You'll need to show your certifier a farm map indicating your field dimensions and where your neighbor's fields are located. In most cases certifiers require a buffer zone of 25 to 30 feet (7.6–9 m), but these specifications can vary, so it's important to ask. If your neighbor uses a ground-driven sprayer, then 20 to 30 feet (6–9 m) could be sufficient. But if an airplane is doing the spraying, you may need a 60- to 100-foot (18–30 m) buffer zone. On the other hand, if you have a thick stand of pine along that field edge, you might get by with just 10 to 15 feet (3–4.5 m).

Farmers are generally loath to give up crop area, but the economics argue in favor of extra

The Conservation Reserve Program

The Conservation Reserve Program (CRP) pays an annual rental fee to farmers who agree to put agricultural land into soil- and water-conserving grasses or native vegetation. This preserves soil, protects water, creates wildlife habitat, and, in nearly all cases, stops the application of agricultural chemicals. The program targets highly erodible or otherwise environmentally sensitive land, such as land next to wetlands or waterways. The contracts are issued for a set number of years, so if you are buying land that is in the CRP, you will not be able to farm it until the contract expires. If you have land you'd like to enter into the CRP, contact your area Natural Resources Conservation Service office.

precaution. Buyers of organic crops intended for human consumption usually test those crops for GMO contamination, and if they find any, as one organic inspector once memorably remarked, "It's now $1.85 corn, as opposed to $4.50 corn."

Buffer zones can be planted along with the rest of the field, or you can put in beneficial organism habitat, such as grasses or flowering plants (alfalfa is nice for bees), or let the edges go to see what sprouts there. The only problem with that is they may sprout a lot of invasive weeds; these should be mowed regularly to prevent them from going to seed and infesting your fields.

If you plant the buffer zone to the same crop as the main field, then you must harvest the buffer zone separately. If you are using the same machinery, it must be cleaned of all crop residues before moving to the main crop. This cleaning must be documented with a record of the date and method(s) used if the main crop is to be organically certified.

If you have streams, ponds, or lakes on your property, these also need to be protected by buffer zones to preserve water quality. The only exception to this rule is in the case of management-intensive grazing (MIG), in which animals are grazed for only very brief periods along a shoreline. This practice can improve rather than deteriorate the shore. (Chapter 11 discusses the effect of MIG on soil retention.)

Tillage Practices

Tillage can aerate and loosen soil, but overuse will destroy soil structure and contribute to erosion and water pollution. Ways to reduce the damaging effects of tillage include the following:

- Minimize tillage (including rototilling), using it only when necessary to prepare seedbeds, incorporate green manures or other soil amendments, or as needed for weed control.

- Avoid tilling wet soils or very dry soils, which destroys tilth and compacts the soil, squeezing out air and water.

- Till as shallowly as feasible.

- Combine tillage operations where possible (such as plowing and disking at the same time).

A buffer zone can be planted to grass, shrubs, and other beneficial habitat, as shown here. Or it can be planted to crops, as long as they are harvested and stored separately from the certified-organic crop.

- Till in spring instead of fall. Spring tilling calls for more precise timing and good luck with the weather, but it helps the soil stay in place over winter.

Contouring

Growing crops on a slope requires extra measures to prevent soil erosion from rainfall. Think of how water always finds the easiest way to move lower, following any natural channel available. In a row crop, the bare ground between rows provides a quick conduit for any rain that doesn't sink immediately into the soil. If rain is falling heavily, quite often there's more water than the soil can instantly absorb, and that extra water starts moving downhill if it can. If the crop rows run straight up and down a slope, the water races downhill between the rows, taking a lot of your topsoil along. If the rows are perpendicular to the slope, there's no easy way for the water to move lower, and it stays in place until the soil can absorb it.

The practice of planting crops perpendicular to a slope, or along its contours, is called contour cropping. On slopes steep enough that water will sometimes flow through even perpendicular crop rows, strips of row crops can be alternated with strips of sod-forming forage crops, which will stop any water that manages to escape the contoured rows. This practice is called contour stripping,

and it's a beautiful sight in the fall to see a hillside striped with alternating bands of ripe corn and green forage. With both methods, it is essential to lay out the contours with the help of a device that will track elevation, to make sure the contours are running level across the slope.

It's much easier to plow and plant straight up and down a hill, but soil losses from this practice can be enormous. If you are planting on slopes that are steep enough to need contouring but strips have not been laid out, contact your local agricultural extension office or the regional Natural Resources Conservation Service office for information on obtaining technical assistance and possibly cost-sharing for properly laying out contours on your land.

Grassed Waterways and Check Dams

Erosion control methods commonly used in more gently rolling land are grassed waterways and check dams. A rolling field usually has a low spot where water collects and runs out of the field. In moderate to heavy rains on bare soil, a gully will form in the low spot, and eventually feeder gullies will emerge as well, as the soil erosion works its way uphill. To halt gully formation, farmers will leave the low spot in grass, rather than tilling it. When working in the field, they simply lift their equipment as they cross this grassed waterway.

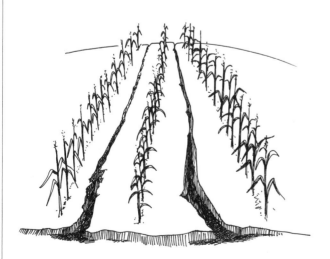

In crop rows that run up and down a slope, the bare ground between rows acts as a downhill channel for rainwater, which takes soil with it.

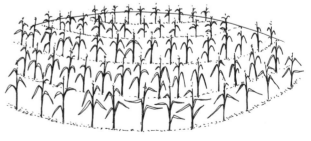

Crop rows that run perpendicular to the slope offer no easy downhill path for water. The rain soaks in instead of running off.

Preventing Soil Compaction

Especially on wet or heavy soils, wheeled traffic or extended use by livestock will compact the soil. Crops on compacted soils yield poorly or not at all, water puddles instead of being absorbed, and soil structure suffers. Remedy this by using low-impact tires and light equipment where possible, driving only on field edges, not working soil when it's wet or very dry, and rotating livestock through different pastures.

On steeper slopes farmers may place an earth berm, called a check dam, across the grassed water- way at one or more spots to further slow water runoff. Check dams are also used in steep pastures, where they not only prevent erosion but can be made big and deep enough to collect water, creat- ing ponds for wildlife.

Conservation and No-Till Systems

Conservation tillage systems stir the soil rather than tilling it, while no-till systems, as you might imagine, don't till the soil at all. These practices leave varying amounts of crop residue on top of the soil. They work best with larger seeds that pro- duce large, vigorous sprouts with the strength to push the residue aside.

On the plus side, conservation tillage usually minimizes damage to soil structure, and leaving residue on top of the soil minimizes wind and water erosion. On the downside, most conserva- tion and no-till systems sprout a lot of weeds. In conventional agriculture, spring weeds are dealt with by spraying herbicides; in organic agriculture they're a real problem, since the weeds are usually up and running before the crop is even planted. However, some growers and organic researchers are evolving successful organic no-till systems, which have the potential to greatly reduce fuel use in field operations and have long-term beneficial effects on cropped soils.

Another erosion-reducing tillage option is equipment designed to create miniature ridges for planting, which decreases wind speed at the soil surface, and equipment that clean-tills only in the crop row, leaving the space between the rows undisturbed for mulches or cover crops.

Seed Sources

To be certified organic, you must use certified- organic seed or have proof that they are unavail- able. Proof means doing your research: check all the possible organic sources you can find for that type of seed and document the phone calls you made or

Where shallow slopes converge and water runoff threatens to create a gully, a grassed waterway can stop the erosion.

On steeper slopes where a grassed waterway isn't enough, a check dam can be built across the waterway to ensure that rainwater is stopped and soaks in instead of running off with a lot of soil.

e-mail correspondence you had with seed suppliers. This makes using the Internet pretty much a necessity in order to find possible seed sources.

The rules are different if you are purchasing plants instead of seeds. If you can't find an organic source for bedding plants (annuals for the garden), then you can't sell the harvest from those plants as organic. There are no exceptions here.

For trees, shrubs, and other perennials, if you can't find an organic source you may purchase nonorganic plants (as long as they are non-GMO), and after a year of organic management you can sell their harvest as organic. (See chapter 8 for more information.)

Open Pollination

Though the Final Rule does not require it, many growers prefer to use open-pollinated varieties. *Open-pollinated* is defined as seed that results from wind, insect, or animal pollination in the field, without human interference. This type of seed is genetically diverse, and the best of each year's crop can be collected and replanted by the grower, so that over the years varieties uniquely suited to that farm's soils, weather patterns, and climate are developed.

Open-pollinated seeds do not produce such uniform plants as hybrid or genetically modified seed, and the plants are easily contaminated by pollen from conventional crops. But using open-pollinated seed opens up an amazing realm of possibilities. The staggering number of available varieties represents not only a long history of home breeding, but an incredible palette of tastes, as well as fine-tuned adaptability to a wide range of environments. For the food connoisseur, or for the market gardener, growing some open-pollinated varieties is the clear choice.

Hybrid and GMO Seeds

Hybrid and GMO seed has been artificially manipulated by humans — hybrids in the field, and GMO seed in the laboratory — to create genetically uniform plants. Hybrids are created by the hand-pollination of two closely related varieties, while GMOs can cross everything from close relatives to entirely different species by directly manipulating a plant's DNA. Hybrid seed is allowed by the Final Rule; GMO seed is not.

Hybrids are usually more vigorous than either of the parent plants, but they often do not breed true, so that you have to buy new seed each spring. The genetic uniformity of hybrids also decreases disease and pest resistance; if one plant develops a problem, they probably all will, since they share identical weaknesses as well as strengths.

Seed Treatments

Have you ever planted sweet corn seed that has a pink color? That's because it's been coated with a fungicide. Seed treatment with chemicals is fairly common, in order to prevent seed from rotting when it's planted in ground that is either too cold for sprouting or infected with a soil disease.

Seed to be used in organic systems must be untreated or have been treated with allowed, certified organic substances. For example, since legumes depend on being paired with the correct rhizobium bacteria to be able to fix nitrogen (as discussed in chapter 2), legume seed is usually inoculated with a specific species of rhizobium in case the right species isn't present in the field. This is acceptable in organic farming as long as the rhizobium is not genetically modified.

Collecting from Open-Pollinated Plants

Vegetables that are pollinated by insects or wind — brassicas, cucurbits, onions, beets, chard, spinach — will cross with each other, so that you'll get hybrid pumpkin-squash and other odd stuff unless you can separate the closely related species with enough space. If this isn't possible, cover the flowers with brown bags until you can pollinate them by hand with pollen from the desired mate. Brushes and other implements are used to deliver the chosen pollen.

Saving seed, especially from open-pollinated plants, is a specialty field, but that's no reason not to experiment. For more information and details, consult one of the good reference books listed in the resources section.

What's Wrong with Genetic Uniformity?

Genetic uniformity means that every plant in a field is exactly like every other, down to the last gene. So, if one plant gets sick or bug-infested due to a lack of resistance or other type of weakness, it's pretty certain every other plant in the field is going to get sick, too.

Genetic uniformity accounts for the virulence of potato blight in Ireland in the mid-1800s; only a single variety of potato was grown at the time in the entire country, and when blight arrived on the island, the result was years of devastating famine. Currently there is worldwide concern among plant scientists about the genetic uniformity of wheat and its lack of resistance to a strain of black stem rust fungus called Ug99. The pace of research to combat this threat has been frantic in recent years.

Sourcing Seed

Fortunately, untreated, open-pollinated, certified-organic seed is becoming more readily available every year, particularly for garden plants. The resources at the end of this book include a list of some well-established national suppliers, and more can be found on the Internet. Ideally, you should find a supplier that produces seed in a climate similar to your own, particularly if you're buying trees and shrubs. If that isn't possible, buy seed or plants adapted to a colder zone rather than a warmer one. Many suppliers offer seed for green manures and cover crops in addition to vegetables, a good thing since these seeds must also be certified organic (or be documented as not commercially available in organic form) if you wish to be certified organic.

If you have contacted several suppliers in search of organic seed for a particular plant and absolutely can't find any that meets the guidelines, then you may be allowed to use nonorganic seed — if you can document your search with phone logs, e-mails, or letters. Even then, you should make a quick call to your certifier before ordering to make sure the alternative seed is acceptable.

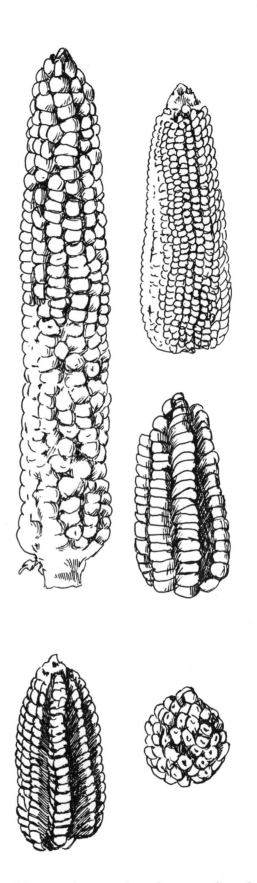

Corn, like many domestic plants, has a complicated genome and can be bred to a variety of growth habits, forms, and colors. The diversity of the species gives it flexibility in adapting to varying regions and pest and disease pressures.

The Final Rule on:
Nonorganic Seed and Planting Stock

The Final Rule includes some exceptions to its regulations for seed and plant sources. Nonorganic sources can be used in the following situations:

- **Nonorganic seed.** If growers can't find a source for untreated organic seed, then they are allowed to use nonorganic seed, as long as it has not been treated with any synthetic substances or prohibited natural substances. But nonorganic seeds may never be used to produce sprouts intended for human consumption (such as bean sprouts). During the transition period, you may use seed that is not certified organic without having to document unavailability of organic seed, but you still may not use any genetically modified seed or seed treated with synthetic substances.

- **Nonorganic seedlings.** Nonorganic transplants can be used to produce an organic crop *only* if a temporary variance has been granted by the U.S. Secretary of Agriculture due to a declared natural disaster.

- **Nonorganic planting stock.** Nonorganic planting stock for perennial crops (trees, shrubs, strawberries, and so on) can be used to produce an organic crop *only* after the plants have been under organic management for at least a year.

In addition, seeds, seedlings, and planting stock treated with prohibited substances may be used to produce an organic crop when the prohibited substances are required by federal or state phytosanitary regulations.

Saving Seed

Saving seed can lead to breeding your own lines of vegetables and — who knows? — maybe to establishing yourself as a producer and retailer of seed for unique, organic vegetables. Saving seed preserves genetic diversity, saves money, and can evolve strains of plants that are uniquely well suited to your soil and climate.

There are two basic rules for seed saving:

- **Save seed only from open-pollinated plants.** F1 hybrid seed won't "breed true" to either of its parents or will be sterile. Seed catalogs that focus on organic growers usually note whether seeds are hybrid or not.

- **Save seed only from the very best plants,** those that are most vigorous, resistant to diseases and insects, and highest yielding. Observe plants in the garden through the season and, if necessary, put a flag or tape on those plants so you can find them when it's time to harvest seed.

It's easiest to collect seed from annuals. Biennial plants, those that don't produce seed until the second season, call for more patience. These include carrots, onions, beets, and chard.

GETTING STARTED

The easiest plants to successfully save seed from are those that are self-pollinating. Because each individual of these species pollinates its own flowers, the seeds nearly always breed true to the parent plant. In the garden these are the beans, peas, lettuce, peppers, and tomatoes.

- Beans: Collect the seeds when the bean pods are ripe and dry.
- Peppers: Collect the seeds when the peppers are red and dry on the vine, or at least shriveled.
- Lettuce: Collect seeds before they start to fall off the plant.
- Tomatoes: Pick the tomato at its peak of ripeness, squish out the pulp into a bowl, add some

water, and let ferment for three days at room temperature; 70°F (21°C) is about right. Stir the mixture gently twice a day, and it should grow some mold that will eat away the protective gelatinous casing around the seed. On the third day pour off what has collected at the top of the bowl. The good seed should be at the bottom.

DRYING AND STORING SEED

Collected, clean seed should be air-dried for at least several days at temperatures of no more than 85°F (29°C) — any hotter and you risk killing the seed. Ideally the seed should be at less than 10 percent humidity; it will feel almost crispy. Once dry, seeds should be stored in airtight containers in a

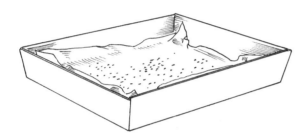

Seeds dry well in an open container on a layer of newspaper.

dark, dry area; a refrigerator works nicely. Label the bag or container with the date, variety, and any other appropriate notes, including planting instructions. Seeds in dry, cool storage can last up to five years.

Pest Management

If your soil is fertile and alive, your plants are well selected and well managed, and your farm is home to a diversity of life, then you should have minimal problems with pests. This is one of the great strengths of organic farming: the holistic approach has the effect of creating such healthy plants and such a well-balanced and resilient environment that problems are rarely able to get much of a foothold. Even if you were willing to use chemical pesticides, you probably wouldn't need them.

But minimal doesn't mean none, and especially in the first years when you are building the system and fine-tuning your planning and management skills, you will encounter pests that have to be dealt with directly and immediately. And there are some

Heirloom Vegetables

Heirloom vegetables are old varieties that have often been propagated by individual families for generations, and they are by definition open-pollinated, so that seeds can be saved to grow next year's crop. As hybrid seed varieties became widely available through commercial seed catalogs in the past 50 or 60 years, the heirlooms began to disappear. Yet heirlooms can offer unique tastes and colors unavailable in mainstream grocery stores, and are important to the organic mandate to preserve genetic diversity.

In response to the accelerating loss of heirlooms, the Seed Savers Exchange (SSE) was founded in Decorah, Iowa, in 1975 as a nonprofit organization devoted to preserving heirloom varieties of vegetables, fruits, and grains. Since its founding, the group's thousands of members have grown and distributed more than a million samples of endangered seeds not available through other catalogs. If you're ever in Iowa, be sure to make the pilgrimage to the SSE's Heritage Farm in Decorah to see the 23 acres (9.3 ha) of certified organic gardens and the 700 varieties of apples in the heirloom orchard, the most diverse in the United States.

Thanks to the Seed Savers Exchange and organic farmers and gardeners everywhere, over the past decade or so more heirloom varieties have become available through commercial seed retailers. Even my local garden store now sells heirloom tomato plants each spring.

pests that seem to show up no matter how good the system. As always, the organic farmer's best defense is a good offense.

Integrated pest management (IPM) is being increasingly used by conventional and sustainable farmers as an effective management method for reducing pest problems, and for managing them with fewer chemical inputs. IPM works great for organic farmers, too, but organic IPM is different from mainstream IPM in that it focuses almost completely on managing for health and preventing problems before they start, instead of conventional IPM's greater emphasis on monitoring for problems and then applying synthetic solutions. Spraying for a pest that's actually present in damaging numbers is a step up from spraying by the date on the calendar or as a preventive, as is widely done in conventional farming, but using allowed synthetics is still clearly required by the Final Rule to be the last thing an organic farmer does, not the first.

An IPM program, according to the brief, clear definition of the Minnesota Department of Agriculture, has six components: planning, setting action thresholds, monitoring and detection, proper identification, action, and evaluation of results.

Evaluation

Let's cover the last point first: if an IPM program is going to work in the long term, it's important to record what you do and what kind of results you achieve from those actions. If you're able to control a problem insect or disease effectively this year, then you know where to start next year, and you'll have some time to research and reflect on how to do even better. If your plan didn't work, you need to evaluate the possible reasons: Did you put on the row cover too late? Were the plants too stressed by dry conditions to fight off mites? Then you can figure out what you're going to try doing differently next season. In organic IPM the first planning question to ask is *not* "What can I use to attack this problem next year?" It's "What can I do to prevent this problem next year?"

As always, remember that what will and won't work for pest control will be unique to your farm and to the organic system you are building there. No one can tell you what exactly will and won't work on your farm, and you can only find out by continual research and experimenting.

Planning

Obviously, then, the most important part of pest management in an organic cropping system of any type is researching, planning, and implementing proactive steps to prevent insects and diseases from getting a foothold. This is done by building an effective crop rotation, timing plantings to avoid pest population surges, growing correctly spaced plants that are healthy and robust enough to ward off disease and discourage insects, and providing a diversity of habitat close to your crops for insect predators.

Crop Rotation Removes Pest Habitat and Food

Rotating crops prevents problems from building up in the soil. Many insects and diseases specific to a crop incubate or overwinter in the soil. When the food they like best is taken away, they tend to dwindle away as well. This is how crop rotation removes pest habitat and food.

Sometimes it's necessary to completely remove a crop and plant residues to make sure there's no place for pests and disease to overwinter. This is sanitation: removing diseased plants or parts from the garden and orchard, and composting them at a high temperature or burning them. So pick up the windfall apples and smut-infested sweet corn and get them out of there. Clean tilling in the fall before planting a winter cover crop will also discourage the buildup of some insects and disease.

Properly Timed Planting Can Mitigate Pest Damage

Timing of planting can be an important tool for the organic grower. Some insect pests and diseases favor specific temperature ranges, moisture conditions, and parts of the season, peaking at fairly predictable parts of the summer. For example, root maggots and vine borers appear early in crucifers and cucurbits; a later planting should help mitigate damage from these pests. But if your main pest problems with these crops are cabbage worms or flea beetles, you'll want to plant as early as possible, under a row cover, as a first line of defense. Early planting also helps against corn borers, earworms, potato leafhoppers, and fall armyworms, while later planting is better when dealing with cucumber beetles.

Proper Spacing Keeps Plants Healthy

Proper spacing, whether for trees, garden crops, or field crops, is essential. As discussed earlier in this chapter, light availability, airflow, and humidity for plants are dependent on plant spacing. When plants are crowded they shade each other, preventing them from reaching their maximum potential, and the ground-level environment has little airflow and excessive humidity, which encourages diseases and insects.

Beneficial Organisms Control Pests

Think of all the bugs you don't want to see in your garden: aphids, mites, thrips, tomato hornworms, potato leafhoppers, squash borers, cutworms, cabbage loopers, root maggots, Colorado potato beetles, corn earworms, and so many more. All of these pests have something in common: during at least one stage of their lives they are all food for predatory organisms, from birds and bats to other bugs like parasitic wasps, flower flies, lady beetles, and more.

Lady beetles (most of us still call them ladybugs, but that refers to only one of hundreds of species) love eating aphids. Spiders will eat about any type of bug, and ground beetles dine on the hated Colorado potato beetle larvae. Parasitic wasps (of which

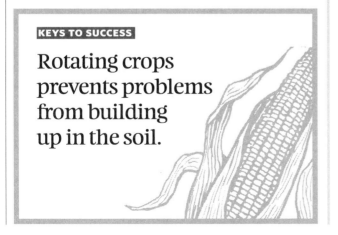

KEYS TO SUCCESS

Rotating crops prevents problems from building up in the soil.

Native Plants for Beneficial Insects

There's no shortage of flowering plants to choose from when designing insect habitat, but many growers start with species native to their region, since these will require the least care and be well suited to the needs of resident bird and insect populations. For specific information about insect-friendly plants in your area, check the Web site of your state university's horticulture department, contact a local Master Gardeners Association, or locate a regional retailer of native plants. The Wild Farm Alliance and the Xerces Society, which promote habitat for beneficial organisms, can also be helpful (see the resources for more information).

there are thousands of species, most of them tiny) lay their eggs right inside pest insects, destroying them. Though lacewing flies are vegetarian, their larvae are carnivores, scarfing up a big menu of moth eggs, caterpillars, scale insects, and whatever else gets in their way. Dragonflies eat a lot of flying insects, including mosquitoes and even deerflies, which don't bother plants but do bother gardeners. There are also pirate bugs, ambush bugs, assassin bugs, syrphid flies (which look like little bees), tachinid flies, and more. They or their larvae are of great benefit to the organic farmer in keeping pest populations under control.

Predatory insects can have a terrific impact on pest insects in an organic garden, as long as you give them a place to live. You can be either quite hands-off or very deliberate in developing habitat for beneficial organisms. In most areas native species of forbs, brush, and trees will sprout on their own if an area is left undisturbed. If you want to attract one or more particular species, though, it's worth doing some research on the types of plants or structures they might prefer.

Many adult beneficial insects need nectar and pollen, so having a lot of flowering plants throughout the season is important. These can be planted in borders around the garden or in rows or beds in the garden. When selecting plants, aim for varieties that will bloom in

Beneficial Insects

Predatory insects can greatly reduce populations of pest insects. You can encourage them to make a home on your farm by cultivating those plants and other habitat features the insects prefer for food and shelter. (Each of the insects in this sampling is shown at about twice its normal size.)

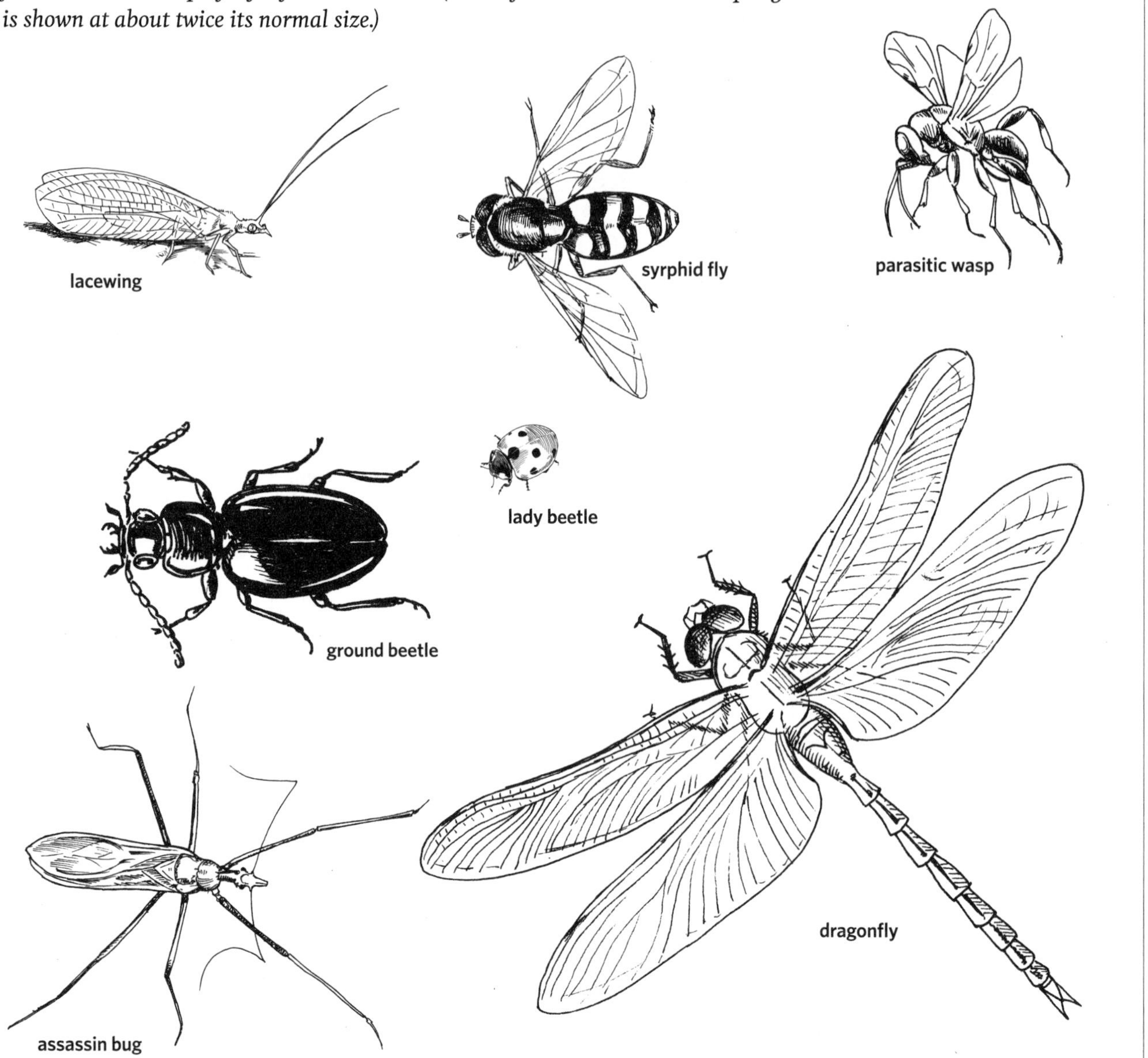

lacewing

syrphid fly

parasitic wasp

lady beetle

ground beetle

dragonfly

assassin bug

sucession from early spring to autumn, so that your garden has flowers in bloom through all of the growing season. Remember that beneficial insects tend to like their flowers small and full of nectar, rather than showy but dry. Yarrow, buckwheat, monarda (bee balm), sweet alyssum, black-eyed Susan, and boneset are all excellent. Asters bloom in the late summer when many other plants are done for the season, and comfrey goes all summer long if you cut it back after each bloom. (Comfrey is horribly invasive if it

escapes your garden, however, so don't plant it unless you can keep it contained.)

Brush piles, log piles, and rock piles provide habitat for many bugs and some larger critters as well, such as insect-eating toads. Buffer zones are excellent places for these larger habitat structures, as well as for brush and trees that shelter birds and bats.

With all these measures in place it's unlikely you'll have a lot of disease and insects in your crops. But you will have some every year. All those birds, bats, and predatory bugs need food to live on, after all. Organic growers prefer to avoid disease but will tolerate low levels of problem pests in order to keep predators fed. When a problem gets to the point where it has the potential to do some serious crop damage, then is the time for direct attack.

Setting Action Thresholds

This brings us to the next phase of the IPM system: setting action thresholds. This means determining ahead of time how much damage from insects and disease you can afford to tolerate before taking action. This calculation will be based in part on what happened last year. If you had corn borers in the field or codling moths in the orchard, then you know you're going to be early and aggressive in dealing with those pests this season. If you don't know yet which pests you'll be facing, then you have to be diligent about monitoring.

Deciding when it's appropriate to take action also depends on knowing the pest's potential for damage and its life cycle. Life cycle knowledge is all about figuring out where the weak points are and at what time in the season they are going to occur. Applying controls at the right time will make all the difference in how effective they are. With insects, in most cases young larvae are much more vulnerable than the adult insects, so find out when in the season you might expect to see them, and whether those larvae are going to be on the leaves, at the base of the plant, or in the soil.

This can get fairly involved. For example, when the first adult codling moth of the season is found in a pheromone trap in the orchard, you can figure that it will take 150 degree-days for the ones that escape the pheromone traps to breed a female who then lays eggs. By then you should have your kaolin clay or horticultural oil applied to keep the females from laying their eggs on the apples. At 250 degree-days the larvae are hatched out, and an organically approved pesticide may be in order.

Monitoring and Detection

Many growers walk their fields or gardens several times a week to look for problems, and orchard

Degree-Days

The life cycle schedule of many diseases and insect pests is determined by degree-days, or the measure of accumulated warmth during the growing season. You can look up how many degree-days must accumulate for a particular insect to emerge from its winter quarters, or to progress to the next life stage, or for a virus or bacteria to multiply. If you are tracking degree-days in your orchard or garden, then you can precisely time disease and insect control measures.

To track degree-days, you'll need a "high-low" thermometer, a device that records the highest and lowest temperatures during a 24-hour period. Add the high and the low together each day and divide by two to get the average temperature for that day. Subtract 50, and that's the number of degree-days for that day. (If the answer is below 50, then it's simply zero degree-days, since negative degree-days are not counted.)

For example, assume that a spring day has a high of 68°F and a low of 36°F.

> 68 + 36 = 104
> 104 ÷ 2 = 52 (average temperature for the day)
> 52 − 50 = 2 (number of degree-days for the day)

Each day's degree-days are added to the running total. The total is compared to pest and disease degree-day requirements, to tell you when to start scouting for problems.

Degree-days can be calculated using the Celsius temperature scale, but a Celsius degree-day is not the same as a Fahrenheit degree-day. Specifically, a Celsius degree-day is nine-fifths of a Fahrenheit degree-day. (In other words, 9 Celsius degree-days equals 5 Fahrenheit degree-days.) So when you're tracking degree-days, make sure you're using the same scale (Fahrenheit or Celsius) as your source of information about a disease or pest's life cycle. Correlations between growing degree days and the life cycles of specific pests can be found at many land-grant university websites.

owners generally put up insect traps and check them regularly. You are looking first for damage, which includes leaf discoloration, holes, bumps, and curling. Second, you're looking for insects and their signs: aphids, spittlebugs, and all their crop-eating kin, as well as egg masses, larvae, and weakened stalks. When you spot a potential problem, canvass the rest of the planting to see how intense and widespread the damage is. The damage level determines whether you're going to take immediate action or simply continue monitoring.

Identification

It isn't always easy to correctly identify a problem. Some nutritional deficiencies can look a lot like a disease, or even some sort of sap-sucking insect damage. If you can see the bug in action that will help; otherwise be careful about leaping to conclusions. Agricultural extension agents can be very helpful in these situations, or find someone in the neighborhood who knows their plant problems and talk to them. Take a sample along. Trying to make a difficult diagnosis yourself from

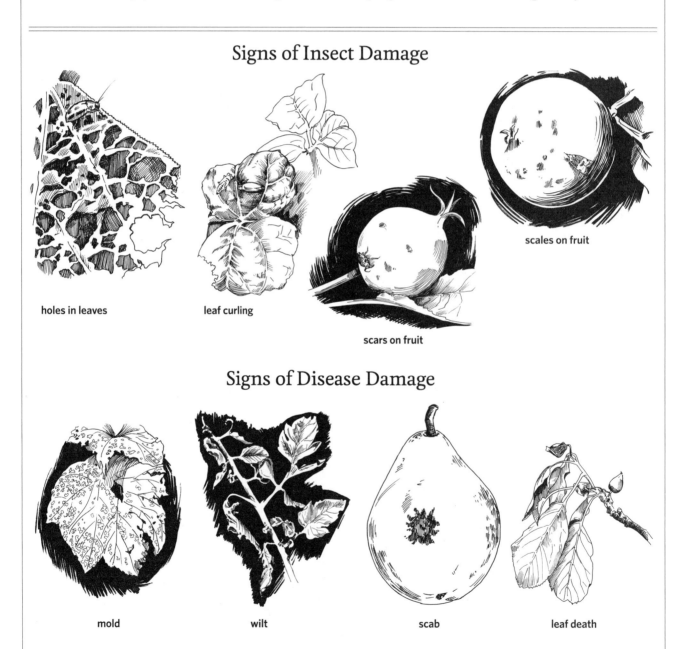

Signs of Insect Damage

holes in leaves

leaf curling

scars on fruit

scales on fruit

Signs of Disease Damage

mold

wilt

scab

leaf death

Damaged plants can signal insect or disease pressure. But sometimes it can be difficult to distinguish insect damage from disease damage. Experience and careful monitoring are key in diagnosing the problem and deciding on a course of action.

photos, Internet resources, or someone not familiar with your area is iffy, especially if, like most organic farmers, you simply don't experience problems often enough to have the necessary experience.

Taking Action

The last IPM step is action. With correct ID in hand and the criteria for action met, it's time to bring out your tools. These tools fall into cultural, biological, and physical categories (all of these categories of practices are allowed and encouraged by the Final Rule), with approved pesticides to be used only as a last resort. Many of these are already in place in an organic system, while others specific to the problem are brought out when the pest shows up.

Cultural Controls

Cultural controls are those pest-mitigating strategies that use plants or remove them, as appropriate. For organic agriculture, there are basically three:

- Providing habitat for beneficial insects, as described earlier
- Removing any plants that are hosts for diseases that afflict your crops
- Planting trap crops

Trap crops are plantings that are more attractive to the pest insects than the crop you're growing for market. Young corn is supposed to draw corn borers away from peppers, for example.

Biological Controls

Biological controls include beneficial organisms that control pest populations, and these are enhanced by the cultural controls of providing habitat. It's possible to purchase beneficial insects for release in your garden, orchard, or field. Some growers swear by this method, while others can't make it work on their farm. Some organic growers are also trying applications of fungi, viruses, bacteria, and nematodes that selectively attack pest insects.

Physical Controls

Physical controls include barriers like screening around the base of young apple trees to stop rodent damage, traps like saucers of beer for slugs, fencing to keep larger pests away, and guard animals. The exact type of physical control varies widely depending on the pest in question.

This category also covers washing, blowing, or vacuuming bugs by hand. Vacuuming, blowing, or spraying with water to physically remove bugs from plants has to be done vigorously enough to remove the bugs, but gently enough to not damage the plant. This takes some practice. Handpicking bugs off plants is another option; most old gardeners at one time or another have picked Colorado potato beetles and will tell you gross stories about drowning them in kerosene or squishing them bare-handed.

Many other clever controls are used by organic gardeners, but most are quite labor-intensive, such as placing collars around the bases of young plants to prevent cutworm damage. These types of controls are best suited to smaller gardens.

ROW COVERS

Row covers made of lightweight see-through material are used to keep adult onion maggots, leafhoppers, flea beetles, cabbage worms, root maggots, and others from laying on your young vegetables eggs that will hatch into chomping, burrowing larvae. It's essential to have the cover in place well before adults are expected to show up, and to close

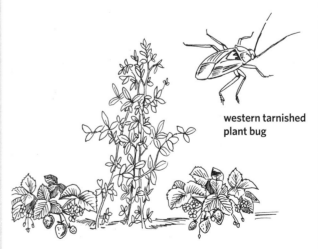

western tarnished plant bug

A planting of alfalfa can function as a trap crop to draw western tarnished plant bugs away from strawberry plants.

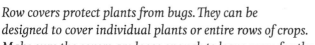

Row covers protect plants from bugs. They can be designed to cover individual plants or entire rows of crops. Make sure the covers are loose enough to leave room for the growing crop.

any seams or gaps. The edges of the cover can be held down with dirt, boards, mulch, or whatever is handy and effective; just make sure you leave enough slack in the material to accommodate the rapidly growing vegetables.

FENCING

If the pest in question has feathers or feet, physical or psychological controls are the answer. Fences may be necessary to keep animals from raiding your garden or orchard. If it's deer, the fence should be at least 8 feet high (2.5 m) to keep them from jumping it, twice the height of the average livestock fence. Rabbits and other small critters will be deterred only by a fence that has openings so small they can't slip through, such as a small-mesh, woven wire. Raccoons are more problematic, since they can climb almost anything, and if sweet corn is the reward, they probably will.

The best solution in many cases may be a mesh electric fence, powered either by a line from an outlet in a nearby building, or by a solar-powered or battery-operated unit. Not only is the cost minimal compared to the expense of permanent fencing, but installation of portable fencing is quick and simple. The biggest drawbacks of an electric fence are the need to check it daily and the need to unplug it before a thunderstorm hits, to make sure it doesn't attract a lightning strike. If the fence loses power and you don't discover it within a day or so, the animals probably will. Also, fur is a fairly effective insulator against electrical current, so the animal usually has to touch the wire with an ear or nose to receive a convincing shock. Tie fluttering strips of plastic at intervals along the fence so

the animal is sure to see it rather than just stumble through, and coat sections of wire with peanut butter to encourage them to sniff it.

GUARD ANIMALS

Another option, if you have the space and the inclination, is dogs. In many cases a dog that takes its guard duties seriously can eliminate the need for a fence. There are just a few wild animals (skunks, porcupines, wolves, bears, and some coyotes) that won't get out of the way for a dog, and those aren't big problems for crops.

In many areas, it is illegal for a dog to chase deer, so you may need to train your dog not to. Most dogs quickly realize they will never catch a deer, and just need a little encouragement to learn to be content with keeping them out of the yard instead of running after them into the next county.

The right cat, one that really likes to hunt, is very effective against mice and voles. Such a cat is useful in both the garden and in storage areas, and should be well cared for but not overfed.

BIRD CONTROLS

Bird damage can be as much a problem as that from four-footed animals. Many birds, especially sparrows and starlings, love berries, tree fruits, and even tomatoes and will either take the whole fruit or leave big holes in it. Some, like crows and turkeys, will pick seed out of the ground or consume whole rows of freshly sprouted corn. But insect-eating birds like swallows and purple martins are tremendous at keeping many types of insects under control, so you don't want to discourage those types from living on your farm. The idea is just to keep the fruit and grain eaters out of the crop. Growers use a variety of devices, from netting draped over plants, bushes, and trees to scarecrows and noisemakers.

Netting works until it gets holes, and it can be a job to put it on and take it off. Though netting also makes harvesting inconvenient, for a small area with a big problem, it's probably the best solution. Other growers have tried things like stringing twine between trees or above bushes, and tying pie tins or other sparkly things to it so they twist in the wind. Scarecrows, fake owls and hawks, and noise cannons all decrease in effectiveness the longer they're out there, since birds quickly become used to anything that is a regular presence. The most effective of scare tactics seems to be a noise broadcaster that emits a variety of noises at irregular intervals.

Eliot Coleman, author of *The New Organic Grower*, says that putting out piles of corn for birds (and deer) until the crop is too big to interest them works well for him, on the principle that if there's other, easier food nearby they won't bother the sprouts.

Many growers' solution to the bird problem is to put up with a certain level of damage. Most farmers can spare a few apples for the birds. But in some areas deer and wild turkeys are so numerous that they can do serious economic damage to field crops, and in this situation control is difficult. Some states have payment programs for wildlife damage, but you must have photos or other documentation to make a claim.

A backpack sprayer is handy for applying sprays — those allowed by the Final Rule — in an orchard or garden.

Approved Pesticides

Although there are some allowed organic pesticides, the rule states that they may be used only when all other control measures, biological, cultural, and physical, have already failed.

The rather short list of approved synthetic and natural pesticides includes:

- *Bacillus thuringiensis* (Bt), a bacteria that occurs naturally in the soil and is effective against some insects
- Pyrethrums, effective against many insects and occurring naturally in chrysanthemum flowers (but do not mistake the synthetic version for the natural one)
- Insecticidal soaps made of potassium salts
- Diatomaceous earth consisting of the silicon-based fossils of certain marine microorganisms
- Azadirachtin, the active ingredient in neem tree preparations
- Plant-based (not petroleum-based) horticultural oils, used to suffocate aphids, mites, whiteflies, and thrips

Bt and pyrethrums break down very quickly in the environment and are often only moderately effective. With the insecticidal soaps, diatomaceous earth, and azadirachtin, best results are achieved when used while insects are still young or before the infestation becomes widespread.

Disease Management

Once a disease has gotten a foothold, disease controls are pretty much nonexistent for an organic grower in the current growing season. For next year, the options are to lengthen the crop rotation, find more resistant seed if it's available, plant cover crops that are effective against that disease if possible, and research allowed organic preventative sprays for that disease, if any are available. In most cases, a longer crop rotation is the best option for gardeners. Sometimes you have to move the affected crop a considerable distance to prevent it from being afflicted again. On a small property this might even entail renting land elsewhere for a few years.

Weed Management

A weed, as many authors have noted, is simply a plant in the wrong place. Corn from the previous year's planting is a weed in this year's soybeans; potatoes missed in last fall's digging will have to be hoed out of the next season's salad greens. And there's no end to plants that arrive on their own to compete with a crop, from the relatively inoffensive purslane and lamb's-quarters to their more difficult compatriots like stinging nettle and quack grass.

Weeds, unfortunately, like precisely the same growing conditions as crops, and it's a rare soil that doesn't have a lot of weed seeds in it, just waiting for an opportunity to sprout. The farmer's job is to get the weeds before they get your crop, crowding and shading out your plants.

Basic Organic Weed Control

Weeds successfully employ various strategies to survive where they aren't wanted. Most have evolved to be more tolerant of less-than-ideal conditions than crop plants, a real advantage. Annual weeds will sprout faster, grow quicker, and produce many, many more seeds than just about any food plant.

But they do have their weak points: most (not all) annual weeds are fairly small and short-lived and don't compete well for light. They are easy to kill with cultivation or smother with mulch, if you get them young. If you keep the weeds under control for just a few weeks, until your crop is big enough to shade the ground, any weeds that sprout after that probably won't have enough light to make much trouble.

There is one caution: Many weedy grasses and some other problem plants are only encouraged by cultivation, since a new plant will grow from each part of the root that you just sliced up. If that type of plant is your problem, heavy mulch might be a better tactic than cultivation.

Dealing with Persistent Weeds

For a really persistent weed problem in a part of a garden or field, many growers will plant and then till under a green manure chosen especially for its suppressive effect on weeds. Buckwheat and oats are frequently used. Planting and plowing under are done several times in succession during the growing season for best results. Another common tactic for tackling a difficult weed problem is the stale seedbed method, in which the ground is tilled several weeks ahead of planting, then cultivated two or three times. Each cultivation kills the sprouted weeds and turns up more weed seeds, which will then sprout and be killed by the next cultivation. This greatly depletes the soil's weed seed bank.

Biennial and perennial weeds are more difficult. These appear not only in crops but also in pastures, orchards, and other permanent plantings. The several varieties of thistle are probably the best known and most difficult to eradicate, but there are many more. Organic growers continue to experiment with control methods such as encouraging the spread of diseases specific to thistle, but for most people the most common control methods are mowing and hand-chopping.

The timing of these operations is important. Mow too early and you won't make much difference; the weeds will come right back. Mow after the weeds have set seed and you won't make any difference at all, since next year's weed crop has already been planted. Mow just as the weeds are blooming but before the seed is mature, and you'll be able to take a break for a few weeks. Some really persistent species, like wild burdock, are really only discouraged by taking a shovel and cutting off the taproot below the soil surface. But mowing at least keeps them from going to seed.

Profile: Jeff and Lori Fiorovich

Crystal Bay Farm, Watsonville, California

From their 3½ hilltop acres (1.4 ha) on the California coast, Jeff and Lori Fiorovich can see Monterey Bay. On the land around their certified-organic Crystal Bay Farm they see acres and acres of strawberries, all well fumigated and sprayed. Lori says:

We're surrounded on two sides by large conventional farmers. Dole is across the street. Thank God people can make a choice about what kinds of products they buy in the store. And then there's folks like us who have the opportunity to take it a step further and choose to do something different. That, for me, is where it's at. Parents come here telling me they used to think their kids were allergic to strawberries until they started buying our berries and found their kids didn't break out in rashes anymore. I had a friend tell me she didn't even like strawberries until she tasted ours.

In addition to strawberries the Fioroviches raise a selection of vegetables, lots of squash, and a large pumpkin patch that is a magnet for school groups each October. Forty trees now coming into bearing offer a wide selection of heirloom apples, from Ashmead's Kernel to Hidden Rose. Too Cute the pony, Sweetpea the goat, rabbits, and chickens add animal diversity and more attractions for kid and adult visitors.

"This is my mom's land. I never farmed conventionally. I'm a building contractor," says Jeff. "My younger brother was in school studying horticulture, and he got me interested. It went from there, and the organic thing was the main driver. We didn't want to spray that close to where we live. And there are all kinds of environmental reasons. The most immediate reason is yourself, and then you start thinking about the rest."

The coastal climate allows a luxurious ten-and-a-half-month growing season, though summers are dry and the farm's sandy soils are prone to erosion. "Farming out here is all about irrigation," says Jeff.

The Fioroviches use a variety of techniques to keep the farm's soil moist and protected. "We have thick mulch under the apple trees, and clover growing in the pathways between. We plant a cover crop in the gardens after each season. Oats, vetch, bell beans, which are like fava beans, and a pea is a pretty common mix out here. It gives you an incredible amount of organic matter and it's a real good thing to bring your fertility up," says Jeff. "Another cover crop I've been really interested in and am starting to use is mustard. There's research coming out that it works as a biofumigant. I need to keep on top of soil diseases, and with the mustard you let it grow chest-high and then till it in. People are also using mustard seed meal, directly applied, and I've bought some this year. It's a by-product of biofuel production, and the meal is an excellent organic fertilizer and soil disease suppressant. This is just in the beginning phases, so we're experimenting."

The Fioroviches let the cover crop grow until Easter. Lori says, "We bring it up about chest-high, and we argue about when to take it down. I take my tractor and trample down a maze in the field to hide Easter eggs. Jeff gets mad because it pushes work out further in the season for him, but I hold fast for tradition."

Jeff now uses a flail mower on the cover crop. "It chops everything up, and that's a big improvement. We have 20-horsepower and 40-horsepower Kubota tractors. The big one I need in order to do my strawberry beds. You want to build them up really high. The top is probably about 2 feet (0.6 m) wide and the bottom 3½ feet (1 m) wide, and I can plant three rows in the bed. I have a rototiller and an A-frame I mount some chisels on to do some ripping and get below the hardpan. It's not a claypan; we have sand, but it turns into a sandstone consistency, and you have to break through that or you don't have any room for roots to grow."

The Fioroviches do not spray their crops. Jeff says:

I'd rather get the soil to the point where I don't need to spray at all. One pest we have is the spider mite, and I release predator mites and they just go to town. After three days the plants perk up. I've always been able to control the spider mites with predators. I usually have an insectary row next to the strawberries, with a lot of native flowers.

We had one back hillside that was really erosion-prone and full of invasive plants. We slowly replaced those with native plants and created an insectary. On the coast here there are specialized native plants: subspecies of fescues, coyote bush, coffeeberry, manzanita, a wild strawberry, lots of native grasses and small shrubs.

Lori says selling their crops, especially the pumpkins, on the farm hasn't been too difficult. "Last year we didn't do any marketing, but I probably see 700 kids from schools in October. We charge $5 per kid. I still have parents coming to my stand whose kids were here years ago. We're located between two state parks with campgrounds, and we're right on the coast, so we're in a heavy tourist area. We offer youpick, and that brings a lot of people to our fields. Our Web site has extended our reach. We live close to Silicon Valley, and those people use the Internet. I'm surprised how many people will come 40 miles (64 km). I'm amazed at the need."

The farm features a small stand selling produce, but the Fioroviches "don't have time to sit there all day," as Jeff says, and so they rely on the honor system, letting customers leave cash for their purchases.

"We're always looking for new ideas to promote our farm. We're planning on having a couple of concerts here this year, just things so people can enjoy the space," Jeff says. Lori adds, "Our little farm is a hoot, and bringing the kids out here is amazing. There's something about getting the kids out of the box and on the dirt."

Garden Crops

Vegetables from Spring through Winter

> *Gardening is not a discipline that can be learned once for all, but keeps presenting problems that must be directly dealt with. It is, in addition, an agricultural and ecological education.*
>
> **WENDELL BERRY**
> *The Gift of Good Land*

A garden used to be standard on the family farm, and so should it still be on the organic farm. Even if you are not growing vegetables for a cash crop, it makes eminent sense to have some fresh, healthy produce for yourself when it's available for nothing but the cost of the seed and some labor. A garden is also the best spot on the farm to run your own small-scale experiments with different vegetable or small grain varieties, soil amendments, crop rotations, tillage, mulches, and whatever else comes to mind.

A garden can be as simple as a couple rows of vegetables next to the back door or as nuanced and complex as any masterwork of art, a dance of vegetable rotations, cover crops, and flowers. If produce isn't going to be your main crop, stick to the simple approach to save time needed elsewhere. If produce is your cash crop, excellent planning and a lot of hours, used efficiently, are necessary.

The organic principles specific to gardening can be applied in different ways, depending on how intensive your system is. None of them are difficult to grasp, and they have the added benefit of preventing or ameliorating weed and pest problems, making the whole growing season more pleasant.

Garden Structure

Designing the architecture of your garden can be as simple as laying out rows or beds or as elaborate as building trellises, framed raised beds, miniature greenhouses, and other structures that maximize space and delight the eye. A deciding factor in how to put together your garden should be what is most efficient for the equipment you will be using and what makes the best (and most comfortable) use of the human labor that you have available.

Deciding What to Grow

If you're planting only for yourself and your family, start with vegetables you know you'll eat. If you're planning a market garden, don't do anything without first researching the market. The market decides what you will grow; do not grow a bunch of stuff and then hope someone will buy it if you intend for your business to survive in the long term. There is no point in growing tons of fabulous sweet corn if everyone else in the county is also growing it. You'll have to sell it so cheap you won't make back the money and time you have invested. But if you can deliver sweet corn a week or two earlier or later than anyone else, produce a delicious heirloom variety, or arrange to be the exclusive supplier for a local restaurant, then you've got a good reason to grow the crop. Marketing will be discussed in full in a later chapter, but until you read it, at least remember that finding what the market wants or needs comes *before* growing the product.

Raised Beds

Raised beds, where the planting area is anywhere from ankle to knee high, offer the advantages of better drainage, an earlier warm-up in the spring, and not requiring as much bending over. They are an excellent idea in a heavy clay soil, but with sandy soil they require more water than ground-level planting and may not be worth the effort. Raising the planting area can be as simple as plowing or hoeing the dirt into a raised row, or you can go so far as to build wooden frames to hold the dirt at the height you want. Just make sure you don't use treated lumber, since the wood is impregnated with chemicals that will contaminate your soil and is prohibited by the Final Rule. (Composite plastic wood is allowed.) Framed raised beds can't be worked with power equipment, but they are easier to work if you're dependent on hand labor, so these tend to be built by small gardeners rather than market gardeners.

Rows versus Blocks

Another structural component to consider in the garden is whether to plant in rows or groups. Traditional rows are the easiest if you're using mechanized equipment, and make planting, cultivating, and harvesting quick and efficient. But rows are fairly wasteful of space, and an alternative is to plant in blocks. Blocks, like framed raised beds, are not amenable to power equipment. The compromise used by many market growers is to plant several rows close together in a bed. The bed is made to fit between the tractor tires, and a cultivator set up for narrow rows can be run through the bed. This saves space and avoids soil compaction in the

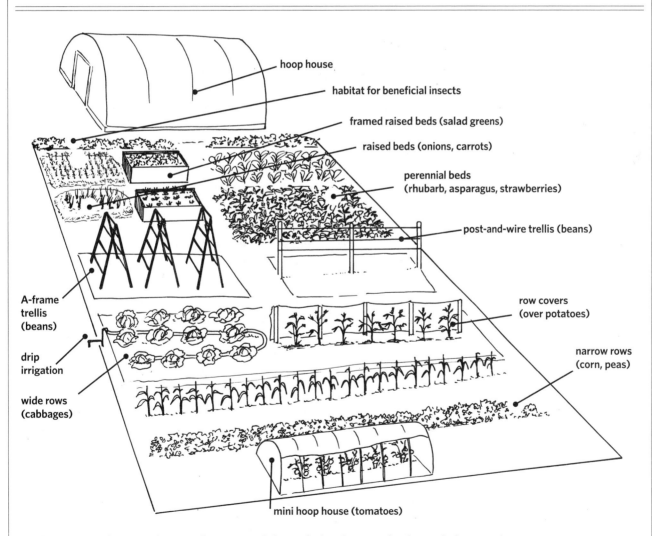

Gardens can combine aesthetics, efficiency, and diversified techniques for diversified crops. The management techniques you employ are determined by local conditions, crop requirements, and your own preferences.

Plan for Rebuilding

Aesthetics, efficiency, and space utilization should of course be taken into account when you are planning fences, trellises, terraces, and other physical structures. One additional consideration: treated wood is not allowed by the organic rule in new construction anywhere it would come into contact with the soil or with food plants. This includes old railroad ties. When you use untreated wood frames for raised beds or to build trellises for the pole beans, they will eventually rot. Construct your wooden garden structures so they're easy to repair and rebuild, since that's what you'll be doing every few years.

planted area, without giving up the efficiency of traditional rows. With this method it's important to make the beds no wider than you can comfortably reach halfway across (or a little farther) so you don't have to step into the planted area for hand-weeding or harvesting.

Vertical Structures

If space is at a premium, peas, pole beans, squash, and other spreading plants can be grown *up* instead of *out* by using trellises. A trellis can be as simple as twine stretched between a couple of fence posts or a teepee of poles, or more elaborately and artistically built of wood. Tomatoes are often grown in individual trellis supports to keep the fruit from resting on the ground, where it is more prone to rot. Trellises can make harvesting easier and reduce vegetable spoilage; they can also be a pain to clean up when they fall over or are completely tangled with vines at the end of the season.

Use your imagination in building trellises, but don't use treated wood. Place them where they won't shade other vegetables (the north end of the garden is a good spot), and above all, build them so they are strong and easy to both clean and dismantle at the end of the season.

The Farm Plan

To be successful in organic gardening and to qualify for organic certification, you must have an ongoing, long-term plan for building and sustaining soil fertility and plant health, as discussed in chapter 5. In the garden, the plan entails two basic requirements: rotating crops and building soil fertility with compost, manures, green manures, purchased amendments, or some combination of these. Putting this plan on paper not only is necessary for certification but also helps you track how well your plan is working.

Crop Rotations

Plants within the same plant family tend to consume the same soil nutrients in the same manner and to suffer from the same diseases and insect pests. If you keep planting members of the same family in the same spot, the nutrients they need most become depleted, so that with each passing season the plants are less nourished and less productive. On the other hand, not rotating crops means good times for the soil-borne diseases and insects that are best adapted to those plants. There is no pause in their favorite food supply, so pest populations and problems tend to explode. The goal of garden crop rotation, therefore, is to avoid planting members of a particular plant family in the same spot where another member of that family recently grew, whether it was last year or within the last few years. For example, you can grow parsley with the salad greens, but then you shouldn't plant carrots in that spot the next year, since carrots and parsley are in the same plant family.

Generally the longer the rotation, the better control you will have over those diseases and insects that spend part of their life cycle (often during the winter) in the soil. A four-year rotation was once traditional for field crops in the British Isles and much of the United States, and it's probably a good place to begin in the garden as well.

To build a rotation, you need to know which plant families each vegetable and cover crop belongs to and how much garden space each will take up on average, and for how much of the season. This information is listed for the most common vegetables and cover crops in the chart on pages 160–61, but it's also a matter of the varieties you choose and the particulars of your site and gardening practices. As you build a record through the years of what you have used for rotations, you'll see patterns, benefits, and problems emerge that will allow you to continually fine-tune where you plant what.

The modern classic on the subject of organic market gardening is *The New Organic Grower* by Eliot Coleman. This thoughtful and thorough discussion by a master of the art should be on the shelf of every organic vegetable grower. "Crop rotation is the single most important practice in a multiple-cropping program," Coleman writes, and follows that with one of the most lucid descriptions ever written of how to build a complex garden rotation.

Coleman suggests starting your planning by dividing the garden into sections, then assigning the various plant families to different sections until each section is full. This requires knowing how much space is needed for each vegetable. Some vegetables will take only a fraction of a section, while others will take a full section or two.

For example, vines, sweet corn, and potatoes are all space hogs and may take one or more entire sections, while the various salad greens (which fall into four different plant families) produce a lot in very little space and can be grown together in the same section, which is convenient for cultivation and harvesting. The other vegetables fall on a spectrum in between. How much space to use for each also depends on your personal eating preferences or, if you're growing for market, on what your research has told you will be most in demand.

Once you've established what you're planting in each section, set up your rotation so that each season each section moves to a different space, preferably farther from rather than nearer to where it was previously grown. Coleman suggests listing the plants to be grown in each section on different index cards, one card for each section. The cards can then be shuffled around until you find a rotation that will work.

The more types of vegetables you grow, the more variables you have to experiment with in

the rotation lineup to find which crops work best ahead of or behind others. For example, heavy nitrogen users, like corn and squash, often do best when following nitrogen-fixing legumes, like peas and beans. Be sure to include cover crops and green manures in your rotation.

Year 1

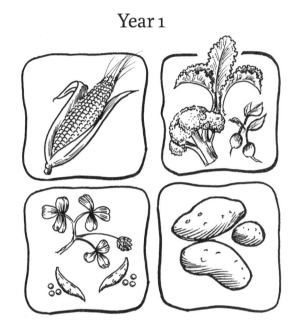

Year 2

Rotating crops quilts your garden every year in a new configuration that builds the soil and discourages diseases and insect pests. Group crops by plant families, and arrange the rotation so that soil-building crops (legumes, cover crops) precede heavy-feeding crops (corn).

Families and Characteristics of Common Garden Crops

Family	Vegetable	Characteristics
Legumes (Fabaceae). All legumes fix nitrogen in the soil and so are helpful to plant ahead of heavy-feeding crops.		
	Alfalfa	Deep-rooted perennial that does well in light soils and brings up nutrients from sub-soil layers. Seed can be quite expensive.
	Beans	Lima beans need a long, warm growing season, while fava (broad) beans need a long cool season. Yellow (wax) and green (snap) beans mature quickly and are good for northern gardens.
	Clover	White, red, alsike, crimson, sweet, and other clovers all have a place as cover crops for adding nitrogen and organic matter to the soil. Before planting, check to see which type is most appropriate for your soil and climate.
	Peas	Mature quickly; if planted as soon as soil thaws, they'll leave time for a second crop or green manure.
Crucifers (Brassicaceae). Do best in cool weather and constantly moist soils. Good early or late crops; kale especially will last late into fall and may overwinter if protected.		
	Broccoli	Many varieties continue to produce good-sized side sprouts after the main head is harvested, for a continuous supply through the season.
	Brussels sprouts	Good as a second crop in double-cropping, since it prefers midsummer planting for a fall harvest.
	Cabbage	Can be planted in spring as soon as soil can be worked, but is a heavy feeder and needs excellent fertility and even moisture to prevent cracking.
	Cauliflower	Can be planted early spring or midsummer for a fall crop. If you're fussy about white varieties being really white, tie the leaves together over the forming head to prevent discoloration.
	Collards	The one crucifer that actually likes hot weather, so it's widely grown in the South for salad and cooking greens.
	Kale	Tolerates some frost, and so tends to last later into the fall than other garden plants. If kept covered it may last through the winter in areas where winters aren't too harsh.
	Kohlrabi	Likes a lot of sun but cool soil, so a mulch is a good idea.
	Radish	Matures in 3 to 4 weeks, one of the fastest-growing vegetables in the garden, and can be planted in early spring, before the last frost. A good candidate for double-cropping or intercropping.
	Rutabaga	Good as a midsummer planting following an early crop. Slow to sprout and doesn't tolerate weeds well.
	Turnip	Same considerations as for rutabagas.
Squashes (Cucurbitaceae). All like warm soil and a reasonably long season. Northern gardeners should select quick-maturing varieties. Squashes are heavy nitrogen users.		
	Cucumber	Viny, heavy-feeding, and productive. Many varieties especially good for pickling.
	Gourds	Gourds are raised for decorative or craft use rather than for eating. They can really brighten up a farm stand late in the season.
	Melons	Most varieties need a lot of heat and a long growing season; northern gardeners should select seed that has been bred for a colder climate.
	Squash	Winter varieties are hard-skinned, late-maturing, and keep well; summer varieties (yellow and green, or zucchini) are thin-skinned, mature early, and don't keep very long.

Family	Vegetable	Characteristics

Grains (Poaceae). In general (with the notable exception of corn), grains are quick to mature and suitable for growing in the garden even though they are not generally considered to be garden crops. Winter grains are planted in the fall, go dormant over the winter, then are harvested the following summer. Spring (or summer) grains are planted in spring and harvested that summer.

	Amaranth, quinoa, sorghum, buckwheat, millet, teff	All are spring grains.
	Corn	A heavy feeder; does best after a legume. Likes warm soil and a long season. Sweet corn is most common in the garden.
	Rice	Needs a particularly long, warm growing season and is feasible in the U.S. only in Southern California and the extreme South.
	Wheat, barley, oats, rye, spelt, triticale	All either come in both winter and spring varieties or will winter over where the weather is not too harsh.

Alliums (Liliaceae). Members of this family are aromatic and easy to grow, but they don't compete well with weeds.

	Garlic	Fall- and spring-planted varieties are available.
	Leeks	Like a longer season and warmer weather.
	Onions	Can be planted in cold soil.

Goosefoot Family (Chenopodiaceae). A diverse family, but all appreciate constant, even moisture and fertile soil.

	Beets	Mature quickly and are a good choice for northern gardens; plant in very early spring.
	Chard	Does better in warm conditions.
	Spinach	Likes cool weather; tends to get bitter and go to seed in warm weather.

Carrot Family (Apiaceae). All like cooler weather and lighter soils.

	Carrots	Can be left in the garden till the ground begins to freeze; some gardeners say carrots are sweeter if they've been frosted a couple times.
	Celery	Prefers lots and lots of organic matter in the soil, so use plenty of compost and water heavily. Demands a long growing season.
	Parsley	Quick to sprout, tolerates cool weather well, and if winter is not too severe will often come back the following season.
	Parsnips	Look like a white carrot and, like carrots, prefer lighter soils and cooler temperatures. Much slower to sprout than carrots.

Nightshade Family (Solanaceae). With the exception of potatoes, this family prefers a long, warm growing season and all share a vulnerability to fusarium wilt, a plant disease that is one of the most common and hardest to eradicate.

	Eggplant	Needs warm soil and a long growing season.
	Peppers	Should be planted in warm soil. Hundreds of varieties available, from sweet to super-hot. Popular items for many farmers' markets.
	Potatoes	Do best in acid soil and cooler weather. Colorado potato beetles are a major insect pest; effective control is difficult.
	Tomatoes	Should be planted in warm soil.

Lettuces (Compositae). Do best in cooler weather; southern gardeners should select varieties that won't bolt (go to seed) in hot weather.

Garden Perennials

Asparagus, rhubarb, and many culinary and medicinal herbs are commonly grown in gardens, but since they are perennials they are not rotated like vegetables. Plant them along the northern edge of the garden, where they won't shade young vegetables, and use compost and mulch as you do with any garden plants.

Strawberries are a special case, since each year the plants send out runners that root, expanding the plot. It's a good idea to allow a strawberry row or bed to move over a little each year, tilling under the parent plants in favor of the runners.

Double-Cropping

Then there's the option of double-cropping, which is planting a second crop in the same place after a first crop has been harvested. For example, you can plant peas in the early spring, then a few weeks later when the soil has warmed up put either beans or summer squash in the same bed. The peas are done bearing by the time the beans and squash run over them, and you have harvested two crops in the space of one. Radishes and many salad greens also work well in double-cropping systems, as do several of the other crucifers. The method does complicate crop rotations if you double-crop vegetables from two different families, but good planning makes this issue easier to deal with.

Companion Planting

Companion planting can provide several different types of benefits, depending on the pairing. Some plants repel or trick the pests of other plants, while other plants attract beneficial insects. Some plants prevent or lessen disease or encourage better growth in certain other plants. For example, prickly squash vines underfoot are supposed to discourage raccoons from eating sweet corn, while marigolds are believed by many gardeners to discourage a variety of insects when planted around the vegetables. And many gardeners plant flowers among their vegetables, not only to make the garden attractive but also to provide nectar and pollen for bees and other beneficial insects.

If you are interested in learning more about companion planting, *Carrots Love Tomatoes: Secrets of Companion Planting for Successful Gardening*, by Louise Riotte, is the standard book on this topic. She covers not only vegetables but herbs and cover crops as well and goes beyond the garden borders to discuss field crops, fruits, and wild plants.

The three sisters of Native American gardening tradition are the classic example of companion planting. The corn supports the beans, which feed the corn by fixing nitrogen in the soil, while the widespread, prickly leaves of the squash discourage weeds as well as — maybe — corn-loving critters like raccoons.

Building Soil Fertility

The second part of your planning is figuring out how you'll build and sustain soil fertility with amendments and cover crops. Because gardens are smaller and more intensively managed than field crops, there are more options for soil improvement. Cover crops and amendments that are not feasible on a field scale due to time limitations, expense, or labor requirements can be very doable in a garden.

Amendments

A gardener can apply organic amendments over a much longer part of the year than a farmer in the field, where wet weather or maturing crops limit application opportunities. More expensive amendments are often used, too, since the overall cost is so much lower for a garden than for many acres of field. Foliar feeding is also more feasible in a garden, where it's fairly simple to mix up a feed and apply it, as compared to having the tank on wheels with a mechanical sprayer, as would be needed in a field.

- **Uncomposted manure** should be applied and worked in before winter, to give it time to mellow in the soil and to comply with the Final Rule.

- **Compost** is easiest to apply in the fall, when putting the garden to bed, or in the spring, a few weeks ahead of seeding. It can also be applied at any time during the growing season, though organic certified gardeners must make sure that this compost is made in conformance with Final Rule requirements.

- **Purchased amendments** that your soil test says are needed can be applied to the entire garden and worked in before planting or dug in along the crop rows during the season, if you want to use less and put more of it where this year's roots will find it (this is called "banding").

- **Slow-release amendments**, common in organic systems, are often best applied in the fall. Sometimes the release rate depends on the fineness of the particles — a coarsely ground limestone will dissolve in the soil much more slowly than a finely ground limestone.

- **Foliar feeding formulas** are liquid mixes of natural fertilizers that can be sprayed onto leaves and are applied during the growing season. Apply foliar feeds in the morning or evening rather than at midday, since leaf pores close in the midday heat. The feed must be absorbed through those pores to have any effect.

Cover Crops

Cover crops are an important part of a garden fertility program. They protect the soil from erosion and, when tilled in, add organic matter. They can be planted in the fall for winter cover, in fallow sections of the garden to build organic matter, or between rows as a living mulch. Planning ahead for these means you can conveniently order the seed along with your main crop order.

LIVING MULCHES

Cover crops are not necessary in the garden during the growing season if you're short on time. Use mulch instead. But if you do have the time, between-row cover crops are definitely worth it for the added fertility and mud-prevention benefits. White clover, for example, is excellent as a living mulch between rows during the growing season. Living mulches can be left to grow until a killing frost and will protect the ground all winter. Clover comes back in the spring, so you'll need to till it in to kill it.

WINTER COVERS

Winter covers are seeded among the vegetables during the late summer lull or in the fall, when the harvest is mostly done and you are working the soil anyway, as you clean up the garden. If you plant an annual or something that's not winter-hardy in your area, it will die in late fall and the dead vegetation will protect the ground all winter and be easy to till under in the spring. A winter-hardy cover crop will begin regrowing in the spring, but the treatment is the same: just till it under in the spring, but be sure you till thoroughly enough to kill it. In either case, it's important to get the cover crop seeded early enough before winter to allow it to make enough growth to cover the ground adequately. Different crops sprout and grow at different rates, and this has to be taken into consideration when making your selections. How wet and cold your fall weather is also greatly impacts growth rates, so talk to your seed dealer and your neighbors to figure out your options.

FALLOW GROUND COVERS

"Fallow" traditionally means leaving land to its own devices for a year or more, to rest from the rigors of growing crops. For an organic gardener, fallowing presents a terrific opportunity to do some major soil-building with cover crops and break up problem pest and disease cycles. In fact, organic market gardeners commonly have half or more of their ground fallow at any one time, since fallowing does so much good for the health of their soil.

Fallow ground-cover crops can be put in while you're planting the rest of the garden. The trick is be sure to till them under before they get tall and rank and go to seed. Buckwheat especially seems to go to seed before you have time to turn around, but it's not the end of the world if that happens — you'll just be weeding out buckwheat seedlings for some time. Try mowing before tilling to make it easier. You could even take off the mowed tops and compost them instead of trying to dig them in.

Season Extenders

Sometimes there just aren't enough days in the growing season to accommodate all the vegetable varieties you'd like to eat or market. Especially in northern states, it's tough to face a long winter with no fresh organic vegetables to eat or to sell to keep cash flow steady. Traditionally, crops that keep well through the winter — carrots, potatoes, squash, turnips, and so on — were stored in large quantities to feed the family through the cold months. Other vegetables were canned or dried. Storage is still an

Hoop Houses and Greenhouses

Traditionally, the term *greenhouses* refers to structures built with glass panes, while the term *hoop houses* refers to structures that are covered in plastic. But hoop houses have become so common due to their low cost and high versatility that they are replacing traditional glass greenhouses in commercial operations across the country, and they are now commonly referred to as greenhouses when used for starting plants. Of course, hoop houses are also being used for livestock shelter, equipment storage, and, in miniature form, to cover rows of crops vulnerable to weather or pests, or to provide additional warmth early or late in the season. To distinguish hoop houses used as greenhouses from hoop houses used for other purposes, this discussion will use the term greenhouses to mean hoop houses used as greenhouses, as well as glass greenhouses.

Glass greenhouses are getting scarce as hen's teeth, since they are expensive to build and maintain and lack the versatility (and portability) of hoop houses. But there is still a place on the organic farm for a glass greenhouse, since, either as a stand-alone structure or a south-side addition to the main house, a greenhouse is lovely for starting plants. A creative farmer with a pile of old windows and some carpentry skills may be able to construct a small glass greenhouse for very little cash outlay.

Put a blanket on your garden for the winter with a cover crop, planted in the fall after the vegetables are done, to prevent erosion, protect soil structure, and hold nutrients. Till it under in the spring to add organic matter to the soil.

excellent idea to ensure a supply of organic vegetables through the winter, but there are also ways to have more fresh vegetables available.

In recent decades it's become common to use various techniques to extend the growing season, so there are fewer months without fresh greens and herbs and all the other vegetables that taste so good but don't keep very well. Extending the season works at both ends of the year, in early spring and late fall. In southern areas, a greenhouse or hoop house may make it possible to have fresh green food on hand throughout the winter.

Starting Seeds Indoors

For annual plants that need a growing season longer than the number of days you can expect between the average date of the last killing frost in spring and the first killing frost in fall, you'll need to start seeds indoors. Small-scale gardeners can do this in any south-facing window in the house, while larger-scale growers will need a dedicated greenhouse or hoop house.

OMRI seal indicates certified-organic qualification

The term organic *does not have the same meaning for commercial soil preparations as it does for food. If you're using a commercial soil mix, be sure it qualifies for use in certified organic production.*

When *Not* to Start Seeds

Vine crops like pumpkins and cucumbers take up so much shelf space that most gardeners prefer to buy short-season varieties and plant them directly into the garden. If you're trying to get those crops to market before the main season, then it might be worth starting them indoors. Crops like peanuts and lima beans, which need a lot of heat in addition to a long season, probably won't produce dependably in the northern tier of states no matter how early you start them, because there just isn't enough heat during the summer.

Getting Started

Seeds can be started in saved cottage cheese containers, newspaper pots, purchased plastic pots, compressed dirt blocks, or about anything you can find, as long as there's a drainage hole in the bottom of the containers and the containers are set in a tray to contain any excess water.

Most growers start seeds in a seed-starter mix, concocted of peat, compost, vermiculite, and other nonsoil ingredients. To produce certified organic plants, you must use a certified organic seed-starter mix. Note that there are plenty of seed starting planting mediums on the market that are labeled "organic," which is *not* the same as "USDA Certified Organic" and will disqualify your plants for organic certification. Only a mix labeled "certified for organic production" or carrying the OMRI seal can be used for production of certified organic plants.

If you can't find a certified organic growing medium, you can mix your own from equal parts of peat, compost, and vermiculite, as long as these ingredients have not been treated or modified in any way from their natural state. The classic Purdue University recipe recommends adding, for each 25 gallons (94.6 L) of finished product, 1½ cups (355 ml) each of superfine dolomite, blood meal, bone meal, and greensand (all excellent sources of mineral nutrients). In a pinch you can use soil from the garden. It's much heavier and seeds will be slower to sprout than with a growing medium, but by midseason there won't be any difference between plants grown in garden soil and in a seed-starter mix.

Different seeds require different soil depths, and sometimes additional steps like presoaking or using a warm potting medium. Slow-sprouting and slow-growing varieties should be planted earlier than the others, so be sure to read the planting directions for each type of seed. All planting dates should be geared to the probable date you'll be moving seeds into the garden, so that your transplants haven't outgrown their pots before you can get them outside. Most growers will transplant at least some of their seedlings (especially tomatoes) once or twice into bigger pots while they're still indoors, but there comes a point in every plant's life where it needs to get outside, and if that happens before it's warm enough in the garden, you end up with some spindly, less-robust plants.

Germinating Conditions

Most seeds prefer a soil temperature of about 75°F (24°C) for vigorous germination. Sprouting will be slower if the soil is colder, and if the soil is too hot or too cold seeds may not sprout at all. Parking the pots next to a sunny window during the day, where an insulating curtain can be closed at night, will work for most small-scale gardens. Larger-scale growers may use heating pads manufactured especially for the purpose under their pots to maintain ideal soil temperatures, and they will need a greenhouse of some sort. You can also use the type of heating pad sold in drugstores that is designed for soothing aching muscles, if you make sure it's waterproof and has a thermostat so you can set the right temperature.

The planting medium should be kept evenly moist until the seeds are up. Some people put the pots in a plastic bag until the seeds sprout to save the trouble of watering every day. This may encourage damping-off disease, however, and limit available oxygen, which is needed for germination.

Once the seeds are up you can remove heating pads and add more light. Even a sunny window isn't enough in northern areas. For best growth hang lights so they're just grazing the top of the seedlings, hiking the lights higher once a week or so as the plants get taller. Though full-spectrum bulbs are the ideal, home gardeners have done just fine for decades with ordinary bulbs and

fluorescent lights, and since the Final Rule is silent on this topic, do what works for you.

Transplanting

Seedlings started in small pots should be transplanted to bigger ones as they grow. Or you can simply start them in large pots in the first place, though it takes up a lot of space and makes heating pads and plastic bags inconvenient. When transplanting the seedlings to larger pots, take the entire block of soil to minimize root disturbance. This applies when putting them into the garden later on, as well.

When moving plants outside for the season, do it slowly. Put them out in the sun in a sheltered spot for an hour the first day, two hours the second day, four hours the third day, six hours the fourth day, and on the fifth day they can stay out all day. This is called "hardening off," and is absolutely necessary for keeping the shock of direct sunlight from killing the delicate seedlings. During these

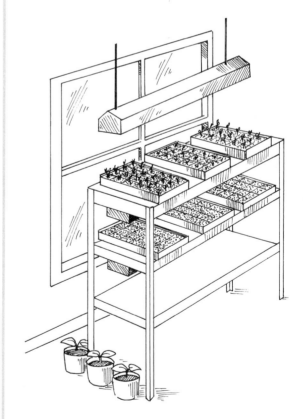

Vegetables that need more growing season than your region offers can be started indoors. A sunny window will do, but for best growth electric lights are needed as well. A lot of plants can be started in a relatively small space with an efficient setup.

Transplanting day is a busy one! When removing a seedling from a pot in order to transplant it, take the plant and the ball of soil it's set in to avoid damaging delicate roots.

days you'll need to water more frequently, since the outside breezes will dry out the pots much more quickly than the indoor environment.

Cold Frames and Hoop Houses

Years ago a bottomless wooden box with an old window for a top was often set over a patch of soil. On warm days the window on this so-called cold frame would be propped open with a stick, and closed at night to guard against frost. Cold frames jump-start the garden season, warming the soil and protecting young plants from cool weather so that seeds can be put in the ground at least a month earlier than normal. Many gardeners now use miniature hoop houses built with a flexible frame and a plastic covering, for the same purpose. Being able to open the cover on warm days is important to keep the plants from being killed by the heat that would otherwise build up inside, just as the inside of a car will get quite hot when parked in the sun on even a fairly cool day. Mini hoop houses can be large enough to cover several rows, or small enough to shelter a single row, whichever suits the crop and your purposes.

Most growers now have given up on the lovely and traditional but expensive and high-maintenance glass greenhouse in favor of the economical, portable, plastic-covered hoop houses. This low-cost alternative to glass has been a boon to market growers, and is available in a vast array of heights, lengths, and structural

A mini hoop house is used in the same way as a cold frame, but it is easier to move and more flexible in design.

A cold frame is an old-fashioned but effective method for giving plants an early start on the season outdoors, or for keeping them growing into the fall.

components. Building it yourself saves even more money. The plastic covering is often doubled, leaving an airspace between the layers for insulation during cold spring and fall weather. Some growers produce in hoop houses throughout the growing season for better insect and weed control, rolling up the sides in warm weather for ventilation.

Hoop houses in all their permutations, as well as the old-fashioned cold frames, can be used at the end of the growing season as well, to get in some late plantings or salad greens, or give those late-season tomatoes a chance to ripen on the vine, or to keep all sorts of other vegetables in production.

Moving through the Garden Season

The steps involved in preparing, planting, tending, and harvesting a garden are simple if it's small, more complicated if it's large. The need for efficient use of labor, effective tools, and regular working hours are much higher with a market garden.

What follows here is merely an overview of a year of organic gardening. If you are seriously considering market gardening as a business, research your markets and local growing conditions thoroughly, and be sure to visit some commercial organic vegetable farms.

Winter

Gardens begin in the dead of winter, when seed catalogs start to arrive. Once you've decided which varieties you want to plant, the next step, while it's still winter, is setting up a garden plan on paper that establishes your rotation and fertility plan.

Early Spring

Four to eight weeks before you are going to plant outside, you can start seeds inside. Whether you do this, and which seeds you start, depends on the length of your growing season, what you have room for, and which crops you want to start producing

Catalog Shopping

Good catalogs will designate whether seed is hybrid or open-pollinated, and whether it is certified organic. For each variety of plant, the catalog will also list which diseases (if any) it has resistance to; the number of growing days it needs to mature; and, where appropriate, the mature height or spread of the plant. A really good catalog will tell you the length of row one packet of seed will plant, which is important to know when you're calculating how much to buy for the amount you want to produce.

For trees, shrubs, and perennials, the catalog should designate the hardiness or climate zone. Don't buy anything that is rated for a warmer zone than the one you live in, unless it's small enough to put in a pot and bring inside for the winter. Plants to be used for certified organic production must have been raised on a certified organic farm and arrive with documentation, or be under certified organic management for a year on your farm before their produce can be certified. For example, if you can't find certified organic strawberry plants, you'll have to wait until their second season to sell organic berries.

In spring, till under any cover crops or crop residues as soon as the soil has thawed and dried. If the cover crops are tall, you can mow or mulch them with a flail chopper first to make the job easier.

early in order to get a premium price. Starting seeds indoors is extra work, but it has the advantage of giving the started plants a big jump on the season and on the weeds, and you get exactly the plant spacing you want when you move them into the garden.

You can start spreading compost in the garden as the pile thaws out. As soon as the soil is ready (meaning it's thawed and dry to the point of being merely moist), you can till under any cover crops and compost, plus any other soil amendments you have added. If you're hungry for green food at this point, you can set up a cold frame or miniature hoop house in the garden to start a few plants.

Plants that like cooler conditions and are quick to mature, such as peas, most of the brassicas, spinach, onions, carrots, parsley, and lettuce, can be planted directly in the garden after it's quit freezing regulary at night, but before the soil is really warm.

Spring!

Prepare the seedbed by turning under any compost, cover crops, or manures, then rake or hoe to break down the soil clumps into a fine, level seedbed. Once the seedbed is prepared and the soil is warm, mark the rows or beds and plant according to the plan you designed over the winter. Main crop planting can be spread over several weeks, beginning with those vegetables that like cool weather and ending with those that respond best to warmer temperatures.

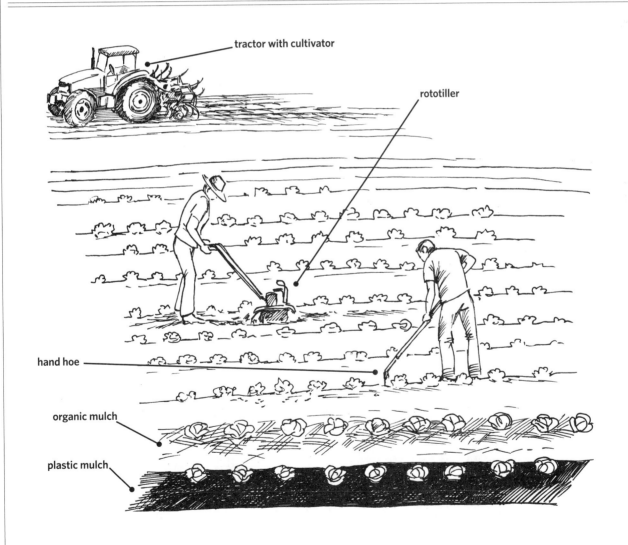

The organic grower has an abundance of tools for controlling weeds. Which you choose will depend on the scale of your garden, the degree of pressure from weeds, the crops you have, and your own inclinations.

Washing and Packing Facilities

Commercial operations will want a dedicated washing and packing facility for efficient operation. You can wash your vegetables in a bucket with a garden hose, but this is slow, messy, and inconvenient when doing more than a handful. For a serious market garden, you'll want a place where vegetables can be conveniently washed and packaged out of the rain and without creating a mud hole from the wash water. A roof and a cement floor are best, while walls aren't strictly necessary since most of the work is being done during the warm season. If you have late-season crops or need a warm spot for onions and garlic to dry or to store root vegetables and squash out of the weather, then you will need a fully enclosed building space. Having your processing facility easily accessible for a delivery truck is a big plus at loading time.

Within those parameters, the ways you can put a produce processing facility together are nearly infinite, depending for the most part on your budget, scavenging and mechanical abilities, and creativity. Old washing machines on a gentle cycle are being used for washing salad greens, used conveyer belts are moving potatoes into bags, and stainless-steel tables from defunct commercial kitchens are great for sorting all sorts of vegetables. The setup that works best will depend on your particular needs, the equipment available to you, and the ways you can dream up to put them together. (See the resources for more information on designing a packing facility.)

The organic payoff: a bountiful harvest of healthy, tasty vegetables. And the garden keeps paying dividends all summer long, and well into autumn.

effective than waiting to tackle them once they've gotten a good roothold.

There are many different hand hoes for this purpose, and all should be kept sharp to be most effective. Rototillers set to a shallow depth or tractor-drawn cultivators do the same job in bigger gardens. But no matter how effective your tools, there are always a few weeds right in the row that have to be pulled by hand. Get them before they go to seed and thus sow next year's weed crop.

Mulch

Mulch is the second line of defense against weeds. Many commercial growers favor plastic mulches, but they must be completely removed from the field at the end of the season to comply with the Final Rule. Plastic mulches, especially black plastic, tend to make the soil warmer, while mulches of straw and other plant materials tend to cool the soil. For this reason, plastic mulch applied early in the spring can accelerate growth, while a plant mulch later in the summer can help to keep cool-loving plants producing longer into the dog days of summer.

A natural mulch usually has to be put on several inches thick to effectively suppress weeds. The moisture-conserving effect of mulches will encourage slugs, so in some cases it may be better to wait until later in the season, when the soil

Once the garden is in, the first priority is weed control. This should start immediately, even before you see any weeds.

Cultivation

Working the soil with a shallow hoeing or tilling will cut off weeds while they're tiny or haven't even emerged from the ground, and that is a lot more

is drier and warmer, to apply mulch. In perennial crops like strawberries, it's best to apply a light layer of natural mulch before the plants appear in the spring if you want clean berries in June. (The plants will come up through the mulch, then spread out and bear berries on straw instead of dirt; the straw also keeps rain from splashing mud onto the berries.)

If you are obtaining mulch from off the farm, examine it first to make sure it's not made from a cutting of mature plants and therefore full of weed seeds. You don't need to import problems. Organic materials (meaning they're natural materials, such as leaves or straw) from off-farm sources are not required to be certified organic, so municipal leaf piles and similar resources can be utilized as mulches by the organic farmer. But do use some caution: Suspect sources should not be used, since they will compromise your organic integrity and residual chemicals may damage your plants. One example of this is grass clippings from golf courses, which routinely have chemicals applied, or composted manure from large poultry operations that feed arsenic as part of the ration.

Summer

As you're cultivating, weeding, mulching, and picking the first peas, salad greens, baby carrots, and green onions, you should also be monitoring for pest and disease problems. As you encounter such problems, you may need to employ floating row covers, vacuuming, spraying with water, and other physical controls (see chapter 5), as well as the old standby of handpicking bugs and drowning them in a can half-filled with soapy water.

Vegetables should be harvested at their peak of flavor and palatability, and the variety in growth

This starter facility for vegetable processing and packing shows the bare minimum for efficient operation. At the left are tubs for washing vegetables, and under the tent top is a table for sorting, bundling, and boxing.

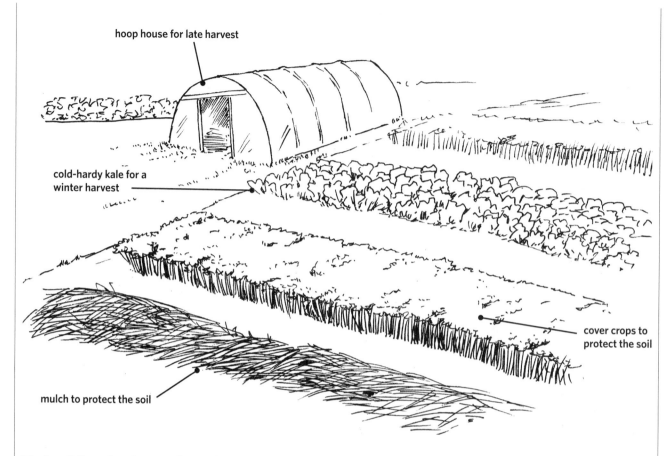

hoop house for late harvest

cold-hardy kale for a
winter harvest

mulch to protect the soil

cover crops to
protect the soil

The late fall garden: Putting the garden to bed properly for the winter not only makes spring work much easier but also preserves and builds soil structure. Hoop houses and cold-hardy crops can extend your harvest into the winter months.

rates and optimal maturity ensures that, in a mixed garden, harvest will begin as soon as a few weeks after planting (with radishes and salad greens) and extend into late fall (with squashes and root vegetables). This means that, with good planning, you will have plenty to sell all season long.

In a market garden, produce should be picked as close to the time of sale as possible. This may mean working into the evening before a farmers' market or CSA delivery, or in the early morning for restaurant and institutional sales. It will all depend on the buyer.

Also important is washing and packing vegetables into attractive, neat bundles or piles, to please the eye as well as the palate of the potential customer. The heart of any business is repeat sales, and the better satisfied your buyers are with your produce, the more likely it is they will return to buy more.

Autumn

As the season draws to a close, more and more areas of the garden are fully harvested, tilled, and sown with a cover crop. Many market gardeners extend their selling season by months or even through an entire winter with late summer and early fall plantings in hoop houses. This also lengthens the season of positive cash flow.

But after the first hard frost, most of the garden is done for the season. If you haven't already, collect any seed you want to save for next year, then remove the dead plants and old mulches for composting. Till the ground (for disease control and seedbed preparation) and plant with a cover crop or, if it's too late to plant, cover the ground with mulch. Turn the compost pile one last time, and take a break until next year's catalogs start arriving.

Profile: **Richard Rudolph**

Rippling Waters Organic Farm, Standish, Maine

Twelve acres (4.9 ha) isn't enough, says owner Richard Rudolph, who manages a sizable staff and a myriad of farming, marketing, and educational activities at Rippling Waters Organic Farm in Maine. Located on the loamy sandbanks of the Saco River, the farm produces 4½ acres (1.8 ha) worth of certified organic vegetables, bedding plants, herbs, flowers, and perennials. These are marketed through an on-farm retail greenhouse and store, a CSA operation, and farmers' markets.

Rudolph puts in a cover crop of oats or rye after a field is harvested and tilled. Which type he plants depends on temperature and time of year. In addition, he sets aside ½ acre (0.2 ha) each year for cover cropping with green manures like clover. Rudolph says:

> Ideally we should have 30 acres (12 ha), enough to do more crop rotation, and to be able to raise straw or hay for mulching. Sometimes crops just can't be rotated far enough away. We're trying to check Colorado potato beetle and other insects, and there's a patch of woods between one field and another. So you move the crop and we can fool the beetles for a few weeks, but eventually they show up on our doorstep. It's a constant problem that we and most farms have.

But the farm's lack of land is offset by the management and creativity of Rudolph and his staff. Four-foot-wide (1.2 m) beds make maximum use of limited land, while an emphasis on greens, which are planted once but cut and harvested several times in a season, allows higher production with less labor. The staff gets an early jump on the growing season by starting plants in a solarium and a 100- by 30-foot (30 x 9 m) heated greenhouse. Tomatoes and similarly heat-loving plants are kept warm, dry, and productive through the cool, rainy summer inside four hoop houses. The hoop houses also extend production well into the fall. The outdoor gardens are bordered and interlaced with flowering plants that attract beneficial insects, crops are interplanted for mutually beneficial effects, and plastic mulches suppress weeds. "Black plastic on the peppers, eggplant, and melons, to increase soil temperature. White plastic for lettuce, broccoli, and other cool crops," says Rudolph.

Since the cost of purchased manure and compost has skyrocketed, the farm is building a 52- by 12-foot (16 x 3.6 m) concrete-floored, covered composting facility to make its own on a large scale. For feedstocks, "we'll be using food scraps from St.

Joseph's College and leaves from a leaf collection program," Rudolph says. "We'll need another tractor with a front-end loader to turn the pile."

Right now the farm utilizes a 22-horsepower Kubota and an old Farmall AV side-mount tractor, plus a lot of hand labor from volunteers and staff. Besides a farm manager, three apprentices, two Americorps workers, and one Americorps/Vista worker, "we have probably 250 volunteers who come to the farm in the course of the summer," Rudolph says. "A lot are school groups, or social service agencies bringing teens with developmental issues or adults with disabilities. We've had a volunteer potato planting day for the last four or five years. We get an acre (0.4 ha) planted in just hours. We did a 'weed and feed' event: Forty people showed up and weeded, we fed them lunch, then we went out on a hunt looking for wild edibles. We have eight or nine events each year."

And that's just the beginning of the education and outreach effort at Rippling Waters. One of the Americorps workers develops lesson plans to be used in the elementary schools, where the farm has initiated school gardens, as well as at the middle school, which has a solar greenhouse. Another works as a community coordinator, doing media outreach and volunteer recruitment. The farm manages gardens for senior citizens, and its Web site offers fact sheets on various aspects of organic gardening.

"Gardening has become a lost art," Rudolph says. "It goes back to the end of World War II and the rise of frozen foods and big grocery stores. Now that's being turned around. With food and gas prices rising, there's a renewed interest."

Rudolph has been in the business long enough (since 1991) to have developed a long-term perspective on the challenges and rewards of small-scale organic farming. He speculates that either a smaller vegetable farm that required little or no outside labor or a larger operation that made it feasible to use more labor-saving equipment might provide a better living than the midsize Rippling Waters Farm. But finding enough affordable land can be a real problem for new farmers, Rudolph says. Land is cheap when the house and buildings are in bad shape, or it's a long way from town, "but we didn't want to spend ten years rebuilding a broken-down farmhouse; we wanted to farm. And we were really interested in finding a place within 45 minutes of a major farmers' market, and within driving distance of a city where there were social amenities."

He learned one of the most important lessons early on, Rudolph says. "The first year we put in something like 600 tomato plants, and I threw my back out just from lugging all those tomatoes around. We went to the farmers' market, and we had to sell those tomatoes, so we lowered our price. So everyone was rushing over to our stand, and a farmer who was there selling, who has since become a good friend, said, 'What are you doing?'"

Not only were they taking business away from others, they were selling at too low a price to be profitable. "Don't grow anything before you know what your market is," Rudolph says. "You have to have a business plan."

The many activities at Rippling Waters Farm are guided by the farm's commitment to supporting greater local food security through education, action, and service. "We want to break down the disconnection between the landscape, the consumers, the food they eat, and the farmers who grow it," Rudolph says. "We try to. I think we are making a difference."

Profile: **Anne and Eric Nordell**

Beech Grove Farm, Trout Run, Pennsylvania

Weed management is a major challenge for organic vegetable production. Weed control typically requires the second largest use of labor on organic vegetable farms, after picking and packing, and weed competition is often the primary reason for crop failure or reduced yields.

This is not the case at Anne and Eric Nordell's horse-powered farm in Trout Run, Pennsylvania. Due to their proactive approach to weed management, competition from weeds is no longer an issue for their high-value vegetables. They typically spend just four to six hours per acre on hand-weeding, primarily to prevent the smattering of remaining weeds from going to seed. Weed pressure is so low at their certified-organic Beech Grove Farm that the Nordells have been able to use living mulches and minimum tillage without compromising weed control or yields.

Their unique and successful system has attracted a fair amount of attention from the organic community, and the couple has authored a booklet and DVD, both titled *Weed the Soil, Not the Crop*, which describe their methods in detail (see the resources).

When the Nordells began their market vegetable operation in the early 1980s, they had three goals, all of which profoundly influenced their approach to weed control. The first was to stay out of debt, so they could maximize creativity and minimize stress. The second was to keep their farm a two-person operation, both to avoid the inevitable complications and expense that come with hired help and because they enjoy working together.

The third goal was to rely on the farm's internal resources as much as possible, rather than depending on purchased inputs. This meant, for example, that weed control would be accomplished with cultural practices rather than off-farm inputs like plastic mulch or hired labor.

Nearly three decades later, the Nordells have achieved all three goals, and the weed control is nothing short of phenomenal. The couple farm their hilltop acres entirely with four draft horses and hand labor, producing consistently bountiful crops on the mellow soil. Disease is rarely, if ever, seen. But they didn't achieve their goals overnight. When they first moved to their farm, the land was an old hayfield, low in fertility and infested with quack grass. Building the soil and the system took many years of observation, experimentation, and adapting old horse-drawn equipment to new functions.

The first problem was to get rid of the quack grass, a tenacious perennial that can sprout new plants from pieces of root. The only way to kill the plant is to kill the root — every bit of it. The Nordells began by shallowly plowing the land, which flipped the sod over and exposed the roots. They then worked the area with a sweep cultivator, which uses V-shaped blades that run under the soil surface, with a flexible pasture harrow or a spring-tooth harrow hooked behind, every two to three weeks for the rest of the growing season. Each pass over the field pulled more roots to the surface and shook the dirt off so they would dry and die in the sun. They worked the field often enough that the quack grass never had an opportunity to green up.

This successful strategy was the beginning of the Nordells' development of their system to "weed the soil, not the crop." The system relies on a deep understanding of various weed types and the effects of different types of tillage and cover crops. It has five key components: bare fallow, shallow tillage, composting, cover crops, and rotation.

The bare fallow, or leaving a field bare of vegetation during all or part of a growing season, is an opportunity to use tillage to intentionally deplete the weed seed bank in the soil. If deep, taprooted weeds like dandelion or dock are an issue, then deep tillage is used to set back these weeds. If perennial weeds with tenacious, rhizomatous root systems, such as quack grass or bindweed, are more problematic, then the Nordells till shallowly and work the ground repeatedly, every two to three weeks over the course of an extended bare fallow, to dehydrate the roots. For annual broadleaf weeds, the couple is more inclined to create a firm, moist seedbed to intentionally germinate the small seeds over a six-week period. As soon as the weeds have sprouted they use shallow cultivation to kill them, then refirm the soil with a roller to germinate another batch of weed seeds.

The timing of the bare fallow is shifted to match the life cycle of the weeds. Initiating the bare fallow in the spring targets cool-season weeds like chickweed and henbit, while a summer bare fallow reduces the seed bank of warm-season weeds like lamb's-quarter, pigweed, and purslane.

Shallow tillage, the second component, is used before vegetables are planted and is important to prevent the planting from bringing up to the surface new weed seeds from deeper in the soil to germinate with the cash crops.

Composting, the third component, kills weed seeds present in the manure used in the compost.

Cover crops, the fourth component, help weed management in two different ways: First, they build soil structure and organic matter, which deters those weeds that are better adapted to poor soil conditions than are cultivated vegetables. Second, cover crops can do their work without increasing soil fertility to the very high levels that give some weeds the upper hand. Cover crops recycle nutrients already existing in the soil, while compost or manure brought in from off the farm can lead over time to too-high fertility.

The Nordells consider the fifth component, rotation, to be the most important, since it integrates all the other components to achieve low weed pressure. Their four-year rotation is composed of alternating years of crop production and fallow. Taking land out of production for a full year before each vegetable crop provides the time and space to optimize the soil-building potential of the cover crops, as well as for a bare fallow period timed to target the life cycle and growth habits of the most pressing weeds.

The rotation also alternates the cash crops between those planted early in the growing season, like onions, peas, spring lettuce, and spinach, and those planted later when the soil is warmer, such as tomatoes, squash, and peppers. Rotating between early and late vegetables prevents cool- and warm-season weeds from getting out of hand. It also makes it easy to select the best cover crop to facilitate shallow tillage. For example, the Nordells intentionally use cover crops that winter-kill, like oats and field peas, before early-planted vegetables because they can easily shallow-till the dead cover crop. Before late-planted vegetables they use an overwintering cover crop, such as rye and hairy vetch, because they will have plenty of time to use shallow tillage to kill and decompose these live covers before the warm-season vegetables go in the ground.

By their third time through this four-year rotation of fallow/late vegetables/fallow/early vegetables, weed pressure was so low that the Nordells could begin experimenting with no-tilling crops like garlic directly into a cover crop, with no hand-weeding or cultivation necessary.

The Nordells say there are tradeoffs to their system. It requires more labor and management at the outset, as well as access to enough land to take half of the market garden out of production each year. On the other hand, the system is flexible, so that growers with a longer season may be able to utilize all five weed control components and still grow a cash crop each year on the same ground. Farmers with a shorter growing season who have some ingenuity may be able to take advantage of some of the rotational principles with annual crop production or simply dedicate a small part of the farm to the fallow-year system specifically for the least weed competitive vegetables, where the savings in weed management may offset the costs of taking land out of production.

"Weed the soil, not the crop."
Anne and Eric Nordell

Field Crops

Grains and Forages on a Large Scale

> *A farmer of deep ecological sensitivity is to the plow jockey on his 200-horsepower tractor what a French chef is to the legions of hamburger handlers at fast food chains. The chef's work is infused with artistic, scientific, and spiritual satisfactions; the hamburger handler's is infused only with the ticking of a time clock.*
>
> **GENE LOGSDON**
> *The Contrary Farmer*

Field crops are in many ways no different from garden crops. They're just produced on a much larger scale. You can grow tomatoes on a field scale, or wheat in your garden. In general usage, though, the term *field crops* refers to grains and forages.

Grains are divided into row crops and drilled crops. Row crops are those species that demand enough personal space that they need to be planted in wide rows, wide enough apart for equipment tires to fit in between. The most important row crops in the U.S. are field corn and soybeans, but there are many others. Drilled crops, so called because they are planted by drilling or broadcasting, are planted in rows so close together you can't drive or walk between them, so that mechanical cultivation is impossible once the crop has sprouted. The most common drilled crops are the small grains (referring to the small seed size), especially wheat, as well as oats, barley, rye, millet, and spelt, among others.

Forages are grasses and legumes that are harvested for animal feed (hay or haylage) or plowed under as green manures. These are generally planted by drilling or broadcasting, like small grains, but they form a sod when mature and can provide several harvests in a single season from a single planting. Grasses and forage legumes will also overwinter and resume growth in the spring, so that they can be harvested through two or more seasons without replanting. Some farmers have perennial hayfields, where they never plant but do harvest whatever happens to come up on its own. With good attention to maintaining soil nutrient levels, this strategy can work for many livestock producers. But to convert a field from tilled crops to permanent forages, it's wise to plant your preferred species. Otherwise, you could have a big harvest of weeds for at least the first couple years.

This list doesn't come close to covering all the species you can grow in fields. There's still sorghum, sugarcane, sunflowers, and lots more. And of course each crop has its individual quirks in terms of ability (or inability) to suppress weeds, use in a crop rotation, planting and harvesting windows, climate and soil preferences, and marketability. Nevertheless, the organic production principles for row crops, small grains, and forages will apply no matter what you're growing.

Fields versus Gardens

The difference in scale between garden and field is the cause of the management differences between the two in an organic system. The first and most obvious difference to anyone driving past the farm is the size of the equipment. The amount of acreage involved in field cropping demands more and larger tractors and implements than a market garden or orchard (though in a small-scale organic system, the machinery lineup will appear dwarfed next to what is used on a big conventional farm). In short, you should have a mechanical bent if you intend to grow field crops.

Field crops usually require much more careful attention to buffer zones than do gardens, which often aren't close enough to property lines for buffers to be an issue. If your field edges lie along power-line corridors or public roads, you should post your property with "No Spraying — Organic Farm" signs. A letter to the power company and the government body in charge of road maintenance (most often this is a county-level department) requesting that they refrain from spraying near your property offers an extra level of insurance. Keep a dated copy of the letters in your farm files. If your field edges lie next to private property that your neighbor does not spray, you may be able to dispense with a buffer zone if you have a signed letter from your neighbor stating that no spraying will be done in that area. Check with your certifier to determine what it requires for this sort of documentation.

Another critical difference from a garden is that a field crop is, by its nature, a monoculture.

Monocultures are contrary to organic principles but inherent in field cropping. In a garden it's fairly simple to have many different species growing side by side in a small area, interspersed with noncrop flowers and forbs, so that biodiversity is maintained. In an orchard or vineyard a diverse ground cover achieves the same purpose. But with field crops, the scale of production demands equipment and work speeds that can't distinguish between different species when planting, cultivating, or harvesting, so achieving biodiversity must be accomplished in different ways. Organic crop producers, like market gardeners, have on many farms built beautifully nuanced rotations that employ complicated combinations of fall-planted crops, spring-planted crops, early- and late-harvested crops, cover crops planted before, with, and after main crops. They also feature a planting sequence that takes into account the specific effects of different crops on particular pests, weeds, diseases, and aspects of soil fertility. In this way their crop

There are three basic types of field crops: row crops, drilled crops (small grains), and sod-forming crops (forages).

The Final Rule on:
Field Crops

Final Rule requirements for soil preservation, pest control, and biodiversity in the garden apply to field crops as well. In the field, some of these requirements have more significance than in the garden:

1. The Final Rule prohibits burning crop residues unless the producer can document that it is the only effective available measure against a specific pest or disease problem or it is done to stimulate seed germination. Burning crop residues is common in some areas, but it destroys organic matter and leaves soils bare through the winter, both contrary to the principles of organic farming.

2. Tillage must be done in a way to prevent soil erosion. In the garden this is not usually an issue. In the field it is always a consideration.

3. Buffer zones in the field, if planted with the same crop as the rest of the field, must be harvested separately from the rest of the crop and the harvesting equipment must be purged between buffer zone harvest and main crop harvest.

4. The crop rotation, as in the garden, must include sod and cover crops. In the field, sod crops are a source of forage for livestock, but if you don't have livestock they can be turned under as a green manure.

rotation accomplishes the goal of biodiversity over time, instead of in a single season.

Maintaining soil life and fertility is handled a little differently in the field as well. High-value vegetable crops grown on small areas make it relatively easy and economical to apply labor-intensive compost and mulch and more expensive purchased soil amendments. Field crops return considerably less profit per acre, and organic farmers are better off focusing more of their efforts on applying green and animal manures, growing living mulches, and buying only those inputs that

soil tests indicate are necessary. Compost production on a scale to cover entire fields is being done, though it requires a high volume of inputs, usually in the form of animal manures, as well as space on the farm for really big compost windrows and the equipment to turn it, load it, and spread it. Unless you have a cow dairy operation or access to a really big municipal leaf pile to supply the inputs, composting on such a large scale is not usually a viable option.

Field Crop Rotations

The first step in field cropping is planning the crop rotation. This, more than in any other type of production, is closely tied to your climate and your soils. For example, the traditional eastern U.S. rotation, that is, row crop/small grain/forage/forage, would be idiocy in the high plains, where the climate dictates no corn and necessitates fallow years to conserve moisture for wheat crops. Besides benefiting the soil and breaking up pest and disease cycles, a diversity of crops has two added benefits for the organic farmer: it spreads planting, cultivating, and harvesting chores through the season rather than concentrating them all into brief, intense bursts, as happens with a single crop; and second, in the event of a crop failure (which happens to everyone sooner or later) you probably have other crops that have done fine. You haven't put all your eggs in one basket, so you still have something to sell.

The traditional rotation of the diversified eastern U.S. farm (row crop/small grain/forage/forage) is now mostly history, as the small grains and forages have largely been replaced by continuous corn and soybeans. However, it is still an excellent basic rotation for an organic system. Here's why.

Row Crops

First up in the rotation is the row crop. In this case, let's say field corn. Corn is a heavy feeder, demanding a high level of nitrogen. That's why for the

When to Plow

Traditionally farmers plow in the fall, when there's more time and the weather in most areas is drier. This also allows the soil to "mellow" over the winter. The crop residues break down, and the freeze/thaw cycles help break up soil clods. But fall plowing leaves the soil open to wind and water erosion all winter, except where you can count on good snow cover (which is almost nowhere in the continental United States these days).

On the other hand, if you have a forage crop planted, waiting until spring to plow down the crop will keep the ground covered through the winter but make the forage crop harder to kill. In addition, with unpredictable spring weather, delays tend to be more common, and there's a good risk of not getting the seed in the ground early enough to mature by the end of the growing season.

So, a good option for the organic farmer is early fall tillage followed by planting of a winter cover crop that will winter-kill. This deprives the farmer of some fall grazing or a final late forage harvest, and it costs money for cover-crop seed, but then all that has to be done in the spring is to spread manure and disc it in along with the cover crop. Or a farmer can wait until spring, plow as early as possible, then cultivate between plowing and planting to ensure a better forage kill and a finer seedbed. If it's a warm, early spring, no problem. If spring is late and wet, this will be difficult.

Whatever the decision, the weather is likely to throw you a curveball, so have a plan B ready. Sometimes this will involve trading in a longer-season variety for one that will mature more quickly, if planting is significantly delayed. Discuss this option with your seed supplier.

previous two years the crop in that field was a forage legume such as alfalfa or clover (often with some grasses mixed in), which fixed a lot of nitrogen in the soil. The forage legume also provided animal feed, which in turn produced manure, which was applied to the field in early spring.

The legume crop should be tilled under in fall or spring, so it won't compete with the corn.

Amending the Soil

Any soil amendments needed for the corn crop (as indicated by soil tests) should be applied before spring tillage. If your soil has good organic-matter content and is not sloped, then amendments (manure, rock phosphate, or whatever) can be put on in the fall with little danger of their washing or leaching away over winter, if a cover crop will be in place. More commonly amendments are applied in spring, just ahead of plowing. This is generally a quick job, since field spreaders cover a pretty wide strip, minimizing the number of trips up and down the field. (Spreading manure is a little slower.) But be sure to order amendments early, since spring is a supplier's busiest season and you want to be sure they arrive on time.

Planting

Many organic farmers like to plant a little later than their conventional neighbors. Corn planting often starts very early, while the soil is still quite cold, simply because farmers have a lot of acres to cover and don't want to get behind. Often you don't get a lot of good planting weather in the spring, either, so the temptation is to seize any opportunity to get the work done, even if it is early. The problem is that corn needs warm soil to sprout, and if it has to sit too long in a cold soil, the seed may rot. Conventional farmers use fungicide-coated

marker disc

A marker disc marks a line showing where you should drive to plant the next set of rows. It takes the guesswork out of the process and keeps the rows evenly spaced.

seed to prevent this problem, but organic farmers can't. Waiting instead until the soil is warm means the corn will be up and growing quickly, getting a jump on the weeds, especially if the field has been cultivated a couple times to kill sprouting weeds before the corn is planted.

Planting is simple enough. Set the planter depth, fill with seed (be sure to pocket the tags from the seed bags for your records), and off you go. The most important part of planting row crops is to have the rows evenly spaced, so that when it's time to cultivate or harvest, the equipment isn't flattening or missing the end rows because they don't fit the row spacings on the implements.

Cultivation

In organic systems, cultivation is essential to successful row crop production. Generally, row crops will be cultivated, using your cultivator of choice, one to three times before they are tall enough to shade the ground completely and discourage weeds. Don't go too fast or get careless on the corners; that's what buries or tears up young crop plants and causes "cultivator blight." If you are cutting your first crop of hay in the next field during this time period, as is usual, you will wind up spending much of May and June on the tractor. If you have a garden planted as well, and maybe some livestock to take care of, you'll be working way more than full-time. The labor crunch during certain parts of

Weed control in an organic system is done the traditional way: by hand-pulling or mechanical cultivation. Mechanical cultivators work in various ways, from uprooting just-germinated weed seedlings to pushing dirt into crop rows to bury weeds.

Purging Harvest Machinery

Harvesting equipment must be free of any non-organic crop residues. If you do all your harvesting yourself and grow only certified organic crops, this is not a problem. But if you have some organic and some nonorganic crops (either in buffer zones or in fields that are not yet certified), or if you hire a custom operator to harvest for you, the equipment must be thoroughly cleaned. Since pressure-washing, vacuuming, and blowing are not likely to get every little bit of grain out of the many nooks and crannies, the accepted method to clean the machinery, called purging, is to run the equipment one length of the field, dump that grain into a separate wagon (or on the ground), and then proceed with the harvest of the certified organic crop. The dumped grain is not certifiable, but there is probably a horse owner or hobby livestock producer in the neighborhood that will be happy to buy it from you. Using the buffer zone for the purge doesn't work. That isn't a certifiably organic harvest, so the combine will still be contaminated. You must purge using certified crop, and you must document that the purge has been done, and the non-organic crop has been stored and sold or used separately from the organic crop. Write down how much of the crop, in bushels, was used to purge the machine. How many bushels will be acceptable depends on the size of the machine and the type and density of crop, so give your certifier a call to get the amount before you go out to harvest.

the season is the primary reason most organic producers don't try to do everything. Diversification is basic to organic farming, but not to the point where you're working yourself to death.

The last cultivation before the canopy closes is an excellent opportunity to seed a cover crop between the rows. In the Midwest, for example, rye, oats, buckwheat, hairy vetch, clovers, or a mix are all good options. In the shade of the corn, these crops won't be able to compete with the cash crop, but they will cover the soil against erosion from hard summer rains, limit the mud factor when harvest equipment moves into the field, and provide additional organic matter when plowed down with the crop residue.

Once the plant canopy has closed — that is, the ground is shaded — you can park the tractor for a while and do other things, maybe even take a nap. The only job in the field at this point is monitoring for pest problems. Take a walk every few days and look for corn borer damage, aphids, potato leafhoppers, or whatever type of pest and disease commonly patronizes your neighborhood. Remember that just because a pest is found, it doesn't mean you have to do anything about it or, in an organic system, that you *can* do anything about it. But it's good to at least know what you're up against; it helps you prepare for next season. Healthy soil, healthy plants, field edge habitat for beneficial organisms, and crop rotations are your main defenses, and they must be in place before the pest arrives.

Harvesting

Corn is harvested either as grain or as silage. A grain harvest leaves most all of the corn plant in the field except for the grain, while a silage harvest removes the entire plant. Corn plant residue in the field can be left entirely or partially in place to hold the soil over the winter. A silage harvest

Unless you have a small crop, a strong back, and a big shovel, grain crops are moved from hoppers to bins to trucks with augers, usually powered by a PTO. Be sure children and animals are clear of the equipment.

leaves bare soil, so organic farmers usually plant a winter cover crop after harvest. (Since silage is used almost exclusively on dairy operations, it is discussed in chapter 12.)

All grain must be dry enough to store without molding. Harvesting grain too early and having to dry it with blowers is expensive, but leaving it too long in the field in the hope that it will dry more risks weather and pest damage, and, in some regions, being shut out of the field until spring if a big snowstorm hits early in the winter. Grain farmers generally test moisture percentages as the fall progresses to guide their harvest decisions. Your agricultural extension agent or your neighbor will be able to tell you what moisture percentage is appropriate for the crop, the locale, and your type of storage facility. Remember, organic crops may not be treated with synthetic fungicides, fumigants, or preservatives to prevent mold or deterioration, so you may want your grain even more dry than is normal with conventional crops.

Storage

Traditionally, small grains and shelled corn were stored in wooden bins inside barns or in a separate building (hence the word "granary"), and this is still a fine way to keep grain dry and out of the weather. Rodents are usually a problem, however, with these types of bins. Only one rodent poison is currently allowed by the Final Rule, and it kills by overdosing the rats and mice with vitamin D. A few cats who are good hunters might be more effective. They won't get everything, but they should be able to keep the rodent population under control.

Corn left on the cob was traditionally stored in wooden corncribs with slatted or wired sides to let airflow through. Corncribs are still used in some areas, and are good storage if you can keep the wildlife away. But even if the squirrels do strip off every grain that shows through the wire, they won't usually get any deeper than that.

These days metal, stand-alone grain bins are the most common type of grain storage. Although they don't circulate air well without the help of fans, they do minimize rodent problems. These types of bins are available in almost any size, and the smaller ones can be easily bought and moved to the farm, while larger ones will need to be built on-site. Be sure to put them on a concrete pad.

Grain is moved in and out of bins and cribs with grain augers, usually powered by a tractor's PTO (power takeoff). You may be able to rent an auger from an equipment dealer. If you do, make sure to purge it, and to document the purge, before using for organic crops. Shovels work for moving grain, too, if you have more time and labor than money for equipment. Combines have built-in augers, and when the combine's hopper is full of grain, most often the grain will be augered from the combine into a gravity box, which is basically a big metal box on wheels pulled by a tractor. The gravity box is pulled back to the grain bin and hooked up to another auger, which moves grain from box to bin. Sometimes the combine can auger grain directly into the bin, but it's a lot of time spent driving to and from the field.

Grain storage bins must be absolutely clean of nonorganic residues. The best way to do this is to own your own bins and store only organic crops in them. Each bin must be numbered and ongoing records kept of all crops that are stored there, and when they were moved in and out, in order to maintain organic certification. If the bins are being used for both organic and nonorganic crops, the record must include a detailed description of how the bin was cleaned (including method and date). Also, no herbicides or other prohibited substances are allowed to be used around the bins, including anticoagulating rat poison (warfarin), the farmer's favorite weapon in the age-old battle against rodents. Having large, obvious numbers and labels on bins makes it easy for truckers to avoid a mistake.

Planting Winter Cover

Once the row crop is off the field, you need to get the dirt covered before winter if it isn't already growing a cover crop. You can seed down a green manure crop immediately after harvest that will be plowed down in the spring to make way for spring planting of a small grain, or you could put in one of the winter grains that will sprout, go dormant over the winter, then resume growth in the spring. A green manure will add to soil fertility and tilth, a benefit after heavy-feeding corn and all the cultivating that was done, but a winter grain saves the field work next spring, a big deal if you have a lot of other things to do. A deciding factor will be whether you have a market for the winter grain.

Either way, the row crop has to come off and the cover crop has to be planted early enough to get a good start before the cold stops all growth.

Small Grains

There are big differences among small grain varieties, and as always in an organic system, it's important to choose one that is resistant to the diseases prevalent in your area. Some growers mix varieties with different disease resistances, so that no matter which disease shows up, they'll still have a crop at harvest.

If your weed control was good in the previous year's row crop, the small grain should come up cleanly. If you're not sure how good your weed control was, you could use the stale seedbed method: run a cultivator or disc over the field several times before planting to kill emerging weeds and turn up new seeds so they will sprout and be killed by the next cultivation.

Small grains are planted with a seed drill. The trickiest part is getting the drill properly calibrated so that it plants the seed at the correct rate. The settings for different types of seed are usually found right on the drill or in the owner's manual. Once the crop is planted, there's not much to do until harvest time in mid- to late summer.

Small grain harvest ideally takes place when the grain is completely ripe and the plant is dead and dry. If a variety doesn't ripen or dry evenly, or if the grain starts to drop off the plant prematurely, then it's necessary to cut and swath the grain (lay it in rows in the field) before threshing or combining. This stops growth and accelerates drying. A grain swather is a dedicated piece of equipment worth the investment if you plan to grow a lot of small grains in a locale where rain is fairly

Definitions

Threshing separates the grain from the stalk
Winnowing separates the grain from the hull and small bits of straw (or chaff).
Combining is so named because it combines all the harvest procedures — cutting, threshing, and winnowing.

Plowing Basics

Plowing is an art. The goal is straight, evenly spaced furrows of a uniform depth. Depth is set by a stop on the plow, and should be chosen to suit the soil and the crop. A shallow soil should not be tilled so deep as to mix subsoil into the topsoil, though in some cases mixing in a little subsoil can be part of the plan to develop the topsoil. A more deeply tilled soil will help heavy feeding crops and deeply rooted plants. Shallower tillage uses less fuel (thanks to less load on the tractor) and disturbs soil structure less.

Let's say you have average soils, and you have checked with the neighbors to see what they do and decided to set the plow depth at 6 inches (15 cm) for your corn. Next you have to lay out the field. The basic moldboard plow throws the dirt to one side, making a furrow and a ridge. Where the tractor changes direction, two furrows will be put together, or two ridges. These are known respectively as a dead furrow and a land. In plowing, you want to minimize dead furrows and lands (to make the field smoother), and minimize the driving you do when the plow is not working. The plow is not working between the time you finish one row, raise the plow to turn around, and lower it to begin the next row. Since tractors and implements can't turn tightly enough to work the row right next to the one you just finished, minimizing turn time takes some planning.

It's actually easier to do this than to explain it. The basic idea is to work the field in sections, a concept that applies to any field operation. Start by mentally dividing the field into strips about 200 feet (61 m) wide (depending on the turning radius of your equipment). Plow down the field edge to the far end, lift the plow, drive 200 feet (61 m) along the end of the field, turn, and plow back to the near end. Then lift the plow, drive back to the first furrow along the field edge, and plow a second furrow next to it. At the far end, turn and put a furrow next to the first one you made 200 feet (61 m) into the field.

Repeat until you've plowed a wide enough section out in the field (as opposed to along the edge) that you can turn and plow the other side of that section rather than driving all the way to the field edge again. Plow up one side and down the other of that section until the field is plowed all the way from the field edge to the far side of that first strip you started out in the field. When that's all plowed, you'll have a land and a dead furrow. One will be about 100 feet (30.5 m) from

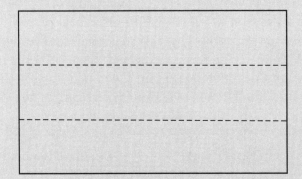

Fields are worked in sections, with the width of the section dependent on the turning radius of your equipment. This method maximizes working time and minimizes time spent turning and driving with the implement not engaged.

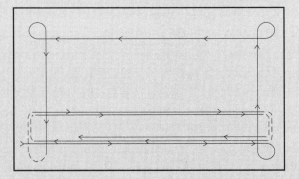

Start by working the outside edges of a section, working your way in toward the center of the section as you make passes down the field.

the field edge, the other about 300 feet (91 m) from the field edge, and you'll have plowed a width of 400 feet (122 m). That's when you open up another strip further out in the field.

Repeat the pattern in the next section of field, and so on until you're done. Then plow around and around the edge of the field. To finish, set the plow depth at half of what it was and plow up and down the lands and dead furrows in the opposite direction, to even the soil surface. If you don't think this is worthwhile, try driving moderately fast across the field perpendicular to the furrow direction — and be careful you don't bounce out of the tractor seat when you hit the dead furrows.

Most farmers use marking discs that run to the side of an implement, or line up trees or other landmarks, to keep their furrows evenly spaced from one end of the field to the other. However, most fields aren't going to be exactly the same width along their entire length. That's okay, because this is an art, not a science. But the straighter and more evenly spaced you can make your rows, the fewer short-tailed rows you'll have and less messing around with turning implements. All subsequent field operations will also be easier if the plowing is done right.

Once the plowing is done, you may want to follow it with a disc, and you can hook a drag behind the disc to create an even finer seedbed. Remember that each pass with an implement over the field costs money in the form of tractor fuel (and often repairs) and your time, so hooking two implements together is a savings. Many farmers will disc and drag at an angle to the furrows, to get a more even soil surface. All other field operations are done in the same direction as the furrows are plowed.

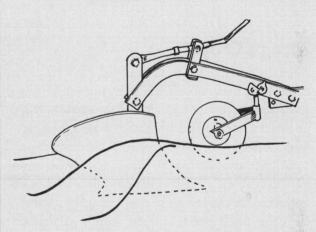

A traditional moldboard plow first cuts the soil with the point. The ribbon of soil rides up the curved share and is flipped over, so the green side is down and the dirt side is up.

When there isn't enough room to turn at the end of the row and start up the other side of the section you are working, start working the adjoining section. Finish the section you started with by working the center on every other pass, until it's done.

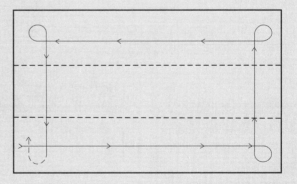

When all sections are plowed end to end, plow around the edges of the field.

frequent at harvest time, since it helps get the crop off the field and under cover faster than waiting until it dries while still standing.

Small grains produce straw in addition to grain. Oat straw makes excellent animal bedding and is a great cash crop in areas with lots of horses or dairy animals, but it's hard for an organic farmer to ship all that fertility off the farm. Maybe you can arrange to have the horse or dairy manure spread on your fields, or use the profits from straw sales to buy seed for cover crops. The straw is baled after the grain is taken off, using a hay baler.

If you leave the straw on the field, you can leave it to cover the soil or plow it under, where it will build organic matter in your soil.

Forage Crops

After the small grain is off the field, the next job is to get your forage crop planted. This can go in as soon as the grain comes off and the straw is baled or turned under, and you may be able to simply disc the field to prepare a seedbed.

Alternatively, if you're planting only legumes (alfalfa is usually the legume of choice in this situation), you can do so in the spring right along with the small grain, and the alfalfa will have a good start by the time you harvest the small grain. Be sure the grain is planted thinly enough that it doesn't smother the legume. If you're putting in a mix of legumes and grasses, skip the grain. (A small grain is a type of grass and so would compete with the grass in your forage mix. You wouldn't be able to harvest the grain easily anyway, since the seed from whatever variety of grass you planted would mix with the grain and be very difficult to separate out.)

Either way, with a spring planting of forage, you may get a hay cutting later that season, depending on where you live. Another option is to plant a forage crop, or oats and alfalfa together, after the summer grain harvest. The oats will serve as a nurse crop for the alfalfa and then winter-kill and protect the young alfalfa all winter. The alfalfa will resume growing in spring, resulting in an early harvest.

Other Options

You can add soybeans to the rotation, since there is usually a good market for certified organic soybeans, though you should first find buyers and find out what variety they prefer. Though soybeans are a legume, the nitrogen benefits for the following

Grain is often cut and swathed so that it will dry more quickly and evenly.

crop are not large since most of the nitrogen is removed with the bean. But soybeans at least supply their own nitrogen and leave very little residue (stalks and leaves), minimizing seedbed preparation for the following crop.

In areas with dry climates, such as the northern high plains, there's not enough rain for corn or soybeans, only enough to grow a crop of small grains every other year. Traditionally, the ground is left bare for a year and then wheat is planted the second year, so it has two years' worth of moisture to grow on. Herbicides ensure that none of the moisture being held in the fallow ground is used by weeds or any other vegetation. A bare fallow field equals a lot of wind erosion. Organic growers in these areas plant cover crops in the fallow years, often deep-rooted alfalfa and similar plant types, which will reach deeper than other crops and pull moisture and nutrients up to the surface. The added organic matter from a plowed-down fallow year cover crop means the soil is better able to hold water, as well.

The Final Rule requires that crop rotations include sod, cover crops, green manure crops, and catch crops, so you don't have to grow small grains (or row crops) if it's difficult in your location and system to store and market them. There's also no rule that a rotation has to be four years long; it can be five, six, seven, or more. Less than three years, however, is unacceptable to most, if not all, organic certifiers.

In short, the possibilities for crop rotation on an organic farm are limited only by your climate, your creativity, and your ability to market the crops. The important thing is to make sure all planning is done according to the organic principles of maintaining a living, fertile soil by rotating crops in such a way as to promote biodiversity, break up pest and disease cycles, and improve soil fertility.

Making Hay

One of the quintessential smells of farming is the scent of fresh-cut hay on a warm summer evening. It warms the soul like the smell of baking bread, or the scent of wood smoke on a cool morning. When the weather holds and the equipment runs, haymaking is a pleasure. When thunderstorms are building to the west and the hay rake breaks and needs welding, it gets stressful.

Hay is alfalfa, clovers, grasses, or any mix of the three that are cut while green and growing, dried in the field, then raked into windrows and baled. Most dairy farmers grow straight alfalfa for its high soluble protein levels and high yields. Many organic dairy farmers and most other livestock producers prefer a mixed grass/legume hay. Timothy and red clover used to be a common combination for a less-rich diet on heavier (having lots of clay) ground, though this mix dries more slowly than alfalfa. Some native grasses like quack grass also make good hay, though the per-acre yield is lower.

If you don't have livestock but still want the benefits of a forage in your crop rotation, you can either grow forages to plow down as green manures or cut hay as a cash crop to sell to certified organic livestock producers. There's always a market for certified organic hay, though it may be some distance away and require hiring a trucker.

When to Cut Hay

Young hay, cut before blossoming, has low fiber and high soluble protein levels, which boosts milk production in dairy animals and fattens young stock, but can cause runny manure, too. Older hay, cut between blossom and seed head maturity, has higher carbohydrate levels, good for maintaining breeding and working animals. Hay that is cut too old, after seed heads are mature or later, will be less palatable to livestock and they may not eat as much. In a mixed planting some species will "head out" or go to seed before others. When to cut a mixed planting is a judgment call.

The number of cuttings made in a season depends on the length of the growing season, the amount of rain, and how hard you want to push your land. Cuttings are generally made from one month to six weeks apart when rainfall is adequate. Dairy farmers in the upper Midwest aim for three or four cuttings each season, beginning at the end of May and wrapping up in early September (though some will cut again in October). Irrigated farms in California may see up to 10 cuttings a year.

Another option is to graze hayfields in the early spring and late fall, taking just one or two cuttings during the summer. This gets manure on the field without the farmer having to spread it, saves the labor and expense of making a cutting, and greatly extends the grazing season.

Harvesting

To make hay, the crop must be first cut and left to dry for a day or two. Then it's raked into long windrows, checked for dryness, and when it's dry enough, picked up with a baler and made into bales. That's quite a lot of equipment, but machinery is what makes it possible to make a lot of hay with just one or two people. A hundred years ago hay was cut and raked with horses or by hand, with

In farm country, the symbol of summer is a field of ready-to-cut hay.

Forages and Grassland Birds

Conserving soils and biodiversity are important reasons for the organic farmer to include forages in the crop rotation. Row crops are good for stirring the soil and achieving weed control, but they are hard on soil structure and organisms and are prone to cause soil erosion. Small grains help improve tilth, their straw can be plowed under to add to organic matter, and many have allelopathic effects on weeds and disease. But forages win the gold star award for their contribution to nutrient levels, organic matter, soil structure, and soil life.

Unfortunately, with the disappearance of horses and the increased use of nonforage feeds for livestock, the demand for hay has dropped tremendously in the past 50 years, and the once extensive areas of cropland devoted to forages have been largely moved into corn and soybean production. While this has had deleterious effects on soil life and structure, it's been disastrous for populations of birds and animals that depend on grasslands.

The earlier and earlier cutting of hay has been especially hard on birds that nest in the grass — bobolinks and meadowlarks are two particularly beautiful and vulnerable species. A bird arrives in the spring, builds a nest in the grass, and has it destroyed by a haybine before the babies can fly or even are hatched. So she builds another nest, lays more eggs, and the same thing happens again with the second cutting. This creates a population sink, where the adults don't replace themselves with youngsters, and the overall population declines.

A generation ago, hay was not cut in many areas until the end of June. This allowed most of the grassland birds to fledge a brood. Grazing the hayfields in early spring will delay maturity of the forage and so delay the first cutting as well, giving these birds a chance.

scythes, then loaded bundle by bundle onto wagons, to be built into outdoor haystacks or stored loose in barns. The process required every available pair of hands, and even then the acres covered were miniscule compared to what can be done with machinery.

Cutting

Cutting is done using the same approach to the field as when planting, though in this case you do the field edges first instead of last, so that you can turn equipment at the row ends without trampling the crop in the headlands (edges). Driving on cut hay is much preferred to driving on uncut hay, though best of all is if you can get the headlands cut and baled a couple days before cutting the rest of the field. Then you don't have to drive on any hay.

Drying

Once cut, the hay is left to dry. This can happen as quickly as a day with a light crop, low humidity, warm temperatures, and a good breeze. It can take as long as two days or more with humid weather,

no breeze, and a heavy cutting. If the hay is cut with a cutter bar instead of a haybine with rollers, before it's raked it may need to be turned with a tedder, a specialized type of hay rake that flips the hay without putting it into tight windrows, where it would take longer to dry.

Raking

Once the hay is dry, it is raked into windrows. If it's very dry you can lose some of it to leaf shattering, so in this case rake gently, that is, not too fast. If it's not completely dry ("tough" is a common term) it may mold or self-ignite after being baled. It's strange to think of wetness causing a fire, but if there's enough moisture, it jump-starts bacterial action, just as in a compost pile. Bacteria create heat. With a bunch of damp bales packed tightly in a hayloft where the heat can't escape, the heat can build up enough to ignite the hay, and your barn burns down. Every year barns burn down due to wet hay igniting in the loft. Technically, hay needs to be below 15 percent moisture to prevent mold, but farmers generally judge readiness by look

When haying, cut the headlands (field ends) first, so you have room to turn when completing rows without driving over uncut hay.

When Rain Interrupts

In the eastern half of the country, making hay is all about catching three to five days of clear weather to get a good crop. A hay harvest interrupted by rain is not the end of the world, but it's not good, either. Rain turns hay brown, gives it a less pleasant odor, and leaches away nutrients. If the hay smells bad from having been out in a lot of rain, the livestock won't eat it. Getting hay dry enough to bale after rain can be a pain, sometimes requiring repeated rakings. If you've cut the hay and suddenly rain is in the forecast, don't rake. The hay will dry more quickly if it's still spread out. If you've raked already, resign yourself to raking again.

and feel instead of a lab test. The color should be a faded green, and it should crackle in your hand. There will be clumps of damper stuff in most windrows. These may cause isolated pockets of mold but aren't generally a problem unless they are big or there are a whole lot of them.

Baling

When the hay is dry in the windrow, baling begins. If you're making small square bales, get a few helpers (if you can) to stack the bales in the wagons and then load the bales into the shed or barn. If you're working alone, you might prefer to make round bales, which are the least labor-intensive.

If you are hiring a custom operator or using someone else's baler and intend to sell or use the hay as certified organic, purging rules apply. Two bales must be put through the baler to clean the machine. These bales must be documented and stored or sold separately from the certified organic bales.

Making Haylage

Instead of or in addition to corn silage, many dairy farmers feed their animals haylage, or fermented forage. They store it in a silo or in big, snakelike plastic bags out in the field. Haylage is cut with a flail chopper (or a flail mower) that chops it into shorter segments. It is blown into the silo or bag at a much higher moisture level than hay, then sealed so that it ferments.

A V-rake, even a small one, will rake two or more rows of cut hay, while a side-delivery rake can handle only one row at a time.

Many farmers apply preservatives to the haylage (and to dry hay, too), but these are not allowed in an organic system. The Final Rule does allow the use of a natural bacteria, lactobacillus, to aid the fermentation process.

Cows love haylage — humans aren't the only ones to appreciate fermentation — and it's pretty similar to hay in terms of nutritional value. For the farmer, haylage is easier to take off the field, since you don't have to wait so long for it to dry and have less risk of rain damage. Also, if you have the equipment, haylage is easier to feed than hay. See chapter 12 for more details.

Planning for Next Year

After the hay is off the field is an excellent time to apply soil amendments like manure, lime, potash, and phosphate, directly on top of the sod. Most organic amendments (except for manure and compost) are slow to take effect, and putting them on a sod, where they are not tilled into the soil, slows down the process even more. Therefore, what you put on now is to benefit next year's crop.

Last, if you're going to leave the field in hay for another season, it's important to give it six weeks of rest from cutting and grazing in the fall, just before the growing season ends, so it can build up enough food reserves in the roots to survive the winter and get off to a strong start the following spring.

"We view our farm as a living organism that is constantly changing and responding to what the farmer is doing to it, as well as to environmental pressures."

BOB QUINN
Big Sandy, Montana

Profile: **Bob Quinn**

Quinn Farm and Ranch, Big Sandy, Montana

North-central Montana is cold and dry, a short-grass prairie environment with an average of only 12 to 14 inches (30 to 35 cm) of rain per year. This is wheat and small-grains country; it's too dry for corn or soybeans. Even to get a reliable crop of wheat, farmers traditionally crop a field only every other year, leaving it fallow and bare in alternate years to conserve water for the next season's crop.

"Our biggest challenge is the lack of water. There's no irrigation available," says fourth-generation, certified-organic grain farmer Bob Quinn. With 3,000 acres (1,214 ha) and no livestock, there's also no feasible method of adding compost or animal manure to boost soil life and organic matter. These constraints have turned Quinn into an artist with green manures and crop rotations. Unlike his conventional neighbors' two-year wheat/fallow rotation, Quinn's four- or five-year rotation alternates between cash crops and green manure crops, with specific crop selection adjusted for weather conditions and weed pressure.

"What we try to do with green manures is to provide enough biomass for the microbes to work once it does rain. The more moisture, the more breakdown and nitrogen release, and it balances perfectly with the increased nitrogen demands in a wet year. In a dry year plants are growing less, so there's less nitrogen demand, but less is being released," Quinn says.

When we do a green fallow we don't allow the crop to go to seed or grow past mid-June. We use primarily field peas. Most of our conventional-farming neighbors are either chemical fallowing or mechanical fallowing. Our green manure crops are planted in early fall or in early spring, then in June we disc and chisel-plow, which leaves most of the residue on the surface, to protect what's going on underneath.

Since I started transitioning to organic about 20 years ago, we've gone through severe wet times and severe dry times, and they call for different crops. I've divided up my green manure crops. Shallow-rooted crops, like peas, are for dry years, and deep-rooted crops like sweet clover and alfalfa are for wet years. Since becoming organic, we have discovered alfalfa not only adds nitrogen but is a significant help with weed control. Lentils add nitrogen, too, and are a high-value cash crop.

Alfalfa's additional benefits, Quinn says, come from its deep roots. They grow in the same soil zone as the roots of Canadian thistle (one of the most hard-to-control perennial weeds in the United States) and outcompete it. In those areas where natural hardpans form saline seeps, Quinn utilizes alfalfa's deep roots to suck up the subsurface water, preventing it from percolating to the surface and bringing the salt with it. This sort of finely tuned problem solving is characteristic of Quinn's approach.

You need to approach every type of weed with a different management solution. You have to understand its life cycle and at what point it's most vulnerable. That's why you have to be aware of what's going on, watch it, and respond quickly. It's a different way of thinking about farming, and you worry about different things.

Most conventional farmers think about yield, not net income. They don't worry about which weed or insect problems are coming, because they believe once these problems arrive they can be solved with chemicals. With organic you have to see them coming. You have to solve those problems before they're full-blown. Most farmers aren't used to walking their fields and projecting out a few weeks what they're going to turn into. You need to go through your fields on foot, or with your ATV or your saddle horse. The chemical guys have sprayers come out and check their fields. They go to the coffee shop more than I do.

Though Quinn grew up on the farm, he left to obtain a PhD in plant biochemistry and to teach for a period. In 1978 he decided he'd rather farm, and returned home to take over the family grain and cattle operation. His first innovations were in marketing. In 1983 he established his own wheat brokerage to gain some control over volatile prices, and in 1985 sold the cattle in order to travel and do more marketing, and to build a flour mill so he could do his own processing. Soon, Montana Flour and Grains (MFG) was processing and marketing for some of the neighbors as well. When requests began coming in for organic grain, Quinn began experimenting with organic methods on his farm, and by 1990 the entire acreage was certified organic. He now grows and markets organic lentils, buckwheat, hard winter wheat, soft white and spring wheat. He also grows Khorasan wheat, a unique Egyptian wheat sold under the brand name Kamut. Quinn travels annually to food shows and clients in Europe and the United States to promote it.

Quinn's crops bring premium prices for being organic and for being high quality. A big part of quality is purity, or a lack of contaminating weeds or grain from other crops. This means meticulous attention to cleaning equipment between the harvesting of different crops, and a similar attention to detail when managing weeds. Though weed-control options are limited by crops and conditions, Quinn has developed, as with soil-quality management, successful methods uniquely suited to his operation.

Small grains are planted too closely for mechanical cultivation, so weed control has to happen before the crop is up. "We can never seed into dry ground like our chemical neighbors," Quinn says. "If I seed into dry ground and it rains then everything comes up at once and we are toast." The trick is to get the crop to germinate before it rains and initiates weed seed germination, he says. "You work the ground so you kill the weeds that are germinating, and then you plant into moist dirt that is dry on top so more weed seeds aren't germinating before it rains. We can't seed if it's too dry." The goal is to give the crop a jump start on the weeds, so the canopy closes and completely shades the ground before the weeds can germinate and catch up with the crop. "A good crop stand with a closed canopy is our best form of weed control during the growing season," Quinn says.

These skills take time to develop, Quinn says, so it's important to go slowly. "Start small, talk to the neighbors who are doing what you want to do, and find out all you can. Take it one step at a time. Don't do the whole farm at once — not more than 10 percent the first year. For our first organic experiment we planted 20 acres (8 ha) out of the 2,000 acres (809 ha) we had in crops that year. The thing you don't want to do is have a wreck right off the bat and get discouraged."

Quinn still makes trying new crops and techniques an integral part of his farming strategy.

I'm now experimenting with dryland vegetables. You just plant them farther apart. We seed corn in 3-foot (90 cm) rows, 12 to 14 inches (30–36 cm) apart. We plant potatoes in 3-foot (90 cm) rows, 3 feet (90 cm) apart. It's amazing how much you can really grow without irrigation. We're trying to figure out which varieties work best on dry land, in the upper Great Plains. We had 42 varieties of potatoes last year, which yielded from 0.9 to 9 tons (0.8–8.1 t) per acre. We took the best half of those and experimented with 24 varieties this year, with the eventual goal of half a dozen of the best reds, whites, and yellows. I'm always doing research on our farm, because that's what I enjoy.

Quinn concludes, "We view our farm as a living organism that is constantly changing and responding to what the farmer is doing to it, as well as to environmental pressures. We can't change those pressures, but we can adapt. But not if we don't know what's out there, what's happening. The farm is not a closed factory, where everything is controlled; it's a living organism, where everything is growing, developing, and going through cycles. You can't farm it by prescription."

Profile: Sam and Brooke Lucy

Bluebird Grain Farms, Winthrop, Washington

Ancient grains — the phrase evokes the scent of a wood fire and baking bread in a village oven, the sound of millstones grinding, and the sight of wind rippling across a field of ripe grain. For Sam and Brooke Lucy, certified-organic ancient grains have become a business and a way of life on their operation in the semidesert of the Methow Valley, where it laps up against the east side of the Cascade Mountains in Washington State. They continue to refine the entire chain of production, from maintaining the balance and fertility of the soil to adapting their grain varieties to their climate and terrain, storing and milling the harvest themselves to ensure quality, and selling directly to chefs and consumers who seek the soul-satisfying taste and high nutrition of emmer, rye, and old strains of wheat. An additional bonus is being able to bring up the two Lucy children on a farm in "one of the most beautiful places I've ever seen," says Sam Lucy.

The main crop of Bluebird Grain Farms is emmer, an ancestor of wheat that produces less per acre but packs a higher nutritional punch and is more easily digested. Emmer is well suited to the arid climate, which is similar to that of northern Mesopotamia, where it originated. It also is easier on the soil than wheat, the Lucys say. There are other differences as well. Sam says, "I've been an organic grain producer since 1999, and the more I grew grains, the more I grew enamored of grains. I sold on the commodities market until we started raising the emmer. We quickly realized that emmer is something different, and we needed to mill and market it ourselves."

Their realization led them to construct on-farm Old-World-style wooden granaries. The moisture-absorbing character of the wood prevents mold, allowing the Lucys to avoid using the fumigants commonly needed to combat the problem in conventional metal grain bins. Since milled grains not treated with preservatives lose their freshness relatively quickly, the Lucys built their own on-farm mill as well. They don't mill the grains until they're ordered, delivering the freshest possible product to customers.

Marketing such an unusual premium product demanded further innovation, so Brooke developed a Web site, built relationships with chefs, and promoted the grains to local food groups. As it turns out, the customers were glad to meet her. "Marketing has been so easy, I never would have believed it," she says. "The local food movement is really happening here in the Seattle and Portland areas, and local food organizations are critical to helping farmers connect with customers. The Seattle Chefs Collaborative hosts events to bring farmers and food buyers together, and that was my jumping-off point for meeting chefs."

The Lucys obtained their first emmer seed from a university that had in turn obtained it from the World Seed Bank about 25 years ago. The Lucys also grow a variety of heirloom rye, called dark northern, and varieties of hard red and soft white spring wheats that were developed about 40 years ago. Since obtaining the original seed they have raised seed for subsequent crops themselves. Brooke notes, "We want our seed to be as adapted to this environment as it can be. That's what is wonderful about these open-pollinated, heritage grain varieties — they will adapt to the soil over time. Emmer grows naturally very well here, and in 20 years we might see our own local variety develop."

The soil is where it all begins, Sam says.

Our goal is to leave land in better shape than we found it. Farming maybe wasn't the best way to take care of the land to begin with. When we plowed up the prairie we created a situation. Here, we know the land is going to be farmed, and I want to be sure the soil benefits from our practices. If we don't balance our soil, we're not going to be raising nutritious food.

The Lucys test their soil religiously, amend accordingly for balance, and use microbes and enzymes from Tainio Technology to ensure soil biological activity. They carefully gear their crop rotation to meet soil needs. Typically they interseed clover with a grain crop and till it under after harvest. They follow two years of clover and grain with a green-manure year of peas and buckwheat and, if testing indicates it's necessary, an additional year of alfalfa.

Sam's background has made him unusually aware of the different characteristics of different soils in different climates. "I grew up on a New England farm, and I came back to farming after college. I helped my brother run a dairy for two or three years, and we put up all our crops. When I moved to the West I went right to work for a crop farmer here in the valley. It's quite a different climate. We have to irrigate, and we bale hay at night, because here it's so dry that baling alfalfa is best when it has a little moisture on it, so it holds together."

When Sam started a land reclamation business, he received hands-on knowledge of how to restore soils:

People would buy old farm ground but wouldn't have the equipment to take out the weeds, and I'd bring it back to a nice sustainable landscape. That's how I began paying off my original equipment loan. That led me to more productive ground that had irrigation, and that's what got me back into farming. I started using grain in my restoration projects, and that's how I got into grain farming.

By and large I think there's better soil in the East than in this part of the West, in the semidesert of the foothills. The East has a little more cantankerous weather, but it doesn't have the extremes you can have in the West, where we've come close to losing an entire crop to hail or to wildfire. We have four distinct seasons, too, and a nicer spring than in New England, and we get on our fields quite a bit earlier. We don't have a lot of spring rain — our rain comes in June.

Though most soils in the area are sandy loams, some of the fields Sam farms have a lot of clay and must be handled differently. And just as he constantly adjusts his farming practices and crop rotations to address soil type and needs, Brooke's marketing program continues to evolve as their customer base grows and the children get older.

"We're definitely coming into a new phase here," Brooke says. "The more direct-to-the-consumer sales we can make, the more money we have in our pocket and the more sustainable we can be." On the other hand, she says, lots of small sales create the need for a sales staff but don't always generate enough cash flow to cover salaries. Working with a distributor to handle part of their crop is one alternative the Lucys are considering.

"Both Sam and I are very independent spirits, so this type of a business suits us well. We can create our dreams and go for it," Brooke says. "At the same time, we have children, so we have to be consistent, and we have to have dinner on the table every night. Balancing it all has been a challenge. But they're right alongside of us as we're putting our ideas into action. I can't imagine not having children, and not having them be a part of this."

Orchard Crops

Fruits, Berries, and Nuts

"**P**lant your fruit trees first." It's old advice and the source no longer remembered, but there is some good sense behind it. Fruit trees take the longest of any crop to begin producing, so the sooner you get started, the sooner you'll have homegrown, organic fruit to eat and market. If you are lucky enough to inherit some old trees, a good, thoughtful pruning may inspire them to produce for many more years to come, and you'll be on your way years earlier than if you had planted young trees.

Bramble fruits, bush fruits, strawberries, and rhubarb will produce a crop at a much younger age than tree fruits and are well worth having along a garden edge or in their own separate patch. Nuts are another crop that add diversity to the farmscape, kitchen table, and market stand. And no matter what your climate, if you can grow vegetables you can grow some type of fruit as well.

There's a world of difference between the casual way you can produce plenty of fruit for yourself with a couple trees in the yard and the attentive management needed if you want your fruit enterprise to be a full-scale orchard or berry patch and a paying proposition. If you plan to have only a few fruit trees, bushes, or brambles for home use, or just enough to bring in a little extra income in the autumn, the guidelines below will put you on your way. If you hope to derive the largest part of your farm income from growing organic fruit or nuts, then you are looking at a whole additional level of necessary expertise and labor requirements. This chapter will give you the outline of what you'll need to know, but to really start learning the business the best option is an internship with an organic grower, preferably of the same type of fruit or nut that you are interested in growing. The alternative is to start with a dozen or two trees, bushes, or plants, learning as you go, and if it suits you, expanding gradually through the years.

No matter what fruits or nuts you decide to grow, the same organic principles apply if you want a healthy crop each year:

- Test and amend your soils appropriately.
- Maintain good sanitation and biodiversity to control pests and diseases.
- Instead of rotating crops (impossible with permanent plantings), manage the ground cover under and around the fruit plants to maximize soil health and moisture retention and to minimize pest and disease problems.

How exactly these principles are put into effect varies according to climate, soils, pest pressure, and type of fruit crop.

KEYS TO SUCCESS

Choose a planting site with deep soil, good drainage, and protection from the wind.

Tree Fruits: What to Plant

Let's start with apples, as an excellent example of what's been happening in the world of tree fruits. Apples are grown by more people in more places around the world than any other tree fruit and offer greater genetic diversity than probably any other crop plant. Unfortunately, much of this diversity has been lost in recent years. Botany professor Don Gordon writes, in his *Growing Fruit in the Upper Midwest*, that in 1905 in the United States there were around 8,000 different apple cultivars, but by the mid-1980s all but 1,200 had been lost. Rare heirloom varieties are well worth seeking out for their wider spectrum of flavors, the higher price they may command in niche markets (but not in wholesale), and in the interests of preserving genetic diversity.

The story is the same for many other fruits. Old varieties particularly well suited to a specific climate and soils have too often disappeared, and only thanks to the efforts of a few interested growers and plant researchers have some been saved. Since public interest in heirloom varieties of

A variety of fruit and nut trees, bushes, brambles, and perennials diversifies the dinner table and the market stand — and adds a serene beauty to the farmscape.

Plan for Preserving

The only fruit that will reliably keep through the winter without very specialized cold storage is apples. The rest will need to be sold quickly after harvest or preserved by processing. Before you plant a lot of fruit, have a plan for how the harvest will be marketed and, if not sold in season, preserved.

domestic plants revived in the 1980s and 1990s, more nurseries have begun to propagate and sell some of these historic cultivars. Heirloom apples are the easiest to find, but a little research and perhaps a phone call to the plant research department of your state's land-grant university, or a regional fruit growers' association, might be the ticket to tracking down old varieties of pears, peaches, or other types of fruits that might do especially well in your locale.

On the other hand, in the past 100 years many new fruit varieties have been developed by plant breeders, which flourish well outside the original species' natural home. For example, there are now many apple varieties that grow much farther north and south than wild apples ever did, including the far south and the coldest reaches of the upper Midwest. Many public universities have strong fruit-breeding programs, and their experimental farms would be one of the first places for the aspiring organic fruit grower to visit and inquire about disease-resistant varieties suited to local soils and weather patterns. Go during harvest season and taste the different types, since taste should be a primary consideration when choosing what to plant.

One word of caution: Many garden catalogs and nurseries that cater to home gardens offer fruit plants for sale in climates that are not suitable. If you want to experiment, go ahead and try the "cold-hardy" apricot in North Dakota or an avocado tree in Georgia. But if you're hoping to sell fruit, then stick to species that are known to reliably produce a crop in your area.

Besides apples, the North American organic tree fruit grower, depending on location, should also consider pears, peaches, nectarines, cherries (sweet and tart), plums, and apricots, as well as some of the more unusual fruits such as mulberries, persimmons, figs, and quince, if the climate is suitable. In the warmest areas, citrus fruits (such as oranges, lemons, limes, and grapefruit) and avocados are on the menu, too. These other fruits add diversity in the orchard and at the table and expand marketing opportunities. All fruit trees are similar in their requirements for site, soil, and care, and the same organic rules apply. No matter what type of tree fruit you decide to grow you will need to pay close attention to site and variety selection, soil type, tree management, and pest control.

Planning the Orchard

Since it doesn't rely on chemical cures for bad decisions, organic tree fruit production requires close attention to appropriate rootstocks, varieties, and planting sites. It differs significantly from conventional orcharding in what is usually allowed to grow beneath the trees and how insects and diseases are controlled. Organic growers themselves differ as to what the best practices are, as they should, since what works for one farmer in one area won't necessarily be the best practice for another farmer someplace else.

As in all other chapters in this book, this discussion presents only guidelines, rules, and tools. Only you can determine how to apply them on your farm, and that will necessarily involve some further research and ongoing experimentation.

Pollination Preparation

Pollen is moved from the flowers on one tree to the flowers on another mostly by bees, though wind and other insects may perform the same task. As is the case for all plants, only pollinated flowers will produce a fruit (or a vegetable or grain). Since most tree fruits can't pollinate themselves, you likely will need to plant at least two trees of two different varieties (also called cultivars) to ensure fruit (the exceptions to this rule are tart cherries and peaches and some specific cultivars of pear

Make a Marketing Plan First

Since you have only one chance to choose, prepare, and plant fruit trees, you must exercise extra care in selecting a growing site, amending the soil, and picking varieties. But before dealing with any of these concerns, if you plan to sell the fruit, you should figure out *where* you are going to sell it, and if those markets will pay enough, year after year, to make the whole enterprise worthwhile. The general options are on-farm sales, farmers' markets, area retail food cooperatives and grocery stores, and selling wholesale to a maker of organic (or conventional) jams, jellies, baby foods, dried fruits, or juices. Or you can build a commercial kitchen, get a license, and make the juice and jam yourself. See chapter 15 for more details.

and plum). Plus, not all cultivars can pollinate all other cultivars, so it's important when picking trees to pick two that are compatible. Pollination compatibility includes overlapping bloom times. If the two cultivars don't bloom at the same time, it doesn't matter how compatible they are, since pollen from one will never meet flowers of the other. If the nursery you are dealing with can't give you knowledgeable answers about pollination compatibility, find another nursery.

If you are going to have an orchard rather than a couple trees in the yard, consider leaving enough

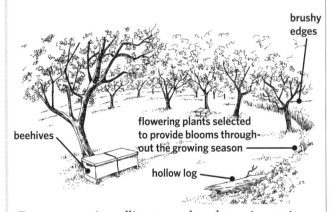

Encourage native pollinators and predatory insects in your orchard by providing homes, habitat, and food sources.

Thinking Beyond Apples

Besides the ubiquitous apple, there's a wonderful selection of other tree fruits that may provide a marketable crop or a tasty addition to the home larder. All tree fruits are similar in their requirements for fertile, well-drained soil, frost and wind protection, and susceptibility to a number of the same or very similar pests and diseases, but they differ considerably in what climates will suit them. Here are some additional notes on the various species:

Apricots. More than 2,000 named cultivars exist in China, and quite a lot less in the United States. It's generally best to plant two different varieties for cross-pollination, even though some cultivars may be advertised as self-fruitful. Apricots have a root system that is shallower than that of most trees, so constant moisture is important, and irrigation is recommended in dry periods. Apricot cultivars are available for most areas of the United States, but reliable fruit production is likely only in southern states, since the trees bloom very early and the flowers (which are the future fruit crop) are easily killed by frost. If a grower in a northern area wishes to try apricots, it's essential to get a cold-hardy rootstock. Most of the same pests and diseases that afflict apples afflict apricots as well.

Cherries. There are two types of tree cherries: sweet cherries for eating and tart or pie cherries, which also taste good in the hand but are more generally raised for processing. The tart cherries are all self-compatible, so only one cultivar is needed for successful pollination, while the sweet cherries are not and two compatible cultivars must be planted for successful pollination. Sweet cherries require a warmer winter than tart cherries, but tart cherries can be grown in most areas of the United States. Minimize pruning of both types of mature trees, as too much clipping can have a dwarfing effect and delay blooming.

Citrus fruits. Oranges, tangerines, lemons, limes, and grapefruit grow where frost is rare or nonexistent, and many varieties of each are available, as well as numerous hybrids. At the northern extremes of their range, growers may wrap insulating material or mound dirt around the trunks during the winter. Unlike most other tree fruits, citrus fruits improve in quality the longer they remain on the tree, so don't be in a hurry to pick.

Peaches and nectarines. Nectarines are really just a variety of peach. Peaches are a southern fruit, since they won't tolerate temperatures below -10°F (-23°C). Unlike other tree fruits, peaches are self-fruitful, so you need only one variety for pollination. More than 2,000 cultivars are known.

Pears. More than 5,000 recorded cultivars are known. Pears generally prefer a somewhat

space to establish habitat on orchard edges for pollination insurance — bees and bugs. Good bee and bug habitat includes brush piles, hollow logs or trees, and as many types of wildflowers as possible. You could also invite a beekeeper to set up hives in your orchard. There are many amateur beekeepers around, and a lot of them don't have land of their own. Local beekeeping associations are an excellent place to inquire. Ask at the agricultural extension office where to find any groups in your area, or call your state beekeepers' association.

Site and Soil

Because trees are so permanent, the grower gets just one good chance to pick a good site and do some deep tillage to incorporate any needed organic matter and soil amendments. Take the time to carefully select a good site, and be sure to test the soil.

Soil

No matter what rootstock and cultivar, all fruit and nut trees prefer deep, fertile soil that is slightly

warmer climate than apples, since they bloom earlier, making them more susceptible to late frosts. Pears suffer from the same pests and diseases as apples, with fire blight and "pear decline" (transmitted by the pear psylla insect) being the most common problems. Though a few varieties of pears are self-fertile under some circumstances, growers generally plant two different (but compatible) cultivars. Unlike other tree fruits, pears must be picked before they are fully ripe and allowed to ripen in storage for best consistency and flavor.

Plums. Plums are either European or North American in origin, and varieties are available that will produce throughout the United States. The European cultivars are self-pollinating, but native American varieties are not. The best fruit and hardiness come from crosses between European and native cultivars, and these types require a native variety for cross-pollination. Pruning should be minimized on mature plants. Wild plums and cherries may be reservoirs for insects and diseases that afflict cultivated plums, and if these are ongoing problems in the organic orchard the grower should consider removing the wild plants nearby.

More unusual types of tree fruits may be worth investigating for market or home production, such as quince, mulberries, figs, persimmons, and avocados.

acid, usually with a pH between 6.4 and 7. A sandy loam is the ideal, as it is for vegetables, but trees will generally tolerate a range of soils, so long as they are not pure sand, heavy clay, or poorly drained. Wet soils stunt growth and eventually kill the trees. Shallow soils, underlain by bedrock, inhibit root development and will limit production and trees' ability to withstand strong winds. Soils a little heavy or light for best production can be improved with an ongoing program to increase organic matter.

Topography

Topography is as important as good soil. Fruit trees should not be planted where they will be constantly exposed to high winds, or where frost will gather on cold nights. This means avoiding hilltops and hill bottoms. A windbreak can shelter a windy site, especially if it's composed of evergreens, but there's not much you can do about frost pockets at the bottom of slopes other than avoid them. A slope is better than flat ground, since it promotes better air drainage, particularly of cold air. In northern areas a northern slope may delay blossoming long enough in the spring that some fruits or varieties may not have enough time to fully mature during the average growing season, while for other species the delay in blossoming may be key to avoiding late frosts that can wipe out early blossoms and so the entire year's crop, as is, for example, the usual problem with apricots in the upper Midwest. A west slope, taking the full strength of the afternoon sun, makes fruit trees more prone to sunscald in colder areas, but that's fairly easy to prevent with a wash of white latex paint on the trunks each fall.

Amendments

Fruit trees require the same spectrum of soil nutrients as other plants, but in general there is much less messing around with soil amendments in orchards, for the simple reasons that you aren't removing as many nutrients with a fruit harvest as you are with annual crops, and the trees have big root systems that reach both deep and wide.

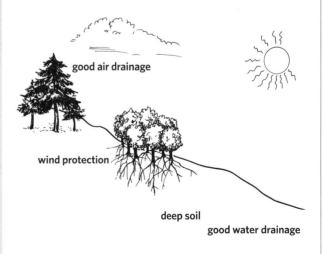

good air drainage

wind protection

deep soil

good water drainage

Plant your fruit trees where there is good air and water drainage, wind protection, and deep soil.

Letting grasses and forbs grow long for at least part of the season under orchard trees will encourage flowering plants (which attract pollinators) and build organic matter in the soil.

They're better at foraging for themselves than annual plants, reaching into the subsoil for minerals and spreading just below the surface far beyond the drip line to find water, air, and nutrients.

A second factor is the strong relationship between tree roots and mycorrhizal fungi that builds up through the years, something that doesn't happen in fields and gardens where soil is disturbed annually. These fungi greatly extend the foraging capabilities of the tree roots, and in return for the favor they get digestible sugars from the tree.

But growers still need to soil test and amend any deficiencies, since having the soil right is the first line of defense against insect and disease problems, and the best insurance for ensuring a good crop. In the ideal sandy loam, this means particular attention to maintaining organic matter and calcium levels. In heavier soils, calcium is less likely to be deficient.

Organic matter and soil cover are your best protection against the stresses of a dry season. Since the buds for next year's flowers develop late in this year's summer, dry weather has the potential to decrease next year's harvest. Soil covers and high levels of organic matter help keep moisture in the soil.

Calcium makes fruit crisp and is important in helping to ward off many diseases. It's also the major nutrient most likely to be lacking in light soils. Calcium is generally added in the form of lime, which also raises the pH. If your pH is where you want it but you wish to add calcium, add it in the form of gypsum.

In an orchard or any other permanent planting where there is infrequent or no tillage, amendments work their way into the soil very slowly. Allow three years for a lime application to have its full effect. This means you can either split the application and apply a little every year, or add what is needed every three years or so. If you are liming fields or the garden also, it's probably easiest to put the orchard on the same schedule.

Organic matter can be added with cover crops regularly tilled under (but only *shallowly*, to avoid damage to tree roots), by using and continually adding to mulch, or by letting a natural mix of grasses and forbs grow in the orchard. These will die back each fall, so that the old roots will become organic matter, and the dead vegetation will eventually rot into the soil, aided by earthworms. Given the workload on most small farms, the latter choice is the most labor-efficient and is the most in keeping with the organic principle of adhering as closely as possible to natural systems. Fruit trees in the wild surely grow in meadows, and less soil disturbance should allow the maximum development of the soil ecosystem.

Compost can be applied on top of the sod at any time. Manure is good, too, but should go on after the trees are winter-dormant since the nitrogen boost could delay hardening off in the fall, making the trees more prone to winter injury. Too many legumes (clovers) as cover crops may have the same effect, but if grasses and forbs are allowed to grow tall, the shade will discourage the shorter legumes. Or manure can be put on in the early spring, if it is tilled in and it's at least 90 days before you expect to harvest the first apples. If you have early-season fruit that matures in mid- to late August, this means the manure must go on before the middle of April.

Choosing Rootstocks and Varieties

Tree fruits do not breed true from seed, which is fun if you have years to spend and you're interested in developing a new variety, but difficult if you want to propagate the variety you already have. For this reason, tree fruits are propagated by grafting a branch from the chosen variety onto a rootstock, a form of cloning. The rootstock controls tree size to a great degree and can be used to add selected traits, such as cold-hardiness or resistance to some diseases, to a variety.

As with site selection and soil preparation, the grower gets one good chance to select the right rootstock and varieties, so it's important to research what's available and how well different rootstocks and varieties suit your site and intentions. Apple trees come in the greatest number of rootstocks and varieties, but with any species of tree fruit a grower will have a choice.

Rootstocks

Most fruit trees come in three basic sizes: small, medium, and large. In the trade these are usually called dwarf, semidwarf, and standard trees. The rootstock controls the size of the mature tree, so you can have the same variety in all three sizes.

As an example, dwarf apple trees mature at about 8 to 10 feet (2.4–3 m) in height, semidwarfs at 12 feet (3.7 m) or so, and standards at 15 feet (4.6 m) and higher. In general, the smaller the mature size, the sooner the tree will come into bearing and the sooner it will get old and die. Most standard-size apple trees will not come into full bearing till they're ten years old (some longer), but they can bear for 100 years and more. In contrast, dwarf trees will be at full production at around five years but may be done at 25 years or less. Other species of tree fruits vary considerably in their sizes and life spans, with peaches being often very short-lived, while pears may live even longer than apple trees. In each case, life span depends to a great extent on the rootstock, climate and soil type, and cultivar.

On the other hand, standard trees must either be aggressively pruned to stay at a reasonable height, or they will require ladders for harvesting and spray units (if you're going to spray anything) that can reach into the tops. Fruit on dwarf trees can be picked with both feet on the ground and a long-handled basket picker, and any old sprayer will do. Big trees take a lot of space; small trees fit in small areas. Dwarf trees may be more susceptible to wind damage and cold temperatures, so much so that in past years dwarf apples were not recommended for Zone 3 or less.

Some of the drawbacks can be mitigated if you're dealing with only two or three trees. For example, you can plant dwarf trees in a more protected area, such as along the south side of a barn. Most commercial orchards favor dwarf and semidwarf varieties for earlier production and ease of spraying and harvest. Many organic orchards prefer the more vigorous standard trees, and there are few farm sights more gorgeous than a huge old standard fruit tree in full blossom.

Fruit tree grafting has been going on for hundreds, possibly thousands, of years, and there are many rootstocks for the different species that are commonly used by commercial growers. Each has strengths and weaknesses. Unless you have the time and space to experiment, stick to rootstocks that have proven themselves through decades in your area. New types are often promising but may develop unforeseen problems years later. As selection criteria, Professor Don Gordon recommends that a rootstock:

- Be winter-hardy for your zone
- Provide good anchorage for the tree in the soil
- Be resistant to disease
- Sucker very little or not at all
- Be tolerant of unfavorable soil conditions
- Promote early bearing and continued heavy production
- Control size without affecting productivity

Also important is to prioritize those qualities most needed in your area; for example, disease resistance is less of an issue in western states. There is no rootstock available that is good in all these categories, so decide what you need most and choose accordingly. A good nursery, especially one that is attuned to organic growers, can be of considerable help in choosing among the confusing plethora of rootstocks.

Varieties

Cultivars, or varieties, of fruit vary not only in their adaptability to different soils and climates, but also in taste, time of harvest, and productivity. Apple varieties, for example, fall into three general seasonal categories: early, midseason, and late. Early varieties have a deserved reputation for having a short harvest window and poor keeping qualities, but as the first apples of the season they taste unusually delicious and bring an excellent price if you can get them sold while still at their peak. Midseason varieties vary in keeping qualities but will have a longer harvest window, while late varieties tend to be excellent keepers and extend the harvest season.

In *The Apple Grower: A Guide for the Organic Orchardist*, Michael Phillips offers a glimpse into just how complicated choosing the right variety can be. Not only do you need to take into account ripening time, and your own and your potential customers' likes and dislikes, you also need to consider whether your soils and climate zone will bring out the best in a particular cultivar, he says. 'Jonathon' shines in the Midwest, while 'Spartan'

The Final Rule on:
Rootstocks

A rootstock from any source is allowed by the Final Rule, as long as the tree is under organic management for a full year before fruit from that tree is marketed as organic.

tastes better when grown in New England. "Many of us easily forget that heirloom cultivars gained their reputations precisely because these apples were adapted to their locale," Phillips writes.

As a general rule, don't put all your fruit in one varietal basket. Plant a selection of choices that will give you an extended harvest season and some choices between sweet and tart, good eating and good juice or jam, and simple and complex flavors. In fact, juice, jelly, and fruit butters are better, in many people's opinion, when made from a mix of varieties. Though there's no accounting for personal preference. In fact, there's an old Midwestern saying: "One man's apple is another man's hog feed."

Layout

Fruit trees are spaced according to their size, to maximize sunlight and airflow but not waste space. Different authorities have different ideas about spacing, but if you're not using labor-intensive wire trellises or specialized pruning, plan on spacing most dwarf fruit trees at least 15 feet (4.6 m) apart, while the standard trees should be 30 to 40 feet (9-12 m) apart. Trees in southern areas will grow bigger than trees in the far north, so spacing should be adjusted accordingly. Again, it's best to ask local nurseries and growers for their recommendations for your area.

Planting in squares, so that the rows run in both directions, means you can mow or run equipment easily on all four sides of a tree. On the other hand, if the trees in one row are staggered so they don't fall on the same perpendicular line as those in the next row, you can get 15 percent more trees per acre, but you'll be able to run equipment in only one direction.

Planting the Orchard

Trees can be bought either as bare-root stock or in containers. Bare-root stock is cheaper and easier to get in the ground, and there is a wider selection of varieties available in this form, but those trees should be in the ground early in the spring, before the tree breaks dormancy (when the buds begin to swell). Container trees also benefit from early planting, but if you keep them watered they can be put in up until late summer in most areas. The later the planting, the less quickly the tree will become established, and the more watering you should do in the first season. One- and two-year-old seedling trees are generally best; older trees are more expensive and usually won't establish any more quickly.

If you can't find a good local nursery to buy from, check the Internet for suppliers. Mail-order nurseries generally do an excellent job, but be sure to use a nursery that offers trees raised in your climate zone. For certified growers, the organic rules require that you use organically raised planting stock, from a certified-organic nursery, or manage plants organically for a year before they are eligible for certification. Obviously this isn't an issue with tree fruits, since it takes several years for them to produce. There are some excellent organic nurseries well worth supporting with your purchases, if they have the variety and hardiness you need.

Identify the Graft Union

With standard-sized fruit trees, the graft union should be planted an inch or two under the ground, deeper than it was in the nursery. The union is visible as a bulge or slight bend at the base of the trunk, and marks where the rootstock was joined to the cultivar. This allows the base of the stem to plant additional roots that are from the cultivar itself. Should the tree ever be lost to wind or lightning (sooner or later this happens to everyone with fruit trees, it seems), then any sprouts from the root should be from the cultivar instead of the rootstock.

This additional rooting is something you don't usually want to happen with a dwarf or semidwarf tree, since then the dwarfing influence of the rootstock could be lost. For this reason, plant dwarf and semidwarf trees with the graft union a couple inches above the soil level. Unless, that is, you wouldn't mind having a different size and variety of fruit in that spot should the original tree be lost. Sometimes you get lucky and a rootstock will produce a fairly tasty fruit, but often they don't.

Dig the Hole

To plant a fruit tree, remove the sod and dig a hole slightly deeper than the longest roots, and at least 2 feet (0.6 m) wide. If you pile the dirt on a tarp or feed sack as you go, you won't have to scrape it out of the grass later. An easy way to judge the depth of the hole is to lay the shovel handle across so it's level with the ground, then hold the tree next to the shovel with the graft union at the correct height. If the roots touch the bottom of the hole, dig deeper. Do not cut the roots!

When the hole is the right size, before putting the tree in it, take your shovel and jab it into the bottom of the hole to loosen the dirt. If it's a heavy soil, do this to the sides of the hole as well, to make it easier for roots to penetrate. Then chop the sod into pieces and throw them upside down into the bottom of the hole, where they will provide a nice boost of organic matter and nutrients.

Set the Tree

Position the tree in the hole and, this is really important, gently spread the roots out so they are not circled around each other or pointing upward. Then refill the hole, starting with the topsoil and finishing with the subsoil. This puts the worst soil at the top, where it's easiest to make improvements. At the half-full point tamp gently and add 2 to 3 gallons (7.5–11 L) of water to eliminate air pockets, and when the hole is full, press the dirt down firmly with your feet and add another 2 to 3 gallons of water. Leave the dirt slightly cupped around the trunk, to hold water. Setting the tree at the right height and straight up and down can be a little tricky to do on your own, so have a helper if possible.

Mulch

The next step is putting mulch around the new tree to smother competing weeds and grass and hold water until the tree's growing roots and increasing height enable it to compete on its own. An effective method is to first surround the tree with six or eight thicknesses of newspaper unfolded to its largest dimensions, then cover that with 6 to 8 inches (15–20 cm) or so of straw, hay, or other natural materials. (Do not use the shiny advertisements or pages with colored ink. They are bad for the tree and prohibited by the Final Rule.) This will effectively control weeds for up to two seasons before melting into the soil.

The mulch should have no direct contact with the tree trunk, in order to discourage the moist, still conditions that might encourage disease-causing bacteria or fungi.

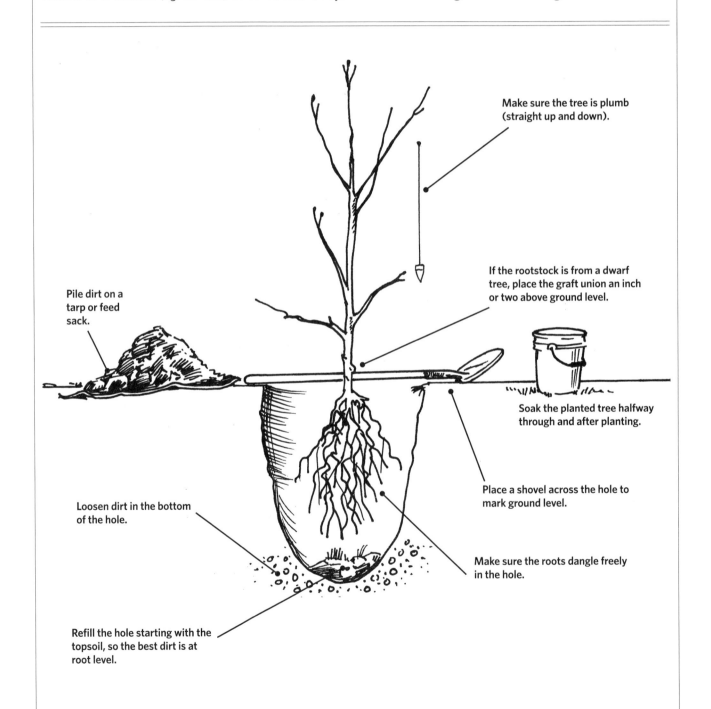

Make sure the tree is plumb (straight up and down).

If the rootstock is from a dwarf tree, place the graft union an inch or two above ground level.

Pile dirt on a tarp or feed sack.

Soak the planted tree halfway through and after planting.

Place a shovel across the hole to mark ground level.

Loosen dirt in the bottom of the hole.

Make sure the roots dangle freely in the hole.

Refill the hole starting with the topsoil, so the best dirt is at root level.

The size of the hole, placement of the tree, and treatment of the roots are all important factors in planting a fruit tree.

Care and Maintenance of the Orchard

Looking after fruit trees can be one of the most pleasant of all farming tasks. You're often out in the orchard in lovely weather, working quietly, attuned to birds, bugs, clouds, and the smells of growing things. The season begins with pruning, continues with thinning fruit (if you're so inclined) and monitoring and control of pests, and finishes with the harvest.

Pruning

Fruit trees will survive without pruning. However, they will have more broken limbs, shorter lives, more problems with insects and disease, and less edible fruit. For those reasons, orchard owners generally prune each tree every one to three years.

Timing

Most orchard owners prune in the late winter, before the trees break dormancy in the spring. Pruning earlier in the winter is possible, but it increases the risk of stimulating bud break during a winter thaw, which will set back spring development and weaken the tree. Sometimes you get behind and are still pruning as buds are breaking. This is a little riskier, since the trees bleed more sap and the fresh cuts can be an invitation to disease. Late-summer pruning is used in special circumstances, such as when an abundance of water sprouts (upright, non-fruit-bearing sprouts — you'll know 'em when you see 'em) decreases the amount of nutrients and water the tree directs to the ripening fruit.

Technique

Pruning is done with several goals in mind. The most important are to select for the strongest junctions between branches and trunk and to shape the tree so that the maximum sunshine and airflow reach each developing fruit. There are different

Pruning New Trees

After planting, many growers cut back a tree's leader (the highest-reaching vertical branch) to encourage it to grow more aggressively. Others don't think this is necessary. In general you'd want a rough balance between the amount of roots and the amount of top growth, and the best course is to inquire at the nursery or of an experienced grower whether pruning or not pruning is best for the type of fruit, age of seedling, and your climate.

schools of thought on how best to shape a tree during its first few years of growth. With apples in particular, the old-fashioned way was to prune so the tree is shaped like a basket, or like an upturned hand, with the fingers representing the main branches and the palm an open area left in the center of the tree. Most sources now recommend a central-leader method instead, where a strong central trunk supports well-spaced lateral branches and the tree overall has more of a Christmas tree shape, so that the lower branches reach out beyond the upper ones, and there's enough sun for all of them.

In practice most fruit tree species are amenable to either method, but there is the occasional tree that refuses to go along with your plans, say for the central-leader system, in which case it might be better to suit the pruning to the tree's ideas instead of yours. A basket-shaped tree is great for little kids to climb.

Once a central-leader tree has reached the height you want, it's pruned in a modified central-leader system, where the leader is trimmed back as needed, encouraging the tree to grow several competing leaders, and so slowing or stopping any increase in height.

Making Cuts

When pruning, consider both how the tree looks now and how the branches will grow in the coming year or two. Walk around the tree and identify the obvious problems: rubbing branches, water sprouts, branches so close together they almost touch along their length. Taking care of these problems first is a good way to get a feel for how the tree is growing, and where cuts should be made to encourage and maintain an uncrowded set of main lateral branches.

Further cuts should be directed at favoring those lateral branches with wide crotches that are evenly spaced around and along the trunk, minimizing growth in the heart of the tree where sun doesn't fully penetrate, and cutting branches that compete with the central leader or with the main branches in a basket tree.

In some cases you may need to remove an entire branch to increase light penetration and airflow. Make the cut at the base of the branch, just above the "collar," or slight thickening that marks where trunk transitions to branch. Cutting closer to the trunk than this makes it easier for disease spores to enter the trunk through the wound. With a large branch, make the first cut from the underside before completing the job from the top. This prevents the branch from ripping off a strip of bark and wood when its weight causes it to begin falling before you complete the cut.

In other cases you may need to cut branch tips to redirect growth. Cut just above a bud that is faced in the direction you want the branch to grow. For example, a branch that has too much of a downward turn can be pruned back to where it is straight and has an upward-facing bud.

Every cut should favor branches with strong crotches over branches with weak ones. Strong crotches are really important. Narrow V-shaped junctures between branch and trunk, or branch and branch, are inherently much weaker than wider crotches of 45 to 90 degrees. A branch loaded with ripe apples or ice is heavy, and if it has a weak crotch, sooner or later it's going to break.

Pruning is also used to keep standard trees to a more reasonable size. However, care must be taken not to overprune. Standard trees tend to retaliate with an abundance of water sprouts if overpruned. And on any tree, don't remove more than a third of the growth at any one time.

Different species and varieties will react differently to different pruning techniques and cuts, and that's why pruning is an art, not a science. But nearly always, a bad pruning job is better than no pruning job at all. Your technique will improve with time.

Training a Fruit Tree

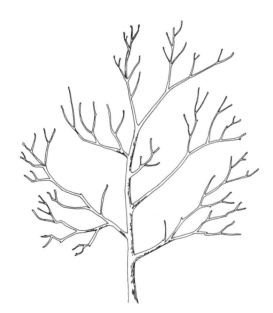

Central leader: Central trunk, with evenly spaced main branches

Basket: No central leader, with main branches all emanating from low on the trunk

Both the basket and the central-leader methods of training a fruit tree accomplish the same two goals: opening up the inside of the tree to light and air and ensuring strong branches.

Cleanup

When you're done, remove all the prunings from the orchard. Either use them to make a brush pile for bugs and small wildlife some distance away, or run them through a wood chipper and add them to the compost pile. Since dead wood can be a reservoir of disease, if you have disease present in your orchard or suspect it, the Final Rule allows you to burn the prunings.

Ground Cover

Organic growers don't agree on everything, and the topic of ground cover in orchards is a case in point. The options are

- Bare ground
- Mulch
- Short grass
- A tall growth of mixed forbs, grasses, and legumes

Bare ground is common in conventional orchards but (besides being kind of ugly) opens the soil to erosion and crusting, contrary to organic principles. If you have high pest pressure or are in a very dry climate and are using irrigation, you may not want anything other than the trees growing in the orchard. In this case the ground must be

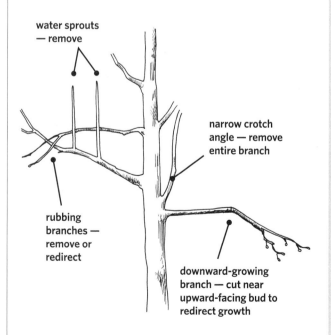

Prune to remove water sprouts, crossed branches, too-narrow crotches, and other problems.

water sprouts — remove

narrow crotch angle — remove entire branch

rubbing branches — remove or redirect

downward-growing branch — cut near upward-facing bud to redirect growth

Encouraging Wide Crotches

In the first few years after planting, trees should need no pruning except to remove broken branches or to correct obvious problems. But you still want to encourage wide crotches, and if they aren't happening naturally, encourage them either by pinning a wooden clothespin on the branch and bracing the other end against the trunk to open the angle, or by hanging a weight on the branch for a season, such as a plastic milk jug half-filled with water.

covered with a mulch. Mulching can be fairly high maintenance, however, since it requires growing or importing the mulch, having the equipment or labor to get it in place, and continually adding more as it decomposes into the soil.

The last two options — short grass or tall — are the lowest labor, though there's considerable debate as to how much water and nutrients the growing ground cover robs from the trees. In a dry climate with poor soils this is more of an issue than it is in wetter areas. Regular mowing will result in more grasses. Mowing restricted to just twice a season — before harvest and before winter — will encourage native flowering plants and other tall-growing species, excellent for beneficial insects and animals.

Which option you choose will depend on all sorts of factors: how much time you have for the orchard and how important the harvest is for your farm budget, what diseases and pests are problems, whether you depend on rain or irrigation, and so on. How all these factors play out on your farm should inform your decision as to what you want to let grow under your trees.

Thinning Fruit

The best per-unit profit in an orchard comes from the fruit that is large, well colored, and unblemished. These are the ones that sell for fresh eating at good prices. The small, damaged, or poorly colored fruits are used for juice, jelly, and similar products, at a much lower price. Fruits that are alone on their few inches of branch will grow much bigger and be less prone to blemishes than a bunch of fruits crowded closely together. This means that, in order to get the highest percentage of big, flavorful fruits it's necessary to thin the crop early in the season.

In addition, many fruit species, and especially most apple varieties, have a biennial habit: they tend to bear a good crop only in alternate years instead of every year. Early thinning encourages flower bud development for the following year, helping to overcome this tendency.

In conventional orchards, thinning is done with a chemical spray. In organic fruit production you can either put up with a lot of low-value fruit (and there's nothing wrong with making your main product juice instead of fruit for eating) and a largely biennial crop, or you can thin by other methods.

With just a few trees, thinning can be done by hand. With more trees, you'll need lots more time or more help, or you can spray salt, vegetable oil, or lime sulfur on the flower blossoms to reduce fruit set, or you can spray fish oil or a citrus extract at petal fall. These sprays need to be used very carefully, since they act by killing or crippling a percentage of the blooms, and the potential for overthinning is large. The pollinated blossom is what produces the fruit, so the more blossoms you kill or disable, the less fruit you will have.

When hand-thinning it's best to have the job done within 35 days of full bloom, since the more

Thinning fruit by hand is slow work, but the reward is more high-value fruit at harvesttime.

you delay, the less benefit there will be to the remaining fruits. With apples, for example, thin to one fruit every 5 to 8 inches (12–20 cm), leaving the biggest and best and getting rid of the worst. Don't leave the culled fruits in the orchard. Put them in a bag or bucket and put them in the compost pile, or feed them to the cattle or pigs.

Thinning is a big job, even with a few trees. Unfortunately, for the diversified farm, it needs to get done about the same time you're working hard in the garden and getting ready to cut the first crop of hay, and you may be calving or lambing as well. In this case a little thinning is better than none, but the world won't end if you don't get it done.

Pest Control

Pest control isn't strictly necessary in an orchard. Even if you don't prune, mow, thin, or spray, you'll still get some edible fruit in most years for many years. But there are two problems with doing nothing: First, if fire blight or some other really devastating disease or insect hits, you could lose every tree before you even realized what was happening. Staying alert and on top of things pays off. Second, if you want more good fruit and less bad fruit, you have to be proactive about it. This includes reducing the chances of insect and disease problems, monitoring so you know when they arrive and when to treat, and having what you need on hand so there's no delay in dealing with a pest.

Fruit trees in general, and apples and pears in particular, are susceptible to damage from an impressive list of insects, starting with codling moths, apple maggots, and plum curculios and down a long list that includes tree borers, European apple sawfly, scale insects, mites, and aphids. The bug list is matched by the number of diseases that can afflict orchards, such as apple scab, fire blight, cedar apple rust, and sooty blotch. Many of these afflict other fruits as well, often with additional problems thrown in. Organic controls for these insects and diseases do exist. But before these are applied, it's important to have the general proactive practices in place. If you can reduce the potential scale of the problem before it occurs, you'll be ahead of the game when the actual disease or insect arrives in the orchard.

Prevention

The prevention of pest and disease problems focuses on three principles:

- Ensuring good airflow
- Making beneficial insects welcome
- Practicing good sanitation

Orchard siting, tree spacing, and pruning really pay off here, optimizing airflow (which in turn decreases humidity) and making it harder for diseases to take hold and flourish.

Beneficial habitat in the form of flowering plants between trees or around the orchard edges gives aphid-eating lady beetles, green lacewings, and syrphids (hover flies) additional food in the form of nectar and a place to lay their eggs. (This habitat also keeps insect pollinators in the neighborhood.)

Insect-eating birds and bats are the top predators of codling moths and eat many other insects, too. Bats need dark, protected spaces to sleep in during the day, such as haylofts, shed attics, hollow trees, or bat houses. The types of birds that eat orchard insects need trees to nest in and brush or tall grasses and forbs for fledglings to hide in and for additional food. Bluebird houses provide homes for bluebirds, tree swallows, and other good species. A birdbath or pond for water and bathing is really appreciated.

Good sanitation is essential in an orchard. This means getting dropped fruit and diseased wood out of the orchard, since both are significant vectors for various pests. Codling moths, for example, crawl out of dropped apples to begin breeding their next generation.

Monitoring and Identification

As discussed in chapter 5, a basic tenet of integrated pest management is being able to identify insects and disease and know when they've reached a level

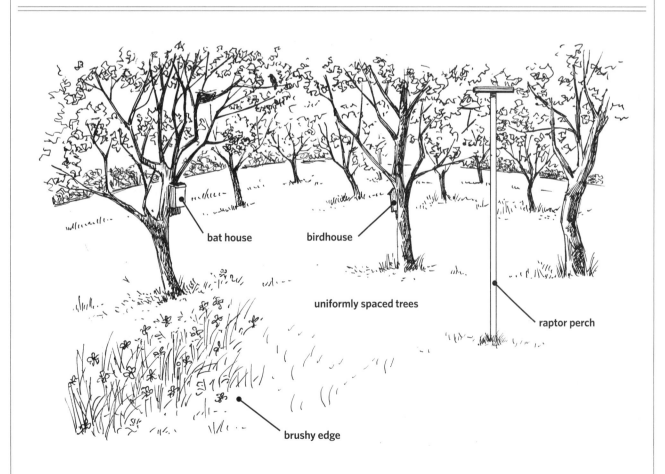

Properly spacing trees, picking up dropped fruits and prunings, and providing houses for birds and bats, perches for raptors, and brushy-edge cover for birds and their fledglings will limit bug and disease damage.

Some Common Orchard Pests and What to Do about Them

Pest	Identification	Life Cycle	Control Options
Apple maggot	Adult fly is smaller than housefly; cut open apple to detect larvae; infected apples have blemishes on skin and corky brown streaks in fruit	Adults lay eggs under apple skin; larvae feed on apple, drop out, and overwinter in soil	Good sanitation; kaolin clay on apples; yellow triangle bait traps during adult feeding stage
Apple sawfly	Brown winding scar on fruit, often a worm in the fruit, and a bad smell	Overwinter in soil; lay eggs in flowers; larvae feed on apples	White sticky traps that mimic flowers in appearance; early thinning of infested fruit
Codling moth	Entry hole in apple and worm inside	Found in nearly all orchards; eggs laid in young apples, can have several generations per season	Mating disruptors; granulosis virus; kaolin clay
Fruitworms, leafrollers	Damage to flower buds; bud leaves tied together; second generation leaves "corking" — areas of tissue that resemble cork	Adults lay eggs in early spring, which hatch just before full blossom	Bt (*Bacillus thuringiensis*) at pink bud stage and again at petal fall if needed; thin apples so fruits don't touch
Plum curculio	Half-moon-shaped scar on apple over eggs; oval exit hole on bottom of apple from larvae	Arrive at or after pink bud stage; lay eggs in developing fruit; larvae feed inside apples, causing them to drop off tree before they can ripen; can be a disaster in an orchard	Kaolin clay spray (most effective); pheromone-baited sticky traps around trunks; jarring bugs loose by hitting the branches with sticks

where they're going to cause economic damage. Of course you can't identify them unless you can find them or recognize the signs they leave behind, and to do that you'll need to spend some time poking around fruit, leaves, bark, and ground covers with some ID books or other aids in hand.

Monitoring traps are very useful in bringing the bugs to you and also giving you an idea of how many are present. A variety of traps are available, from simple white cards coated with a sticky substance to traps baited with species-specific pheromones (sexual hormones that attract adults looking for a mate). Suit the trap to the problem species, and follow directions from the manufacturer for placement, rebaiting, and cleanout. Close monitoring of problem species gives you an idea of how many are present, and of where they are in their life cycle, which tells you when you'll need to apply sprays for maximum effectiveness.

Many organic orchardists have refined their insect-control techniques to the point of monitoring humidity and temperature levels, since life cycles of some of the worst orchard pests are closely attuned to these environmental cues. For example, codling moths don't begin mating until

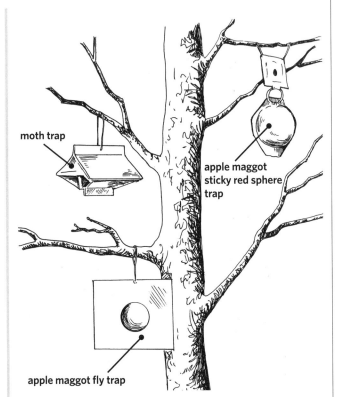

moth trap

apple maggot sticky red sphere trap

apple maggot fly trap

Insect traps are used to monitor or reduce pest levels in orchards. They are colored and baited differently, according to the type of insect they are intended to trap.

Some Common Orchard Diseases and What to Do about Them

Disease	Identification	Life Cycle	Control Options
Apple scab	Corky scabs on fruit	A fungus that overwinters on fallen leaves; releases new spores in wet weather	Remove fallen leaves or dust with lime; prune for good airflow; spray sulfur
Fire blight	Wilted leaves, browned or blackened branches, leaves that stay on tree in winter; pears are especially susceptible	A bacterium that spreads through the tree, killing it; transmitted by insects and rain	Plant resistant cultivars; prune out affected branches and burn
Rust	Yellow spots on fruit; misshapen fruit	A fungus that overwinters on alternate host plants	Remove alternate hosts (juniper); plant resistant cultivars
Sooty blotch	Black sooty blotches on fruit; harmless to people or animals	A fungus spread by rain and humidity; overwinters on bark and twigs	Prune for good airflow; wash apples after harvest to remove blotches

evening temperatures stay above 60°F (15° C), and the first eggs will hatch after 243 degree-days have accumulated. The day the eggs hatch is the day you do the first spraying.

This level of monitoring requires a good "weather station" setup that will record highs, lows, and humidity levels, as well as extensive knowledge of insect and disease life cycles. It's a bit much for the home orchard, but essential for the commercial organic orchard.

It's unlikely that the organic orchard will ever be completely free of insect damage. Keeping it to a reasonable level is the goal, and for the small-scale home orchardist it's also important to distinguish between those bugs that make fruit inedible, and those that just cause blemishes. Blemished fruit is fine for pies, jelly, and juice.

Dealing with Larger Pests

Deer will travel a long way for most fruit and are the major animal pest in orchards. There are two ways of dealing with this problem. The most expensive is building an 8-foot (2.4 m) fence around the orchard. This can be either woven wire or high-tensile strands, but it has to be at least 8 feet high or they will jump it. An electric fence is a possibility as well. To baffle the deer and keep them from jumping over or creeping under it, run a one-wire fence about 3 feet high outside and 3 feet from a two wire fence, with the wires at about 2 and 4 feet high.

The second method is to plant standard-sized trees and fence them individually for a few years until they get big enough to withstand some damage. This is done by setting 6-foot-high (1.8 m) circles of woven wire around each tree, held in place by a fence post (a metal T-post works fine and can be pounded in quickly). Take the fence off each fall for harvest and to trim ground cover short around the tree for the winter, then replace it. This method is not cheap and doesn't work with dwarf trees, since they won't grow large enough to get above deer feeding, but it is very effective.

Woven-wire hoops around trees are fairly expensive and time-consuming to install, but for a small orchard they may be the most effective method of protecting trees from deer damage. Be sure to anchor each hoop with a fence post to prevent it from blowing over in high winds.

Once the standard trees are large enough so that most of the crop is above the reach of deer, you can put up with some damage. This works if the deer pressure is not heavy (dogs are a real asset here if they are free to roam the orchard at night). On the good side, having a few deer in the orchard can greatly reduce the job of picking up the dropped apples, and they will keep the trees pruned up to a height of about 3 to 4 feet (about 1 m), which will make it easy to mow or apply mulch under the trees.

The second biggest animal problem in the orchard is rodents. When other food is scarce, such as in winter, is when mice, rabbits and all their sharp-toothed kin delight in chewing the bark off young fruit trees. If the bark is chewed in a complete circle around the trunk, the tree will die. Prevent this disaster by placing nylon or metal screening (such as is used on screen doors) *loosely* around the base of each trunk from ground level up to 8 inches (20 cm) above the anticipated snow depth. Big old trees with tough bark aren't

A loose cylinder of hardware cloth or metal screening around the base of trees protects the bark from rodents, especially in the winter when they are hungriest. Make sure the protection extends several inches above the expected snow depth.

attractive to rodents, so once your trees are big you can skip this precaution.

Birds are major problems with cherries and often other soft fruits, eating or damaging much of the crop in some orchards. For a few small trees, netting can be placed over the tree to keep the birds from getting at the fruit. In a larger orchard, various devices are used to scare away birds, ranging from recorded noise and music to pie plates strung on twine stretched between trees so they rattle in the breeze. Studies have shown that these methods are not very effective in most situations, since the birds quickly become accustomed to the noise and motion. Only if the noise or motion were frequently changed to something new would the birds stay away. If small birds are doing a lot of damage to your crop, installing nest boxes and perches to encourage raptor species that prey on smaller birds may be the best option.

Harvest

This is absolutely the best part of the whole year: picture yourself picking apples on a cool, crisp day, with plenty of friends and relatives, the cider press going strong, and a bonfire set for the evening. It's farm life at its best. But when should you schedule the picking days?

Different varieties of fruit ripen at different times and hold on the tree for shorter or longer periods. With apples, if most early varieties aren't all picked within a window of a week and a half, they get mealy and oversweet and are good only for deer food. Later varieties, on the other hand, may hold on a tree for a month or longer. Fruit on a single tree doesn't ripen all at once, either, so start by picking the ones on the outside that got the most sun — those will be ready first. Work your way inward on subsequent pickings.

There are four clues for telling when fruit is ripe:

- First, the color deepens.
- Second, the stems break easily — you don't have to work too hard to break them off.
- Third, mature fruit will start to drop from the tree.
- Fourth, and best, the fruit will taste ripe — it'll lose that puckery green taste.

Fruit Storage

Apples keep best of all fruit. They will last through the winter if they are kept cool (between 32 and 38°F or 0–3°C) and at 80 to 85 percent humidity. An old-fashioned cold cellar with a bucket or clay pot of water may work well, but do not put the apples next to the potatoes or onions, since the apples emit ethylene gas that encourages the potatoes to turn green and the onions to sprout. Most other fruits will not keep long, and will need to be canned, jellied, jammed, juiced, frozen, or dried to preserve the crop. This offers many opportunities for being creative in your own kitchen and for making value-added products to sell during the winter, when cash flow generally dries up for growers.

Waiting till peak ripeness to pick tends to give the best flavor (with the exception of pears, which should be picked before they are fully ripe). However, for storage, fruit picked a little early will generally keep better, as fully ripe fruit goes soft in storage. If you're making juice, jelly, or jam, it's usually good to mix in some underripe fruit for an added flavor snap.

As you're picking, clean up the drops. At harvesttime many of the drops fall simply because they're ripe, and they are still good fruit. If there have been no animals in the orchard, these are safe to use for juice and jelly. If there have been animals grazing there (including deer), compost the drops or feed them to the livestock, to avoid any possible bacterial contamination.

This is also a good time to examine fruit for the telltale signs of various insects and disease, like the smudges of sooty blot, the half-moon scars of plum curculios, and so on. Most types of blemishes left from bugs and disease are just that — blemishes. The fruit is still perfectly good to eat (except ones with the worm still in) but may be hard to sell to a customer. If you don't want to eat it yourself, put it into the juice bucket or jam pot. More important is assessing the overall level of damage. This will tell you if you're doing okay with your pest management or need to address some big problems in the coming season.

Putting the Orchard to Bed

Once the harvest is done and the drops are cleaned up, it's time to call it a season and tuck the orchard in for the winter. If your ground cover is more than a few inches high, mow it short to discourage rodents from taking up winter residence in the orchard. If you have hawks and owls in the area, not many rodents will be brave enough to cross a big expanse of short grass, even with snow cover, for a meal of fruit tree bark.

If you have mulch around the trees, pull it back 2 feet (0.6 m) or more from the trunk as an additional precaution against rodent damage. Wrap screening or hardware cloth *loosely* around the base of all tree trunks with bark young enough to invite chewing. Plastic spiral guards or wrapping the trunk with tape are not generally recommended,

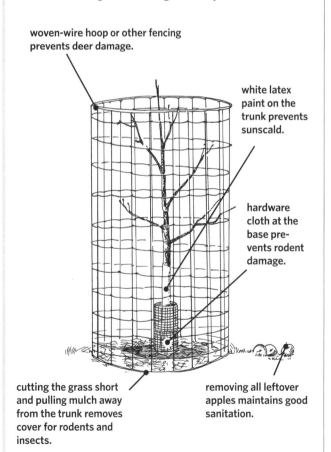

woven-wire hoop or other fencing prevents deer damage.

white latex paint on the trunk prevents sunscald.

hardware cloth at the base prevents rodent damage.

cutting the grass short and pulling mulch away from the trunk removes cover for rodents and insects.

removing all leftover apples maintains good sanitation.

Tucking in a tender young fruit tree for winter involves giving it protection from pests and the weather.

since both tend to hold in moisture and prevent air circulation, leading to rot, infection, and injury.

Nut Trees

Nut trees are generally bigger and take longer to come into bearing than fruit trees, but they are beautiful in the landscape and a couple of species have wood with commercial value. In general the requirements are the same as for fruit trees: choose deep, fertile, well-drained soil and plant species and varieties appropriate to your climate zone. Nuts are afflicted by their share of pests and disease, and again, the same rules apply as for organic fruit production: provide good siting, airflow, and sanitation, and get to know the life cycles and preferences of specific pests. Nut harvest is a little simpler than harvesting fruits, since you can either wait for the nuts to fall off by themselves, or shake the tree gently to make them drop. Organic requirements are the same as for fruit trees.

Almonds

Almonds are restricted to warm areas where their February blooms won't be killed by frost, and they prefer a drier climate and more sandy soil. Commercial almond production in the U.S. is limited to Southern California, and bees are critical for pollination.

Butternuts

Also called white walnuts, butternuts are native to most of the northeastern United States, but unfortunately butternut canker, a fungal disease, has killed or is killing the trees throughout their range. No resistant cultivars are available, but research is ongoing to identify wild trees that are resistant.

Chestnuts

The native American chestnut was wiped out nearly a century ago by a blight, though sprouts still shoot up briefly from old stumps before being killed in their turn. Blight-resistant Chinese varieties are now used for commercial production in the United States, and most varieties are hardy from Zone 5 south. The Badgersett Research Farm in Minnesota has developed hybrid varieties for more northern areas.

Filberts

This is the only nut produced commercially in the United States that grows on a bush instead of a tree. Native varieties growing in the wild are called hazelnuts, even though they are the same species. This nut grows farther north than any other, and it will tolerate some shade, as well as a wider range of soils than other nuts, but not dry conditions.

Pecans

These trees need a frost-free growing season of at least 200 days, so production is limited to southern states. The nuts are in high demand, and the beautiful wood has commercial value for cabinet making and other high-end uses.

Walnuts

Black walnuts produce both nuts and extremely valuable wood, so much so that the tree is more often grown for the wood and not the nuts. Pruning the branches to a height of 10 feet (3 m) or more from the ground reduces knots and increases the value of the wood. Black walnuts are fussier

Nuts don't bruise, so instead of picking them by hand you can shake the tree to make them fall, greatly speeding up the harvest.

about having deep, rich soil, but are tolerant of a wide range of climates, growing naturally throughout most of the eastern United States.

Brambles and Bushes

Not all fruit grows on trees. Many good fruits come from shrubs, such as blueberries, currants, and bush cherries, while brambles are well-known producers of berries, including raspberries and blackberries.

As with tree fruits, good organic management of bushes and brambles begins with researching the particular climate, soil, site, and management preferences of the species you intend to raise, and with choosing a variety adapted to your area.

Blueberries

Blueberries are native to much of the United States, growing in the wild as low, spreading bushes. These lowbush blueberries are used for commercial production in the northeastern United States, but growers in the rest of the country generally plant highbush blueberries, which grow 3 to 10 feet (1–3 m) in height. In the southern tier of states rabbiteye blueberries are grown, and these bushes can be more than 30 feet (9 m) high.

All varieties of blueberries need a particularly acid soil, with a pH between 4.0 and 6.0. Sulfur is a common soil amendment for decreasing pH, and elemental sulfur, though a synthetic product, is allowed by the Final Rule. Mulching with naturally acid materials such as pine needles or oak leaves will also help to lower pH. In northern areas be sure to buy cold-hardy varieties.

Currants

Like blueberries, currants prefer an acid soil, though they're less fussy than blueberries and will tolerate a pH between 5.5 and 7.0. They prefer a cooler climate than blueberries and are mostly produced in the northern tier of states.

Other Native Berries

Other native berries, such as gooseberries, serviceberries, huckleberries, and elderberries, can be picked wild or grown in the garden, and they may be a good way to expand a line of organic jams, jellies, or pie fillings. Cranberries, another native fruit, are grown where the summers are relatively cool and wet, in low areas that can be flooded at harvesttime to float the berries off the plants for easy gathering.

Some areas prohibit the cultivation of black currants, since they are an alternative host for white pine blister rust, a serious disease of white pines. Red currants are the variety generally grown commercially, while yellow currants are more of a curiosity. The berries are very small and picking enough to work with can be tedious, but they are superb in jellies and sauces and could be an excellent niche market.

Bush Cherries

Cherry trees, discussed earlier in this chapter, are not the only source of cherries. Bush cherries include:

- Sand cherry, a native bush from 4 to 5 feet high (1.2–1.5 m) that is cold- and drought-tolerant and a good choice for the northern plains. All varieties are self-sterile, so plant two different cultivars.
- Nanking cherries, another bush cherry, that is native to Asia and produces bright red cherries with very large pits and an excellent taste. This variety also needs two cultivars for pollination.
- Cherry-plum hybrids, available as small trees or big shrubs.

Brambles

Brambles are low, thorny plants whose shoots live for two or more years, producing leaves the first year and fruit and leaves in following years, before dying back to the ground. Blackberries and raspberries are the two bramble-type plants used for commercial berry production (loganberries,

Minimize thorn-to-thigh contact in bramble crops by maintaining mowed paths between the rows. These paths will also ease harvest and improve airflow to discourage disease.

boysenberries, and dewberries are varieties of blackberry).

Blackberries are native to North America (though cultivated varieties are of Eurasian natives) and are bigger, more tangled, tougher plants than raspberries. Their thorny canes will fruit for two or three years in most varieties.

Raspberries are not native to North America. Like blackberries, they bear fruit on thorny canes, but they are smaller and more susceptible to diseases than blackberries. Summer-bearing raspberries bear fruit on their two-year-old canes, while fall-bearing raspberries bear on first-year canes, so that a spring-planted patch will bear fruit that year. Again, plants whose fruits you intend to sell as certified organic in the same year that you obtain them must come from a certified organic source. Plants that don't bear until their second year do not have to be sourced from an organic grower, as they will have been under organic management for a full year before harvest. Most raspberries have canes that die after the second year.

Removing old canes from raspberries and blackberries helps control disease, though with good soil and airflow, pest problems are usually minimal in these plants. Birds may be a problem, but fortunately canes are small enough to cover with netting if necessary.

Grapes

Grapes are cultivated in more places worldwide even than apples, and grape production has a culture and history all its own. There may be as many as 10,000 named varieties around the world, and every region has its own specialties, along with its own palette of problem pests and diseases. The many native North American species are generally quite resistant to problems and good for jelly, but most often make poor table or wine grapes. Fortunately, hybrids of European and North American varieties have been developed, many in just the past 50 years, and good wine grapes are now being grown as far north as Zone 4.

In general, disease problems are minimal in the West, making organic grape production fairly easy there. In the higher-humidity eastern states, growers must pay more attention to choosing a suitable variety as well as to sanitation and other disease- and pest-control measures. Birds are a problem in many areas and netting may be the only effective way to deal with their depredations.

Grapevines are as permanent as trees, and are trained to trellises. Trellis construction, vine training, and vine pruning are quite specialized with

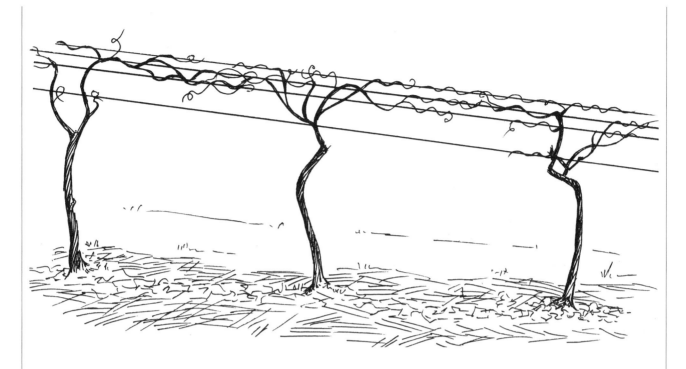

Grapes are grown on trellises, which should be stoutly built of long-lasting materials. In conventional vineyards, the ground under the vines is usually sprayed with herbicide to keep the ground bare. In organic vineyards, mulches or ground covers (mowed regularly) control weeds.

grapes, and represent a substantial investment of time and money before a crop can be harvested. Growers interested in organic grape production should visit other producers in their area and do some thorough research before launching themselves into this very specialized and competitive niche. The wine market in particular is extremely crowded with small vineyards, so be sure you will have a unique product and a large enough market before making a big upfront investment.

Strawberries

A perennial favorite for "you-pick" operations, strawberries offer a quicker return for a smaller investment than perhaps any other fruit crop. New varieties are being continually created, but basically there are three types: June-bearing, everbearing, and day-neutral. Spring-planted June bearers will not produce berries until the following year, while the other two types will produce berries

the same year they are planted. As is the case with grapes, selecting the right variety of strawberry for your area and farming system is crucial for best production and best pest control.

Commercial growers often manage everbearing and day-neutral strawberries as annuals, planting, harvesting, and plowing under all in the same season. In more traditional production, plants are kept for several years and generally grown in

Strawberries renew themselves every year by sending out runners that root and establish new plants. Straw under the plants keeps the berries clean and the soil evenly moist for best production.

The Final Rule on:
Berries, Brambles, and Grapes

As long as plants are under organic management for a full year before their fruit is sold as organic, they meet the requirements of the Final Rule and do not need to be purchased from an organic source. But if you are buying older plants from a nursery and plan to sell fruits that season, then the plants must be certified organic.

There are various ways to manage strawberries, each suited to different varieties, soils, and systems. But weeds, pests, and disease are always waiting to move in and, as always in organic production, are held at bay with proactive measures. With strawberries, this means starting by planting your new bed in clean-tilled ground free of weed seed, but not where tomatoes, peppers, or potatoes have been grown in the past several years. Strawberries are very susceptible to verticillium wilt, a common problem with those vegetables, which can survive in the soil for several seasons. Putting a mulch down in early spring before the plants are fully up is advisable, as it's much easier than trying to stuff mulch under the grown plants. In addition to protecting soil and controlling weeds, mulch keeps the berries from getting muddy. A dry mulch helps control slug damage as well.

wide rows. The plants propagate by sending out runners, which take root a little way from the parent plant and produce a new plant. A single plant of most varieties will produce berries for a few years. The new runners can then take over berry production.

"Farming . . . is a routine in tune with the seasons and markets. It's diverse and dynamic and I love the creativity it brings out in me."

CYNTHIA CONNOLLY
Ladybird Organics, Tallahassee, Florida

Profile: **Cynthia Connolly**

Ladybird Organics, Tallahassee, Florida

Grapes are notoriously susceptible to a whole range of pests and diseases that thrive in humid climates, so north-central Florida would seem to be one of the worst places in the United States to establish an organic vineyard.

"Pests and disease present no problem at all because we grow *Vitis rotundifolia*, the native muscadine grapes, and we have a very sound organic system in place," says Cynthia Connolly, owner and operator of Florida's only organic vineyard and winery, Monticello Vineyards and Winery. But while the native grapes are resistant to a wide variety of grape problems, they have a reputation for making poor wines. But the reputation is undeserved, she says.

> That comes partly from the Old World wines, where they have a strong reputation from centuries of winemaking. Our wines aren't from the European grape varieties, and they've always had to prove themselves. The two things that affect the wine are how the grapes are grown and the winemaker. I think organically grown grapes are superior. The old-timers who made wine from muscadine grapes used to put sugar in it for a syrupy sweet wine. I like only dry wines myself.
>
> The first year I made wine, a friend asked me if I wanted to go to the state fair. He asked me to bring some of my wine, and I won first place in the hobby winemaking category.

Connolly both grows the grapes and makes the wine, and that is only part of what she produces on her 50 rolling acres (20 ha) of sandy loam soil near Tallahassee. Her farm, Ladybird Organics, also produces fresh grapes for eating, persimmons, pecans, pears, satsuma oranges, Meyer lemons, and Marsh grapefruit. Wheatgrass, pea shoots, sunflower, and buckwheat sprouts are grown to order, and farm livestock consists of a flock of laying hens and a couple batches of broilers each year, which she butchers on the farm herself. And then there are the red worms, grown in a half dozen or so large worm beds, where they turn farm wastes into castings that fertilize the entire farm, with enough left over for sale to gardeners and local nurseries.

"The worms were just serendipitous," Connolly says. "I bought a cup of worms, and it turned into more than I ever imagined. The system is intuitive and absolutely marvelous. One thing has led to another and it's all connected. It's amazing. The worms have been a foundation for recycling and for all the life on the farm. The castings support everything I grow, they support the baby chickens, and they're a product themselves. As

226

"Pests and disease present at all because we grow *Vitis rotundifolia*, the native muscadine grapes, and we have a very sound organic system in place."

Cynthia Connolly

the largest of the microbes in the soil, the worms have helped me to see the connection of all the life-forms on this land, the web of life."

Farming is Connolly's chosen career. "It wasn't anywhere in my family. I was an English, French, and physical education major. I took French to get abroad. I pursued English in graduate school at Florida State University, which got me to Tallahassee, where I fell in love with the land and the community. Agriculture was like a calling for me. I started out with the resources at hand. There was a land-grant university in my backyard, and that was my point of entry. The importance of agricultural education is something I hold dear. It was the first door that opened for me."

After earning a degree in agriculture summa cum laude from Florida Agricultural and Mechanical University, she went on to the doctoral program in agricultural education and agricultural engineering at Iowa State University. Her specialization in agricultural mechanization and construction included course work in welding, diesel and small engines, electricity, and construction and maintenance of farm buildings and equipment. Her doctoral research in agricultural education included setting up a vegetable production program for women at Ahfad University College for Women in Sudan, the only secular college for women in Africa.

"I didn't know it at the time that I applied to Iowa State, but there weren't any other women in the doctoral program in ag education. I checked around the country, and there were no other women," she says. "I finished that degree on May 24, 1980. When the department chair walked me across the stage for my PhD, he said, 'You know, you're the first.' I said, 'Yes, but I'm not going to be the last.'"

Ten years in international work following her doctorate included living and working in a variety of countries on several continents. Connolly worked first for the USDA in Washington, DC, and then the United Nations Development Program (UNDP) in New York, and the UN's Food and Agriculture Organization in Rome. She says:

I spent a good amount of time in that career. You travel, you go from one country to another, you land in another world with two suitcases and your whole community and language changes. After so many years, even though the work was professionally very satisfying, I got to feel too disjointed. I wanted to feel rooted in a community, with friends and family, and I wanted to farm.

My heart was in Tallahassee, the land and the community I had come to love when I went to FSU. I wanted to buy land here and farm. I did not have any requirements laid out, except I didn't want swampland; I wanted arable highland and good water. I wanted to grow organic food, and I didn't have any limits on it. Organic wasn't really a codified way to farm when I studied or worked administratively in agriculture, but I always saw a logic to working with natural systems and without synthetic chemicals. When I was still with the UNDP they sent me to the International Fertilizer Development Center in Muscle Shoals, Alabama, and there were at least three people I worked with there that died of cancer. That was a lot, in my experience.

Connolly bought her farm in August of 1989. "I was 39 years old. My first purchase was a 3910 Ford tractor, and I bought one implement with the tractor — a very heavy-duty mower. Everything else I found used or put together with the welder, on the farm. I found all kinds of parts in the junkyard in Cairo, Georgia."

Fortunately Connolly became acquainted with the local Butler Building representative, Al, who had rebuilt the building, which needed some repair work when she acquired the farm. She and Al became fast friends. "The basis of our friendship was that he could come anytime and pitch in and help, and then we'd go out and get a meal and talk about farming. We had a synergy. He had the skills of someone who grew up on a farm, and I had the theory down, and we built all kinds of things together."

Her first venture was small grain production. "We can't grow hard red winter wheat or even red spring wheat here, but we can grow the soft white wheat," Connolly says. "I had the combine, I had the drill. But it was always a challenge because grain has to be cleaned, and I had to have it cleaned twice because it was organic. I had to carry it up to Georgia to a cleaner, and when they came under new ownership they charged me so much that it was a losing proposition. So I moved on to vegetables."

But the leftover grain was handy when she began raising organic chickens, since organic feed was scarce at the time. There were some pecan trees on the farm already, and she planted a variety of deciduous and citrus fruit and nut trees. In 1992, grapes and worms were added to the mix.

"Farming for me has been more intuitive than charted out," Connolly says. "Yes, I keep a running spiral notebook and jot down stuff so I don't forget, but the earth and the plants and animals drive me, and I know what to do and when to do it. It is a routine in tune with the seasons and my markets. It's diverse and dynamic and I love the creativity it brings out in me."

With so many crops and so little help, there have been some rocks in the road, she says. "It's a bit of a juggling act, but I've got my schedule down, and my seasons. The hardest is the grape season, because you have the fresh grapes and the wine. I hire workers from the agriculture industry in the area on their days off. I speak Spanish, which helps, and I feel a real link with these guys, who work side by side with me. I also have a supportive customer base, which is a pleasure and very affirming. And I have the very best farm dog anyone could have."

Al died in 2000, and Connolly still misses him. "I had the good fortune to have my friend Al for ten years," she says. "There aren't a lot of farmers in this area anymore, and I don't have a partner. Farming is much more isolating than I ever imagined. Socially, it's been a challenge. You farm from can to can't, the saying goes, or from dawn to dusk, and are too tired for much else."

Marketing is another ongoing challenge, she says. "I'm making a living, but marketing has always been challenging. I learn new things every year, and I've had to find new ways and places to sell products. As soon as I found a niche market or specialty product, others who saw my success would start to compete with me, driving me to push the envelope in new ways. There haven't been a lot of support structures or help for organic growers, either extension- or research-wise, though there are more now."

Connolly is doing her bit to remedy that situation. She currently serves on the board of directors of the Organic Farming Research Foundation, where she has a particular interest in the OFRF's new grant program for organic education, and she also offers consulting and short courses on various organic production topics.

The next issue will be how to pass on the farm. "How do you retire? The joke is that you 'retire' every night and then get up again in the morning. Retirement is a question that I'm struggling with. I don't have an answer right now," she says. "The only way farmers retire is to sell the land. But I'm still visioning a way to transition my farm to something that I can keep a hand in without being so overworked, maybe teaching."

Farming, Connolly says, has not been easy, but it has been right. "I think this was my calling and meant to be my path," she says. "I think we have to live our passion. Find a way."

Profile: Richard and Linda Byne

Byne Blueberry Farms, Waynesboro, Georgia

The most important pieces of equipment at Byne Blueberry Farms are a selection of pruning shears and saws. Says owner Dick Byne, "As soon as the season is over (in late July), we start pruning. We should be finished by August, but this year we didn't finish till November. I'm addicted to pruning. I think you're mismanaging when you let the bushes grow high. You need to keep them below 6 feet (1.8 m), or you can't pick them. It shocks the plant, but it really invigorates the plant."

Dick and Linda Byne and their four daughters own and manage 20 acres (8.1 ha) of organic native highbush blueberries near Waynesboro, Georgia, possibly the oldest organic blueberry operation in the state, Dick says. The blueberry patch is part of the 400-acre (162 ha) family farm where he grew up, helping his dad raise cotton and his grandfather take care of the dairy cows.

Dick graduated from college in 1978 and began planting native blueberries after installing drip irrigation in three fields on the family farm. There was never any question in his mind about using organic production methods. "I got into organics not to make money but just because it made sense. In '76 when we were sitting in agronomy class we kept talking about how the organic matter needed to be raised up. They were taking DDT off the market at the time, too, and that raised a lot of red flags. Organic just made sense."

Both Dick and his wife, Linda, work off the farm; he as a high school substitute teacher and bus driver, and she as a school librarian. But during the picking season in June and July they are on the farm full-time, and their four daughters help Linda grade and pack the berries.

"It was like a hobby between 1980 and 1995, and we really had a good time, though we weren't making any money," Dick says. "In 1995 people started saying that they wanted blueberries. Now we work with five different processors to make 13, soon to be 15, different products." The final products, from jam, preserves, jelly, and syrup to chutney, chowchow, and a blueberry-jalapeño jelly, are wholesaled to two grocery chains specializing in organic foods, Whole Foods and Earth Fare. (Byne's Blueberry Salsa won the Taste of Georgia Award in 2007.)

Picking at the peak of ripeness and flavor is critical. "All the blueberries are hand-picked," says Dick. "A machine really beats up the plants, and half of the berries fall on the ground. Your cost with the machine is 20 cents per pound (0.5 kg), but you're losing half the crop and you're getting all the green ones. For handpicking we pay 70 to 75 cents per pound, and, I figure, about a person and a half per acre (0.4 ha) per day during the season."

Careful picking pays off on the on-farm grading line, where a conveyer belt carries the berries first past a fan to blow off leaves, sticks, and small berries, then by a crew of inspectors, usually the Byne daughters, who pick out the soft berries before the rest are packaged and delivered to the processors. "We never pick them red or green," Linda says. "The flavor will never be there if you don't pick them blue. I think that's the reason for our success. Our berries taste good."

Careful picking and ample rain help produce great flavor, and so do healthy soils and healthy and well-pruned plants, Dick says. "We forget the basics sometimes, the fundamentals. Pruning lets the light in, gets your plant full of sunshine and air, and keeps the fungus down. I have very little disease, and I think it's because of the pruning."

Insects aren't much of a problem, either. "If you're organic, you're not spraying, and you're not killing the good bugs either." And though there are plenty of berry-eating birds around, "we're out there picking at six in the morning, because it scares the birds off," Dick says.

He has been testing soil and leaf samples since he began farming organically to guide his fertility program and to track pH, a critical factor with blueberries. "Our pH is 5.5, probably on the high end of where blueberries need to be. I keep dropping it, but it's kind of like holding one of those bobs in the water — it keeps popping back up. The pH is something you have to be working on all the time, like fertility. I have used sulfur, though not many times, and I use peat humus, which is acid."

Instead of applying lime to supply necessary calcium, Dick uses land plaster, or gypsum, which provides calcium in a form that plants can take up, but without raising soil pH as lime does.

Dick has also replaced the original drip irrigation with overhead irrigation, for frost protection. "In 2007 we lost everything that was outside the water to 23 degrees (-5°C) on April 8. You don't forget the date of an event like that."

The Bynes also raise honeybees, which provide raw material for their blueberry honey, but more importantly, for the bees' pollination services. Blueberries are not self-fertile; they must cross-pollinate with a different variety in order to produce berries, and the bees serve as the catalysts for that process.

The Bynes raise seven of the more than 20 varieties that have been developed from native berries of the region, and are working on creating their own variety. "We just keep watching plants, tagging the ones with the sweetest berries. We've mated two different kinds and hope to come up with our own in maybe two years. I want to have all the varieties eventually. Someone may not like a variety grown on a different farm but may like that same variety grown on your farm, because of different weather and different soils."

Dick Byne's comprehensive approach, commitment to improving his farm, and quality products were recognized in 2009 when the Alumni Association of the Abraham Baldwin Agricultural College in Tifton, Georgia, chose him for its Master Farmer Award.

Though Dick is continually experimenting with different varieties, fertilizers, and farm practices, he doesn't do things arbitrarily, and he keeps good records of results. "You can tell what is positive, and what doesn't work. Take your soil samples, take your leaf samples, and manage what you're doing out there. Find out at what point you have diminishing returns."

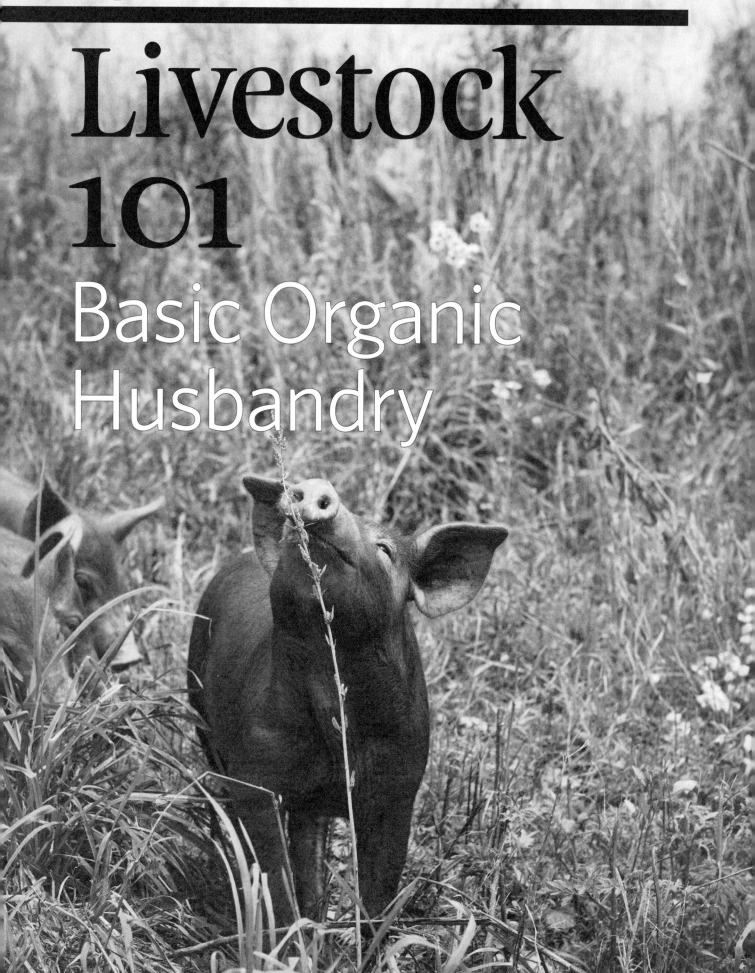

Livestock 101

Basic Organic Husbandry

> *Big domestic mammals were crucial to those human societies possessing them. Most notably, they provided meat, milk products, fertilizer, land transport, leather, military assault vehicles, plow traction, and wool, as well as germs that killed previously unexposed peoples.*
>
> **JARED DIAMOND**
> *Guns, Germs, and Steel*

Livestock do wonderful things for the small-scale organic farm. Among other things, they add diversity, activity, and manure. Manure is the gold standard for organic fertilizers, full of organic matter, nitrogen, and all sorts of other nutrients. It is to soil what chicken eggs are to people: about the most nutritionally complete natural food there is.

Owning grazing livestock means having pastures and forage crops, which work their own magic on soil quality and conservation. And while that grass and clover is improving your soil, it's also being converted by your livestock into high-value organic meat and milk that can add to your diet and your farm income.

Organic Livestock Basics

The Final Rule (and common sense) requires that a farm operation "must establish and maintain livestock living conditions which accommodate the health and natural behavior of animals." Specifically, this means that all animals must have:

- Clean water
- Appropriate shade, shelter, and bedding
- A certified organic diet suitable to the particular species
- Access to the outdoors for sunlight and exercise
- Access to green, growing pasture during the growing season for ruminants (grazing animals)

In addition, any certified organic livestock producer must ensure the following:

- All agricultural products in feed must be certified organic; all feed additives (such as vitamins and minerals) must be approved.
- Species and types of livestock should be suited to your conditions and resistant to diseases and parasites common in your area.
- Animal identification and records must be maintained.
- Preventive health care practices must be established.
- Manure must be stored and handled so that it doesn't contaminate crops or water.

Since the natural behaviors, diets, and suitability of each type of livestock are somewhat different, the details of how you follow these rules are a little different for each species. In this chapter we'll discuss the general requirements in detail, and we'll cover how these requirements are applied for different types of livestock in subsequent chapters.

But the very first thing to understand is that organic livestock production is based on grass. Whether you're talking about pigs and poultry, which need a relatively small proportion of greens in their diets, or sheep, goats, and cattle, which require forages, the organic farmer has to understand good pasture before thinking about livestock. Here's why.

Pasture Comes First

Domestic livestock species are either ruminants, with multiple-chambered stomachs adapted to a diet of all forages, or nonruminants, which benefit greatly from some grazing but need a substantial amount of more nutritionally concentrated food, such as grains, to grow properly and fatten enough to suit modern human tastes.

Cattle, sheep, and goats are ruminants, and the Final Rule requires that these species have access to green pasture during the growing season. Poultry and pigs are nonruminants and so are not currently required under the Final Rule to have access to green pasture, though they still must have access to the outdoors. But pigs and poultry love green vegetation as much as the next animal,

and putting them on pasture (instead of dirt) not only will make them happier but will decrease their need for other feed and improve their overall health.

All the other types of livestock raised for meat, eggs, or dairy products on small-scale farms fall into either the ruminant or nonruminant category, including ducks, turkeys, guinea fowl, bison, elk, red deer, rabbits, emus, and ostriches. So, if you have organic farm animals, you'll want pasture before anything else.

Green, growing pastures don't happen automatically. In an organic system, pasture can't be taken for granted, or abused. Good pasture management has to be understood to be able to build a sustainable farm system, to produce certifiably organic livestock, and to comply with the Final Rule's requirement that pasture be managed to "provide feed value and maintain or improve soil, water, and vegetative resources."

But first, let's dispel some old myths about grazing.

creep feeder gives piglets access to food without competition from the sows

enclosure protects pigs from drafts

deep bedding allows for rooting and nesting

The Final Rule requires that livestock be kept in conditions that accommodate their health and natural behavior. Pigs that have plenty of space and the materials for rooting, nesting, and living in social groups are happy pigs.

Grazing Is the Solution, Not the Problem

The most common notion of grazing among both farmers and nonfarmers is that it means fencing off a pasture, turning out the animals, and leaving them there. Under most circumstances this system degrades biodiversity and soil quality, reduces the productivity of the pasture, and encourages noxious weeds.

Management-intensive grazing (MIG) is a system that transforms grazing animals from a burden on the land to the best means available for improving degraded land and soils and maintaining a sustainable agriculture. Also called rotational grazing, MIG simply means moving livestock to new pasture on a regular basis, leaving the just-grazed area to rest long enough for grasses and legumes to regrow. Rotational grazing works with nonruminants as well as it does with ruminants. In arid regions, rotational grazing may be the only practical means of restoring degraded land to its presettlement fertility and diversity. In any region, it is a low-cost, natural, and extremely effective method of improving open land.

Grazing Myths

This last point is extremely important, and little understood. The idea that livestock destroy land has been enshrined in professional training programs for land management, in the media, and in the public mind. It is completely wrong, as any intelligent livestock farmer of 80 years ago could have told you. It's also dangerous, since it so often in the misinformed mind leads to the naive and unsupportable idea that we should get rid of domestic livestock and all go vegetarian if we want to have a truly sustainable agriculture and enough food to feed a growing world population. This erroneous conclusion ignores the fact that without forages and animal manures, crop-based agriculture is deprived of its most sustainable source of soil fertility.

Manure was the cornerstone of soil fertility on traditional diversified farms, and while you can replace manure with green manures and compost, it requires more management, labor, and expense. Nitrogen in particular, the soil nutrient that is most difficult to maintain in an organic system, is always a challenge to find in adequate amounts, and on the small-scale farm manure is the best source of nitrogen. Many small-scale organic farmers who don't raise their own livestock bring in manure from neighbors who have animals but no gardens or crops, converting one farmer's waste disposal problem into another farmer's treasure. Of course, if you're a conventional farmer, nitrogen is easy to obtain. Synthetic fertilizer plants around the world pour out massive amounts of soil-structure-destroying synthetic nitrogen, made with massive amounts of fossil fuel. Take your pick.

The idea that humans should rid themselves of domestic livestock entirely and revert to vegetarian diets to free up land for crop production also ignores some basic facts. A good deal of the land used for livestock production around the world is far too poor, dry, rocky, steep, cold, or otherwise unsuitable for crops. Without livestock that land would produce nothing at all for human food, and the world would actually have a lot less of a food resource.

Another popular myth about domestic livestock that needs to be dispelled is that grazing destroys land. This fairy tale arose from the practices of farmers who dump their herds into a pasture or rangeland for an entire season and let them graze it right into the dirt. This relentless chewing stunts the grasses and clovers above and below the ground, letting opportunistic weeds move in and opening the land to wind and water erosion. Root systems shrink, and the soil cycles fewer nutrients and is held less securely in place. This type of poor grazing practice has degraded huge swaths of land worldwide, and the blame is wrongly pinned on the livestock instead of the farmers. Before farmers ever arrived on the scene, wild herds of grazing animals following their natural instincts kept grasslands healthy around the world. Grasslands become gullies, deserts, or weedy brush not because of the animals but because of how they are managed by humans. The livestock are just a tool, and as it turns out, they are really the finest tool available when it comes to restoring the same land they are accused of degrading.

Grazing Facts

The true nature of the problem, and its MIG solution, was worked out in Africa by Allan Savory, now head of the Holistic Management International Institute in Albuquerque, New Mexico. As a game manager in his home country of what was then Rhodesia (now Zimbabwe), Savory's goal was to reverse the ongoing desertification in the country's grassland game reserves. The working theory then was that too many animals were on the land, and they needed to be removed for the land to return to its former lushness. But when the animals were removed, the condition of the land worsened instead. The breakthrough came when Savory recognized and applied knowledge of the natural behavior of herds and the biology of grasses to game-management practices.

When not managed by humans, grazing animals of all species, whether they're buffalo in North America or wildebeest in Africa, tend to graze a small area intensively for a short period of time, then move on to another area. The herd doesn't return to the first area for weeks or months, giving it time to regrow. While the herd is grazing, it's also breaking up the soil surface with hooves, and fertilizing with manure and urine. Buried seeds are exposed to sunlight and quickly germinate. The plants that have been grazed are now fertilized and sprout new stems. When human game managers began moving herds to mimic that natural behavior, the land returned to its original productive state. This may seem a little overdramatized; it's not. Seeing it with your own eyes and doing it on your own farm makes you a believer.

Though many farmers and herders around the world have long been aware of the value of rotating pastures, Savory's documented studies and research, as well as decades of practical experience working with more and more farmers and

Managed versus Unmanaged Grazing

managed grazing   **unmanaged grazing**

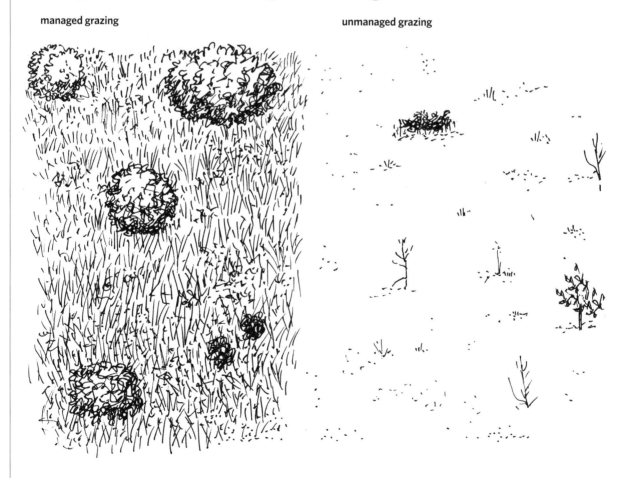

Side-by-side field trials have shown that management-intensive grazing produces a higher volume and a higher quality of forage for grazing livestock and favors legumes and grasses over unpalatable forbs and brush.

ranchers on increasing numbers of acres world-wide, have provided the foundation for a new appreciation of the real causes of desertification and land degradation. Savory's work and the work of many others is slowly revolutionizing grazing practices and reclaiming degraded land on every farmed continent. You may even be able to find MIG operations in your neighborhood by driving around at the end of the grazing season. Their pastures will still look green and lush, while most others will be grazed to dirt level, except for the tall thistles and other weeds the animals won't eat.

Though the Final Rule does not specifically require management-intensive grazing, understanding and using MIG is the most effective way of complying with the rule's requirements, and it makes sense even for livestock producers who aren't certified organic. For farms where livestock numbers are low and carrying capacity of the land high, rotational grazing is not so necessary. But in some areas, especially those with little rainfall, MIG is essential to being able to ecologically and economically sustain a small-scale operation. And for anyone interested in a dairy operation, using MIG considerably reduces the start-up and running costs of what is a very capital-intensive type of farm. When done well, MIG increases the amount of forage grown per acre, making it possible to sustain more livestock on less acreage.

Grass Biology

Have you ever wondered why you have to mow the lawn every week, but you can prune a shrub and it stays pruned for a year? Most plants regrow from the tips of branches and stems. If a deer bites off the tip of a low-hanging branch in the apple orchard, it's biting off the regrowing point, and the tree will have to establish a new one (through a hormonally regulated process) before the branch can grow again. In contrast, the growing points of grasses and many legumes are at the base of the stems. When a grazing animal bites off the stem, it exposes the growing point to a burst of sunlight. The plant quickly responds with new growth.

Now consider what happens if you don't mow the lawn. In an area with high rainfall, the grass gets tall and rank, and if it's left long enough it begins to thin out as the ground becomes more

Holistic Management

Allan Savory eventually broadened his thinking on management-intensive grazing into an entire system for decision making, which he named holistic management. HM, as it's called, is simple to understand, effective, applicable to any aspect of human endeavor, and completely congruent with organic thinking. So much so, in fact, that it's common for organic farmers of all kinds to use holistic management techniques to manage their operations for both profitability and sustainability. HM is a touchstone concept for many in alternative agriculture, and the basic principles are great tools for developing an organic farm plan. For more information, consult Savory's organization, Holistic Management International Institute (see resources).

shaded and the growing points are stifled by the tall foliage. Forbs that tolerate shady conditions for sprouting move in, and over the years the grass turns to brush. In dry regions, as soil dries it crusts, making it impossible for any but the toughest and most invasive plants to sprout. Eventually the grass disappears altogether and what was grassland becomes more and more desertlike, covered sparsely with tumbleweed and other invasive, inedible species.

What if you mowed the lawn every day? The grasses wouldn't be able to grow enough foliage to power their root systems, and they would dwindle away or hang on in a stunted state. That would leave plenty of room for weeds to move in.

Grazing animals are nature's lawn mowers. Constant grazing has the same effect as constant mowing. No grazing has the same effect as no mowing. But by using grazing animals to periodically trim the grass and then leaving it alone for a rest period, a farmer stimulates development of a thick, lush sod that holds the soil and develops an extensive root system that is constantly adding organic matter and structure to soil. It's just like mowing the lawn on a regular basis, except that livestock are even better than a lawn mower, since they also fertilize and break up any crust on the soil surface.

Guard Animals

Grazing sheep and goats suffer many casualties from predators in some areas, especially the western plains and mountains. Though bears, mountain lions, and wolves will all take lambs and goat kids, by far most losses are to feral dogs and coyotes. In many European countries donkeys and special breeds of guard dogs have been used to cut predator losses, and llamas have done the same in South America. In the past three decades the use of guard animals for sheep and goats has become increasingly popular in North America as well. Guard animals can dramatically reduce losses, sometimes preventing any losses at all.

Dogs

Guard dogs, bred to bond to the flock, may deter even the bigger predators such as bears and mountain lions. Research done by Colorado State University Extension wildlife specialist William Andelt found that the most effective guard dog breeds are the Akbash, Great Pyrenees, Komondor, and Anatolian Shepherd. The Maremma, Kuvasz, and extremely rare Šarplaninac all have their advocates as well. The smallest of these breeds matures at a whopping 75 pounds (34 kg); some get as big as 140 pounds (64 kg). All must be properly trained and handled in order to be effective as guard dogs; the first step is to

Properly trained guard animals can dramatically cut livestock losses from predation. Coyotes are a constant problem for sheep producers, and guard dogs an effective deterrent.

have the dogs live with the flock beginning at six to eight weeks of age. Once trained, the dogs don't need humans around to direct their guarding activities. Many will be unfriendly toward strange humans, as well, so guard dog owners should caution visitors. Guard dogs need to be fed dog food and should be vaccinated against rabies (this is legally required in most areas) and other common canine ailments.

Llamas

Llamas, unlike dogs, will eat what sheep eat and share the same vaccination schedule. You can shear them at the same time as the sheep, too, and trim their hooves. For these purposes, llamas should be trained to wear a halter, to be led and tied, and to lift their hooves for a human. Though they aren't usually fierce enough to confront bears or mountain lions, they detest canines of all sorts and will chase them with the intent to kill. A llama of any age has the potential to bond with a flock, unless the llama shows a preference for human society. If a male llama is used, it should be castrated.

Donkeys

Donkeys, like llamas, take great pleasure in chasing and if possible killing canines, though they can learn to recognize and get along with your farm dogs. Donkeys should be raised with the livestock they will guard from a young age so they bond with the herd; otherwise they may not be too diligent about protection. They also eat what sheep eat, but they need horse vaccinations instead of sheep vaccinations. They don't need shearing, but their hooves need to be trimmed every two months or so. Donkeys should be trained to wear halters and to be tied and led, and to have their feet picked up. Donkeys, unless they've been mistreated, generally like people very much and can be quite affectionate. They also have the great added advantage of loving thistle flowers and may greatly reduce a pasture's thistle problem.

The inevitable conclusion is that grazing animals are necessary for the health of grasslands. Remove one, and you lose the other. But also necessary is giving the grass an adequate rest period after a grazing. That, in a nutshell, is MIG.

Setting Up a Grazing System

The first step in beginning managed grazing is to split the pasture into smaller paddocks. This is easily done with cheap, portable electric fencing. Some producers who don't want to mess with maintaining an electric fence put in permanent fencing, which is plenty of labor and expense up front but saves time later. Once the pastures are divided into paddocks, the herd is moved as necessary until it has rotated through the entire set of paddocks and is back to the first one, which by

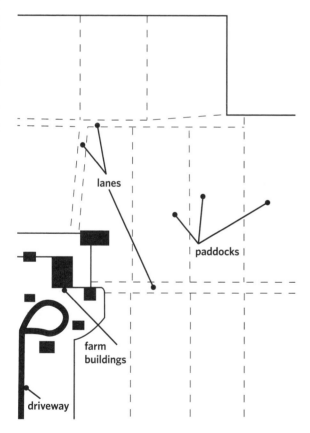

In an ideal world, paddock layout for a management-intensive grazing system has evenly sized paddocks, lanes on higher ground to minimize mud, and an efficient design that will minimize the distance animals must travel between paddocks and the number of lanes needed. In reality, you have to work around your farm's topography and field layout.

Stockpiling Paddocks

Stockpiled paddocks are essentially standing hay, or hay that is being stored as it stands in the field for later use. Paddocks are "stockpiled" by giving them time to regrow during the growing season, then saving them for grazing after the growing season ends. In some areas good stockpiling management can get you through an entire winter without having to feed any harvested hay.

now should have regrown to lush, palatable forage. The size of each paddock is geared to providing just enough grass for the period the animals will be in it, which can be anywhere from 12 hours to a week.

How quickly you move a herd through the paddocks changes with the class of livestock and how fast the grass is growing. In the Northeast and Midwest, where the pastures are dominated by grasses and legumes that grow fastest in cool weather, there is a tremendous burst of growth in the spring and again, though not quite as intense, in the fall. In the late summer, when it tends to be dry and hot, pasture growth often shuts down for a month or more as plants respond by going dormant. In these conditions a common strategy is to set aside half the pastures in the spring for a first cutting of hay, while the herd is rotated through the other half. As the summer progresses and growth slows, the rotation is lengthened by expanding it to include those paddocks that were cut for hay. In the fall the expanded rotation is kept, "stockpiling" paddocks to extend the grazing season well into late fall and even early winter. With variations for your climate and rainfall patterns, this system works anywhere, and it is especially effective in dry areas such as the high plains. Arid regions generally require much longer rest periods, while hot climates may require a combination of paddocks with grass species that grow best in warm weather and paddocks with grasses that grow best in cool weather.

Certified-organic operations may not use buffer zones for grazing, so be sure to plan your paddocks and fencing to exclude these areas.

Converting Cropland to Pasture

If you are in the position of needing to turn cropland into pasture for your grazing animals, you will in most cases have to seed down the area. When the fields are going to be used for hay as well, many growers plant pure alfalfa since it produces more tons per acre of high-quality feed than any other forage. On the other hand, pure alfalfa pastures can be dangerous for livestock, since grazing when the foliage is especially lush or just after a frost is notorious for causing bloat, which can kill animals within hours. (Bloat occurs when the fine foliage forms a mat at the top of the rumen and prevents the animal from belching. The gas accumulates, the rumen inflates so much it keeps the lungs from working properly, and the animal dies.)

A mix of alfalfa or some other legume and grass is a less risky proposition, and gives your animals a more balanced diet than pure alfalfa. If you're on heavier ground red clover might do better for the legume component than alfalfa, though it's slower to dry when you're making hay. Other clovers are available to suit many different climates and soils, along with an excellent selection of cool- and warm-weather grasses, ranging from timothy and orchard grass to the Bermuda grass popular in the far southern states. In fact, the options can be a little mind-boggling, and calling your extension agent and neighbors to chat about their experiences with various species in your area is immensely helpful in making seeding decisions.

Converting Brush to Pasture

If you're dealing with an abandoned area that is overgrown with brush and various forbs, you'll have to mow the field, using a brush mower if there's a lot of woody growth. Often just a few mowings will encourage an excellent growth of grass and get rid of the unwanted species. Any grass that livestock will eat — and that's most of them — is fine as forage, though the amount of forage produced per acre varies considerably among species. You may also want to drill in some legumes with a no-till or grassland drill to add additional protein to the pasture diet, or, if the ground freezes in winter in your area, try frost seeding. This involves broadcasting seed on to an area during that brief window of opportunity in the spring when the ground is bare of new growth, the days are above freezing, and the nights are below freezing. The freezing and thawing of the soil surface tends to pull the seed into good soil contact and so increase the "catch," or amount of seed that successfully germinates and grows.

If you have the time, equipment, and money at your disposal to plow down the field and replant it this can be done as well, though it's not necessary in many situations. And don't worry too much about weeds in a pasture — the livestock will eat a lot of them, adding more balance to their diet. You wouldn't want to eat the same thing at every meal, and neither do livestock. They like a few weeds. The ones they won't eat can be dealt with by clipping

With lightweight step-in posts and reels for the wire, moving portable electric fencing is quick and simple.

pastures after grazing, or running another type of livestock through that does like that kind of weed.

Natural Waterways in Pastures

If your livestock have to cross a stream or wet spot to get to pasture, set gravel or cement at the crossing point to prevent mud formation and soil erosion. If you also need to move equipment across the wet spot on occasion, make the lane wide enough to accommodate your tractor. For livestock a 10-foot-wide (3 m) lane is fine; for a medium-sized tractor pulling an implement the crossing should be a minimum of 16 feet (4.9 m) wide.

You can use a pond or stream to water livestock, but to preserve water quality and soil you must be be quite careful in how you do it. Hooves constantly traveling along streambanks and shorelines break down pasture sod, leaving mud and loose dirt that washes into the water, eroding the banks and degrading the water quality. As a general rule, put a fence between livestock and water. To allow livestock to drink, there are two basic methods: First, you can set up a watering tank on the livestock side of the fence and run a pipe from the water to the tank. Second, you can build a cement ramp into the water, like a boat landing, so

that livestock can walk up to the water and get a drink without stepping on the sod. Put a fence around three sides of the ramp, leaving it open at the pasture end, so that livestock can access the water from the pasture only via the ramp, and they can't go beyond it. Make the ramp wide enough that they can turn around and walk back to the pasture.

If you are using management-intensive grazing, you may not need to fence off the water. Livestock allowed onto a streambank or shore for a very short period of time will have a beneficial effect in many situations, reinvigorating the sod just as they do in the rest of the paddock. But you do

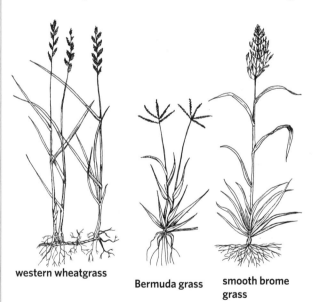

western wheatgrass

Bermuda grass

smooth brome grass

Forage species should be suited to the region, just as crops and animals are. Western wheatgrass thrives in dry areas, while Bermuda grass can tolerate the high heat and humidity of the Deep South, and smooth brome grass prefers cooler temperatures and good rainfall.

Noxious Plants

North America is home to several dozen plants that can cause illness and death of livestock, but fortunately most of them taste bad and have to be consumed in fairly large quantities to cause problems. Most are not usually a problem unless the animals are very hungry and can't find anything else to eat. All the same, it's good to be aware of what plants might cause a problem in your pasture, and pull or mow them as you have time. Some of the more familiar troublemakers are jimsonweed, buttercup, black nightshade, black locust, curly dock, locoweed, lupine, pigweed, and white snakeroot.

Other plants can be invasive, displacing palatable pasture with their inedible selves. Purple-spotted knapweed is a particularly troublesome species that continues to spread east, while thistle, which comes in several species, is probably the most universally despised of all pasture weeds. In an organic system, chemical weed killer is not an option. Mowing or hand-chopping (quite feasible on a smaller farm, less so on a big one) are effective ways of preventing these plants from going to seed and spreading. If it's a weed that spreads through the roots, maintaining a good sod cover with rotational grazing can be effective, though in severe cases it may be necessary to plow down the pasture and reseed.

The Final Rule on:
Pastures

Obviously, paddocks and pastures must be managed according to certified organic guidelines for animals that graze them to be certified organic. The same holds true for any field that produces hay for those animals.

In some cases producers may need to rent additional pasture, such as when they are having a parasite problem in their own pastures. Under the Final Rule, if you can document that the rental land has had no prohibited substances applied for the past three years, you can include it under your farm's certification. This means the landowner does not have to go through the certification process in order for you to rent the land for organic use, greatly simplifying the arrangement.

need to watch them carefully, so that you can move them out of there before they have grazed the bank short enough for their hooves to begin doing damage. And during wet periods or in spring when the ground is very soft you should keep livestock off banks and shores entirely.

Fencing Pastures and Pens

Fences prevent livestock from wandering off and getting lost, stolen, or hit by a car. Fences should be built so that the fenced animals won't jump over, crawl under, squeeze through, or push them

Prevent livestock from damaging shorelines by fencing off any body of water. Provide drinking access by constructing a blind lane that allows limited access or piping water to a water tank nearby.

over. Though a fence is a significant investment in labor and materials, it's worth every penny if you never have to get up in the middle of the night to answer a sheriff's call to come get your cows off the highway.

Nonelectric Fences

A wooden fence is aesthetically pleasing and effective, but too expensive for most operations. For cattle the old standby is barbed wire, often called barbwire. This is strung on metal or wood posts generally spaced 8 to 10 feet (2.5–3 m) apart. A fence 4 feet (1.2 m) high with three strands of barbed wire will hold cattle; four strands will hold cattle and calves.

Sheep don't need as high a fence as cattle, but they need more barbwire strands to prevent them from wiggling through the fence. Woven wire, which has vertical as well as horizontal strands, is considerably more expensive and harder to keep tight than barbwire, but it's safer and more effective for sheep.

Chickens can be fenced with chicken wire, similar to woven wire except the openings are much smaller and the wire itself is much lighter. Laying hens don't really need to be fenced to keep them from wandering off, since they will return to their roost on their own each night. But meat chickens aren't as good about coming home, and all chickens are easy targets for predators, so containing them in permanent pens, portable pens, or with portable electric chicken wire may be necessary.

Goats will climb or crawl under barbwire or woven wire if they are hungry or just bored, and pigs will certainly root under either type of fence and probably not even notice that there was a fence in their way. For them the best option is electrified fencing.

Electric Fences

Electric fencing will hold horses, cattle, sheep, goats, and pigs. You can even find portable electric poultry netting for chickens. An electric fence works by shocking an animal that touches the fence. The live wires on the fence hold the electrical charge, and when an animal touches the wire the charge passes through its body to the ground to complete the circuit. If the ground is very dry or snow-covered it will not function as a ground, and

Keeping Fence Lines Cleared

Whatever type of permanent fence you choose to build, site it so that you can reach it with a mower or trimmer, preferably from both sides. The common practice of herbiciding fence lines to keep weeds down is prohibited under the Final Rule, and mowing is your best option for keeping fence lines clear.

so a ground wire (a nonelectrified wire that, like the ground, completes the electrical circuit when an animal touches the fence) is put on permanent fences as insurance for all kinds of conditions.

Permanent electric fence is mounted on pounded posts and, for maximum effectiveness, includes a ground wire in addition to the electrified strands. A permanent fence can be strung at a relatively low tension, as with barbed wire, or, with the use of special high-tensile wire, can be strung very tight. A high-tensile fence requires fewer posts but stronger corners. Electrified or not electrified, high-tensile wire will bounce back instead of stretching and breaking after being hit by a running animal. The running-foot cost of high-tensile fencing is comparable or a little more expensive than that of barbed wire, but the fence lasts longer and is almost a necessity for some livestock such as bison and elk.

Portable electric fencing is what makes rotational grazing so simple. "Step-in" fiberglass or plastic posts (so called because you can push them into the ground with your foot) and a reel of lightweight plastic wire allow a farmer to string a single wire just a couple feet (about 60 cm) off the ground to fence off a paddock in 15 minutes or a half hour. Paddock sizes are easily adjusted, and the fence can be quickly removed for haying or clipping. Portable electric fence is a psychological rather than a physical barrier and is not reliable enough for a perimeter fence, which should be of permanent, multistrand fencing materials.

Any electric fence must be monitored regularly to find shorts and breaks in the wire. Animals learn to sense when a fence is shorted out or turned off and will walk through the wire.

Fencing Options

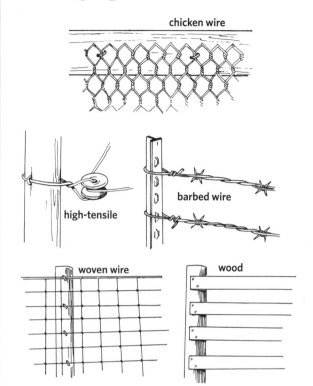

Permanent fencing appropriate to the livestock species should be used on the outside boundaries of your pastures. Wood is traditional but expensive; woven wire is pricey and difficult to install but will hold small livestock; barbed wire is inexpensive and will hold larger livestock but can injure animals; high-tensile wire is reasonably priced and effective and doesn't cause the injuries that barbed wire does; and chicken wire is appropriate for small enclosures and small livestock.

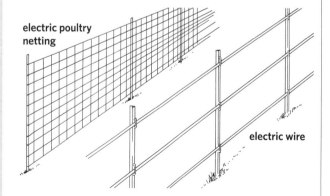

Portable fencing simplifies the management of grazing livestock. A single strand of electric wire will hold livestock that are trained to it, while multiple strands may be necessary for young stock. Electric poultry netting is a little more cumbersome, but perfect for keeping poultry in place.

The Rest of the Organic Livestock Diet

When the grass isn't growing, hay or silage must be bought or made. The rest of your animals' diet — the concentrates — must also be either grown or purchased. Concentrates are higher in fats and amino acids (the precursors of protein) than forages. On an organic farm, grains are the most commonly used class of concentrates.

Poultry and pigs require a majority of concentrates in their diets, while concentrates are not a natural part of the ruminant diet. But you'd never know it from the animals' behavior. Once they've tasted grain, most of them would trot over hot coals to get some more. And there is some benefit in giving it to them: though pasture-based organic farms emphasize forages over grain in the ruminant ration, feeding some grain to ruminants helps them attain a more consistent rate of growth and more consistently tender meat.

Prohibited Feed Additives

Sad to say, but conventional livestock operations routinely add antibiotics, growth hormones, urea, poultry litter, animal slaughter by-products, and plastic pellets to animal feeds. The Final Rule specifically prohibits all these things.

Whether it is homegrown or purchased, certified-organic feed can contain no synthetic ingredients except those specifically allowed in the Final Rule, which are limited to FDA-approved trace minerals and vitamins and inert ingredients allowed by the Environmental Protection Agency. Rather than try to track down yourself what's allowed or not allowed, check with your certifier or check the OMRI brand-name list when buying commercial mineral mixes or vitamin supplements. Complete certified-organic feed mixes are available, and though they're more expensive, they're also very convenient.

Feed additives that aren't prohibited may be used only if they are certified organic. Dried molasses, for example, is commonly added to feeds to make them more palatable, and it must be certified organic to be used on a certified farm.

Grain that has been contaminated by GMO pollen from nearby crops but is otherwise certified organic is sometimes sold for animal feed. This is allowed under the Final Rule.

Obtaining Organic Feed

To be certified organic, animals must have certified organic feed, and this can be hard to obtain if you aren't in a livestock- or grain-producing region where there are other organic farms. The problems with organic feed for most producers are first, finding a source, second, figuring out how to pay for it, and last, if you can't find any, how to grow or mix your own.

Mixing a Feed Ration

Mixing a feed ration yourself is simple for ruminants and more complicated for nonruminants (pigs and poultry). Cattle, sheep, and goats eating a diet of mostly forages harvested from healthy soils are getting all the balanced nutrition they need. Any grain in their ration is more or less just a dessert aimed at making them fatten faster or (if they are dairy animals) produce more milk. You can feed them straight grain or a mix of grains, with perhaps a little molasses mixed in for palatability. Corn is the most common grain fed to ruminants, but you can feed other grain just as easily if it is locally cheaper. Many cattle farmers like to feed their animals a mix of two parts corn to one part oats, on the theory that it makes their coats shinier.

Nonruminants like pigs and poultry, on the other hand, receive most of their nutrition from concentrates and need to have a carefully balanced ration. You can either buy a premixed complete ration or mix your own according to a recommended recipe. (For more information, see chapter 10.)

If you plan to be certified organic but not grow your own grain, find a feed supplier before you get any animals. If the closest organic feed supplier is still some distance away you may be able to split trucking costs with other organic farmers in your area (another reason to participate in farmer networks).

The cost of certified organic feed is usually higher, and often much higher, than that of nonorganic feed. And if it's sourced from some distance away, trucking costs increase, not to mention the general inconvenience. In fact, the price of organic grain has been a significant disincentive for some producers to become totally organic.

Growing your own feed requires owning a bigger land base and having the time and equipment to do it. Some noncertified producers follow all the organic rules on their own land but end up buying conventional feed; many customers are okay with that, even though it means you can't label your product organic. Others have the land, the equipment, and the expertise to produce their own grain and hay or silage.

Whatever you feed for a concentrate, be sure that your herd or flock always has salt and a mineral ration available.

Choosing a Livestock Enterprise

The ease of producing or obtaining certified organic feed is an important factor in deciding what sort of livestock and how many of them you can support on your farm. Also important are the amount of land you have, the amount of time you have to care for the livestock, and your facilities for handling the livestock when they need to be sorted, vaccinated, or loaded for shipping.

Space

The types and numbers of livestock you choose to own must be suited to your land base, which determines how much feed you can grow and pasture you can provide. The more acreage you have, the more options you have for livestock, and the greater the number of animals you can comfortably and cleanly care for.

At the small end, a few laying hens can be kept just about anywhere, including in a small backyard. Meat chickens for market require a little more space; if you're going to put them on pasture in portable pens you'll need perhaps ¼ acre (0.1 ha) of level, grassy ground per pen of 50 to 75 chickens.

In many European countries a pig or two were traditionally kept in a shed behind the garden, as a handy disposal for spoiled and overripe vegetables and garden waste. But the Final Rule requirement that animals have room to exercise and express their natural behaviors eliminates the tiny-shed-in-the-yard option. For a pair of pigs to be bought in spring and slaughtered in fall, ¼ acre (0.1 ha) is probably plenty for them to have room to root and wallow. More pigs need more space. You also want them far enough from the house that the particularly pungent smell of their manure isn't in evidence when your city relatives are visiting.

The amount of pasture and hayfield needed to feed ruminants varies considerably around the country. In the Northeast and Midwest, for example, on average raising a cow and a calf for one year requires 2 to 2½ acres (0.8–1 ha). In dry parts of the West, it can be 20 to 40 acres (8.1–16.2 ha). In the South, where the grazing season is much longer, you can get by with more pasture and much less hayfield. If your pastures and hayfields are on poor soil, are very wet or dry, or have other disadvantages, you'll need more land. If you're interested in

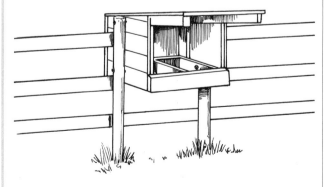

A salt and mineral feeder built into a fence line is out of the way and handy to keep filled.

Let the Processing Plant Come to You

In some areas innovative poultry producers have put together "mobile processing units," using a flatbed trailer or even an old school bus as a portable chicken plant complete with plucker, gutting table, and water connections for cleaning the carcasses. The unit is then shared among a group of processors, traveling from farm to farm as needed.

raising just a few meat animals for yourself and a couple friends or neighbors, you need less land. If you're planning a bigger enterprise, you need more land and better equipment.

How Many

With any livestock venture, long-term success is more likely if you start with just a few animals. The inevitable mistakes will be caught sooner and cost less, and the close observation and contact that is possible with a small group of animals provides invaluable knowledge and experience. Once you have the hang of things, taking care of a large flock or entire herd doesn't take too much more labor than a couple hens or a pair of calves.

But, as a rule of thumb, don't start with a single animal. All farm animals are by nature social, and having two or more satisfies their natural instinct to live in a flock or herd, which in turn satisfies in part the Final Rule's stipulation that livestock living conditions accommodate their natural behaviors. A single animal is a lonely animal.

Having more than one species of livestock, like chickens and sheep, or pigs and cattle, is a fine idea for a diversified farm. It may take a little more chore time, but rotating different species through pastures is a useful means of breaking up parasite and disease cycles. And of course it greatly expands the farm's product line.

Time

Next consider whether you're interested in a year-round enterprise or in just having animals around part-time. Meat chickens, yearling beef calves, and

pigs can be bought in the spring and butchered in the fall, so that there's no need for winter shelter or feed. Just-weaned four- to six-month-old beef calves are even cheaper and easier to find, but they will take another year or more to reach slaughter weight, depending on the breed and your feeding program.

Dairy operations and breeding animals of any type are a year-round proposition, requiring good winter quarters, feed storage, and a water supply that won't freeze. Hay for the winter can be bought, but that rapidly eats up a profit margin. Owning your own hayfield is a real advantage for a year-round operation.

Regional Infrastructure

A final consideration in choosing a livestock enterprise is having good access to animal breeders, processing facilities, and other services such as veterinarians and feed dealers. The most common problems are with pigs and sheep, which aren't so common anymore that it's easy to find other producers in the area to sell you animals. Most breeds of sheep also need to be shorn at least once a year, and sheep shearers are a rare breed in some neighborhoods. On the other hand, sheep shearing courses are occasionally offered through sheep producer associations and agricultural extensions, so you could learn to do this yourself — if you have a strong back.

The bottleneck in organic poultry is processing. If you are fortunate enough to have an organic poultry processor within a half-day's drive, or any poultry processor that will deal with small-scale producers, thank your lucky stars and be sure to send them a Christmas present each year to encourage them to stay in business.

On-farm butchering of poultry is feasible and fairly common, but it's a big job, and one that many people don't want to tackle on a market scale. It's also a job that, unlike, say, picking apples, few friends are going to volunteer to help with. Pigs, sheep, and goats can also be butchered on the farm, but it's an even bigger job. Finding a facility that will do it for you can be difficult in many areas, especially for sheep. Butchering cattle generally requires more work and equipment than a small to midsize farm would want to take on (though of course somebody somewhere is always making a go of it).

Before making any plans for on-farm butchering, confirm your state's processing regulations, because many states don't allow meat processed on-farm to be sold. Poultry is the exception, but usually states set a limit on the number of poultry you can butcher and sell on the farm in a year.

The Right Breed for the Right Place

Once you've decided which species of livestock you want to begin with, the next task is quite interesting: finding the right breed for your climate, terrain, and personal inclinations. The Final Rule specifies that species and types of livestock be selected "with regard to suitability for site-specific conditions and resistance to prevalent diseases and parasites." In practical terms, this means you should select a breed suited to your region.

The Origin of Breeds

Different breeds of livestock arose naturally through human history as various populations of farmers moved to new locations, taking their livestock with them. Those individual animals best adapted to the new region's combination of available feed, climate conditions, and disease and insect patterns survived and reproduced. Their descendants increasingly showed characteristics that made them recognizably different from their

Spanish goat

Cotswold sheep

Sussex chicken

Pineywoods cattle

If you're interested in rare breeds, do your research first to find the one best adapted to your climate and farming system. For example, Pineywoods cattle are adapted to hot, humid weather and coarse feed, while Spanish goats excel in hot, dry climates. Many rare breeds are smaller and better adapted to foraging than the common commercial breeds of livestock.

ancestors, so that each little group of farmers, isolated in different regions, wound up with distinctive types of poultry, sheep, goats, pigs, and cattle.

By the eighteenth century some progressive farmers, especially in England, realized that they could take an active part in breed development by establishing which characteristics were most desirable and selecting for breeding those individual animals that best exemplified the desired traits. Often these traits centered on color, coat or feather type, height, and other visible traits, but qualities related to production were favored as well, such as ability to grow quickly, tolerate the local parasites, put up with extreme heat or cold, or have a skeletal structure that held a lot of good cuts of meat. By the early twentieth century, researchers had identified thousands of distinctive breeds of cattle, poultry, goats, sheep, and pigs around the world, each particularly well suited to the purpose of the animal (meat, milk, eggs, wool, labor), the climate, and the farm practices of its place of origin.

Rare Breeds and Organic Farming

As mechanized, chemically dependent agriculture gained traction through the last century, many breeds began to disappear in favor of the fastest-growing, highest-producing breeds that were most amenable to confinement and unnatural diets (that is, with minimal forages). The most visible example of this trend is probably with dairy cattle, for which the black and white Holstein seems to have taken over the country. It's uncommon now to see any of what used to be the "big six" milking breeds: Jersey, Guernsey, Milking Shorthorn, Ayrshire, and Brown Swiss. And it's almost unheard of to see any other type of milking cattle at all.

But organic farming is emphatically not about "one size fits all" production systems that rely on controlled environments, unnatural feeds, chemical feed additives, and a full arsenal of modern veterinary medicines. Organic farming *is* about fitting the breed to the region and the system, and the farm system to the soil, the terrain, the climate, and the farmer. It's also about preserving the genetic diversity and biodiversity that are so essential to the long-term health and survival of all species and ecosystems. If you can find them and afford them, using one or more of the now rare traditional breeds of livestock can make terrific

good sense in an organic system. It's an excellent first step toward meeting the organic requirement of building a holistic system that fits the natural environment.

But preserving rare livestock breeds isn't just about economics and following rules. (In fact, if all you're worried about is following rules, you're not getting this at all.) It's also very much about hanging on to the joy of farming.

On the downside, rare breeds can be a tough sell in the marketplace. Their cuts of meat may not look the same and often aren't the "right" size. The solution lies in educating your buyers. That, too, can be part of the joy of farming.

Whether you buy a rare breed or a common one, you're usually better off buying them in your own region, so that you get animals that are acclimated to the weather patterns and prevalent parasites. Animals from other regions may take a year or more to become physiologically accustomed to conditions in your region, which can result in somewhat lowered production and more health problems.

Animal Identification

Once you have your animals, you'll need to keep track of who is who. Because organic certification relies on being able to track a farm product from beginning to end, all crops must be documented and all animals be identifiable. The Final Rule states that a farmer "must maintain records sufficient to preserve the identity of all organically managed animals."

Numbered ear-tags are the most common method for individually identifying cattle, sheep, and goats. Pigs can be ear-tagged too, or you can go the traditional route and notch their ears according to a code that identifies their mother and year of birth. Neck chains may also be used, or even photos or drawings of animals that have clearly unique markings, such as Holstein cows. Ear-tags and the tool for putting them on can be purchased from any farm store or catalog. The tags come in a variety of colors and are either prenumbered or come with a special pen for writing your own numbers. Read the label before buying ear-tags, since some are impregnated with insecticides and so are prohibited for organic production.

For each tagged animal or ear-notched pig you'll need a corresponding paper record that tracks parents, birth date, vaccinations and other procedures, any health problems, and production records. These should be kept with the rest of your farm records, and available for the organic inspector.

Attaching an ear-tag is simple; the trick is accurate placement and speed. The animal has to be either restrained by a headgate or pinned by a helper or barrier. Grab the ear smoothly and firmly, without any tugging, twisting, or squeezing. Slip the ear into the tool, center the tool, and give it a quick, firm squeeze to attach the tag. There will be a bellow or squeal, but fortunately the whole incident will be quickly forgotten — at least until the tag is lost and you have to put on a new one.

Poultry of course have no visible ears, so it's tough to put on a tag. Fortunately it's allowable to maintain poultry records for entire flocks rather than for individual birds. Any group of poultry that you purchased or hatched as a group can be documented as one entity. Be sure to record how and when the group was acquired, along with the details of how they were raised, fed, and processed.

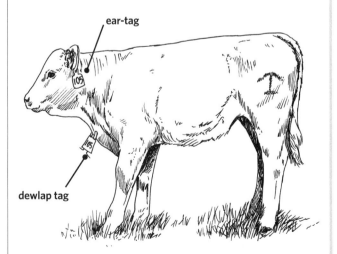

Except for poultry, animals must be individually identifiable to meet Final Rule requirements. Ear-tags are most commonly used for identification, though some dairy cattle owners use neck chains or dewlap tags.

Shelter

The Final Rule (and, again, common sense) requires that animal shelters be safe for their inhabitants and designed to provide temperature control and good ventilation. Shelter for most classes of livestock can be quite simple, so long as it provides good ventilation without cold drafts. As far as safety is concerned, the shelter should keep out predators and be free of protruding nails, broken boards, and junk. For practical reasons, the shelter should be easy to clean and bed, since these are constant chores. For organic purposes, make sure that no treated lumber (prohibited by the Final Rule) is anywhere that animals might come into contact with it or be able to eat plants that grow around it.

Many beef cattle and sheep never see the inside of a building, but poultry, pigs, and goats need a place to get out of cold, wind, rain, and hot sun. And in areas that see a lot of cold rain or bitter winds, beef cattle and sheep will do much better if they can get under a roof or behind a windbreak. That said, livestock of any type must have access to the outdoors; they can't be kept in total confinement.

Poultry are normally shut indoors at night to minimize predation, but all other animals should be able to go outdoors at will unless there is heavy predator pressure. The Final Rule allows keeping animals completely confined for a brief period only in bad weather, when birthing or ill, for their own safety, or if there is a risk to soil or water quality. For example, if your flock is lambing and there are a lot of coyotes in the neighborhood, it might be a good year to lamb indoors. Or if you're experiencing an unusually wet spring and the cattle hooves are badly cutting up the pasture sod and destroying soil structure, maybe the herd should be fed in the barn or barnyard for a week or two until everything dries up.

Pasture Shelter

For grazing livestock, a three-sided shed, with the fourth side open to the south or southeast, is economical to build and easy to clean. Putting the

Finding Plans for Livestock Facilities

Feeders, sheds, chutes, and other essential live-stock equipment can be purchased from numerous dealers, bought used, or in many cases built cheaply by the livestock owner, often from scrap wood. In general the less the animal weighs, the easier it is to construct adequate equipment. Chicken equipment is simple to build; cattle need to have facilities that won't be pushed over by a three-quarter-ton (0.7 t) cow.

Having the right size and spacing is also important. Animals of different species and ages require different amounts of floor space. And a chute doesn't work so well if it's a few inches too wide and the animals can turn around inside and head out the wrong end.

Years ago the Midwest Plan Service of Iowa State University in Ames developed a plethora of well-designed, easy-to-build plans for small and midsized operations covering everything from grain and hay feeders to sheds and chutes. These have been superseded by more modern designs aimed at bigger operations, but fortunately the old plans are still available (see the resources). These plans will save you a lot of guesswork and trial-and-error rebuilding.

The gold standard for livestock handling facility design has been set by Temple Grandin, a professor of animal science at Colorado State University who is revolutionizing livestock handling in the United States. She designs systems attuned to animals' natural behaviors and instincts, so that when it's time to handle livestock the animals go smoothly and willingly. Grandin's book, *Humane Livestock Handling*, is an excellent resource, as is her Web site (see the resources).

shed on skids allows a producer to move it with the animals through a pasture rotation.

Enclosed Shelter

Ease of cleaning is a primary factor in designing enclosed shelters such as coops, pens, and barns for livestock. The type of livestock will determine the type and size of shelter necessary; the following chapters provide specifics.

If you own an old dairy barn, the old barn cleaner may still be there, or you can install one. This automated cleaner sweeps the gutter behind the stanchion stalls and dumps the manure into a waiting spreader. If you retrofit the barn with pens for sheep or pigs, consider keeping the cleaner to make manure moving quick and simple. Otherwise, make sure gates and doors are wide enough for a wheelbarrow or, in a larger operation, for a skid steer.

Retrofitting Shelter

Many old farms have an oversupply of barns and buildings. If any are in decent shape, an economical option is to convert them for livestock instead of building fresh. A shed can be easily converted into a chicken coop or a pigsty, for example, while a pen can be fitted into a corner of a larger building. Larger buildings can be used for larger livestock. Just be sure to place boards in front of any low windows to prevent breakage by curious noses.

Shelter to protect livestock from extreme temperatures, winds, and precipitation can be as simple as a three-sided "run-in" shed or as snug as a well-built barn.

Bedding

In cold or wet weather livestock shelters should be well bedded with straw, leaves, old hay, sawdust, or other organic materials to keep the animals warm and dry. Farrowing (birthing) sows should be provided with deep bedding several days before their due date to allow for their natural nesting behaviors, and any animal giving birth indoors rather than on pasture should be provided with deep, clean bedding. If the bedding is of a material that is typically consumed by the species, such as straw, cornstalks, or old hay, then it must be certified organic. It's okay to use newsprint as long as it doesn't include any of the shiny or colored sections, and sawdust is also allowed as long as it does not come from treated wood.

Pens and stalls should be cleaned regularly; to conserve bedding just pick out the dirty parts. If you're cleaning the pen of a cow and new calf, keep an eye on the mother. It's rare, but every so often one will decide you're a threat and come after you. Be sure you can leave in a hurry if this should happen.

Many producers use a deep bedding system, where clean bedding is piled on top of soiled bedding through the winter, with a total cleanout in the spring. The heat given off by the underlying manure and bedding as it begins to compost helps keep livestock warm, but the spring cleanout is a huge job. Consider renting a skid steer if you don't own one and the shed is big. Another idea, pioneered by Joel Salatin of chicken tractor fame (see

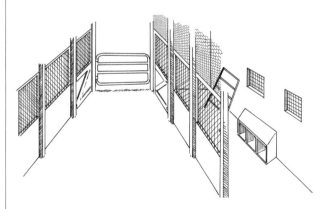

A chicken coop can be retrofitted into nearly any farm building. Here an old stall has been turned into a coop by running chicken wire up to the ceiling and filling any holes or cracks big enough to admit rats or weasels.

chapter 10), is to sprinkle corn over the soiled bedding before putting down fresh material. In the spring when the cattle move out to pasture, put some pigs in the pens. In their digging for the corn they will turn and compost the bedding pack, so that when they're through it's ready to spread on the fields.

Handling Manure

If you have livestock, you're going to have a lot of manure. If the animals are on pasture, they will spread it themselves and you don't have to worry about it. When they're in pens or sheds, the manure must be removed, whether every day or, as in the deep bedding method, once a year. The collected manure can be spread (by hand or mechanical spreader) on garden or fields, or it can be stored or composted for later use.

As detailed in chapter 3, the Final Rule has strict guidelines about how and when raw manure can be spread on fields (see page 79). If you are going to store manure, do so with the aim of conserving nutrients. Ideally this would be under a roof and on a cement slab, but that isn't always affordable or practical. If you have to store it, at least mix it with enough bedding to start the composting process, and pile it high and tight to preserve as many nutrients as possible.

The Final Rule prohibits spreading manure in any way that might pollute water. If you spread manure on frozen ground where it can possibly run off in the spring with the snowmelt into a nearby stream, swamp, lake, or pond, not only will you lose your certification, you'll also probably get a hefty fine from the local authorities. In many areas adjacent to wetlands and waterways it's now illegal to spread any manure from off-farm sources, and on-farm manure is often closely regulated as well. For some producers this has eliminated the use of poultry litter from large poultry production operations, which is a good source of organic fertilizer.

If you are able to spread poultry litter, you must first document that it is free of arsenic, which is becoming a common ingredient in chicken feed for large poultry producers. Spreading poultry litter that contains arsenic residues will cause you to lose organic certification, plus it's a highly toxic chemical that no one wants in a farm's water or soil.

Proactive Health Care for Organic Livestock

What saves more human lives: modern medical procedures, drugs, and technology, or vaccinations, proper diet, cleanliness, and exercise? Statistically it's the latter — the proactive approach pays off. In the past century or two vaccinations, hand-washing, cleaner water supplies, and a better understanding of and access to good nutrition have allowed the majority of humans in the world to survive past the age of five.

It's the same with animals: the most effective way to have healthier livestock is by being proactive about nutrition, clean water, clean buildings and pastures, pest and parasite control, and vaccinations. These measures will save you far more in vet bills, animal stress, and lost production than the best that reactive medicine can provide. Though there will always be some level of illness and injury among your livestock, producers who follow these guidelines as a general rule have far fewer vet bills than conventional operations.

The Final Rule quite properly makes proactive health care a requirement of organic production, stating, "The producer must establish and maintain preventative livestock health care practices." Specifically, the rule requires that:

- Species and types of livestock be selected for site-specific conditions and resistance to prevalent diseases and parasites
- The feed ration meet nutritional requirements
- Housing, pastures, and sanitation practices be designed and managed to minimize disease and parasites

- Animals be kept in conditions that allow them exercise, freedom of movement, and reduction of stress appropriate to the species
- Any physical alterations (castration, dehorning, tail docking) be done only if they promote the animal's welfare and in a manner that minimizes pain and stress
- Vaccines and veterinary biologics be administered as appropriate

Practicing Good Sanitation

Internal parasites and external parasites can be kept to a minimum by a few common-sense sanitation practices. Each pest species has its preferred niche, that combination of food, housing, and safety from predators that allows it to survive, multiply, and harass your livestock. But most of them agree that their eggs and larvae like to be moist, warm, and sheltered. These conditions can be found inside continually wet dirt, fresh manure patties, tall vegetation, or piles of dirty, wet hay or bedding.

Keeping manure-based compost piles hot will kill most parasite eggs, and keeping weeds trimmed along fence lines and around buildings deprives many adults and some larvae of favored sheltering areas. Clean, dry stalls and pens minimize parasite reproduction and survival. When livestock have been wintered outside, it's important to get the clumps of dirty waste hay from the feeding areas scattered or cleaned up early in spring, before fly breeding cycles begin with the first warm weather.

Liver flukes, microscopic worms that feed on blood, can be a problem anywhere there are snails, which are a necessary intermediate host in the life cycle of these internal parasites. Snails are found in water, so keep livestock away from low, wet areas, especially swamps. Hard as it might be to give up that pasture ground, if you have liver flukes (which your processor will be able to identify in the liver of a butchered, infected animal), then fence off the wet areas. Check your watering tanks periodically for snails, and clean them out if you find any. Cleaning out tanks regularly is a good idea anyway, since animals will drink more water if it's clean, and staying well hydrated is important in staying healthy.

Water tanks placed on dirt provide a constant supply of fresh mud, which encourages flies and other pests. The mud will also keep your animals dirty. Place tanks on cement pads or use portable tanks and waterline to move them frequently as animals rotate through paddocks.

Animals kept in clean, dry conditions will not only have fewer fly and parasite problems, they'll be more resistant to disease and less likely to pass it around to each other. Cuts and scratches will be less likely to get infected as well.

Cleaning Up Junk

The old tires, spools of wire, decrepit machinery, and piles of junk that are too often found on old farms are great spots for pools of stagnant water, noxious weeds and overgrown grasses, and sharp edges. Cleaning up these problems, or at least keeping livestock away, eliminates more pest and parasite breeding and nesting areas, as well as a significant source of cuts and scratches.

Getting rid of junk also eliminates a significant source of "hardware disease." This problem happens when a grazing animal accidentally (or on purpose — you never know with cows) swallows a nail, bit of wire, or other small piece of metal. Often it just sits in the animal's stomach for the rest of its life, but if it puts a hole in the intestine, you'll have a dangerously ill animal on your hands and be calling the vet for surgery to remove the offending object. This is a common enough problem that most farm stores sell "cow magnets," which are fed to the afflicted animal in the hope that the magnet will collect the metal and keep it from causing any harm.

Livestock can also easily cut themselves on protruding nails and other sharp edges inside buildings. Clean these up, and always keep an eye peeled for shiny objects in pens, pastures, and barnyards. You never know what's going to fall off a piece of machinery or work its way loose from a board, so be alert and grab it before an animal punctures a hoof, scratches an eye, or eats it.

Preventing Falls

Like humans, livestock can slip and fall on icy or slippery surfaces. Cement itself can often be very

The Final Rule on:
Parasiticide Use

The one exception to the Final Rule prohibition against conventional wormer medication is when an approved preventive management plan for parasite control has not worked and a *dairy* animal has become seriously infested. The parasiticide ivermectin may then be used as an emergency treatment. The milk from the treated cow must be kept separate and not sold as organic for a period of 90 days, to allow the drug to fully clear from her system. After that her *milk* may be sold as organic, but her *meat* may never be sold as organic. If she is in the last trimester of her pregnancy or if her calf is nursing, her calf may not be sold as organic either.

slick; you can minimize this type of hazard from the outset by grooving any cement surface when it's being poured. If it's too late for that, you can keep a little straw, sawdust, or manure scattered on the surface when it's icy.

Controlling Parasites

Since the Final Rule prohibits the use of most synthetic parasiticides (sometimes called wormers), the organic producer has to use other methods, such as herbal remedies, to control these pests when good sanitation isn't enough. (There are some circumstances when wormers are permitted, but for all practical purposes the average organic producer is not going to use them.) And sanitation isn't enough. Parasites and pests are ubiquitous in most environments, and though sanitation will reduce their numbers in pens and around buildings and waterers, it doesn't have much effect on those that breed out in the pastures.

Parasiticides are chemically formulated to kill both internal parasites (called worms) and external parasites. They're typically given to livestock on a regular schedule, usually in the spring before they're put on pasture and in the fall after the grazing season ends. There are good economic and health reasons for this: Internal parasites,

Proactive Health Care for Organic Livestock

Dragging Pastures to Kill Parasites

Some producers drag pastures to break up manure pats after they've been grazed and so kill the larvae and eggs of worms, flies, and other parasites by exposing them to hot sun. Others find that this practice also destroys many beneficial insects, such as dung beetles, that if left on their own would have just as much of an effect in reducing pest populations in manure.

especially roundworm (a nematode) and liver flukes (a trematode), can cause significant losses for livestock producers. Infested animals are thin, don't grow or fatten well, won't conceive as easily or produce as much milk, and are more susceptible to other health problems. A very heavy infestation of internal parasites can cause diarrhea, anemia, coughing, and, in some cases, death. External parasites such as cattle grubs, fly maggots, lice, mites, and ticks all cause discomfort, affect growth, and can transmit disease.

Controlling a parasite problem using organic methods instead of conventional parasiticides requires proactive pasture management and an understanding of parasite life cycles.

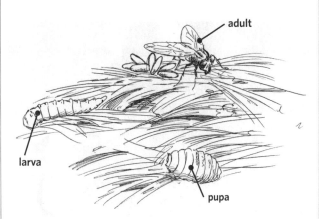

At each stage of its life cycle, a fly requires appropriate environmental conditions for best survival. Many species prefer moist, still areas such as a pile of wet, manured bedding, for laying eggs, since there the larvae will have plenty of food and be protected from cold and from drying out.

Managing around Worms

Internal livestock parasites spend only part of their lives on or in the animal; the rest of the time they're in the manure, dirt, or grass (or, in the case of liver flukes, they're in snails). In general, the adults living inside the animal lay eggs that are pooped out in the manure, where they hatch and turn into larvae. The larvae hang out on grasses and legumes until an animal eats them, and the cycle begins again.

The precise number of days from egg to larvae ready to be ingested depends on how warm and moist the environment is. Parasite numbers in the grass generally peak in the moist, cool spring months and bottom out in hot, dry weather, but larvae of many species can go dormant and survive up to a year in the grass.

The one problem with rotational grazing is that it can increase internal worms, since the livestock often arrive back in a paddock just as the worm larvae have moved up the grass stems and are waiting to be ingested. If your pastures are heavily infested, rotationally grazing them may be a bad idea until the worms can be controlled. The best way to do this is to keep animals off those pastures for at least a year, leaving them fallow or cutting them for hay. This means finding other pastures to rent or keeping the animals confined, something most easily done if you start in fall and let them linger in the winter feeding area until late spring, then start the grazing rotation on those paddocks that have been out of use the longest. For example, taking one or two hay cuttings from a pasture before using it for grazing late in the summer after the weather turns dry will considerably reduce the parasite load on that pasture.

A second good method is to have two types of livestock using the pasture, since most parasites have preferences about which animals they like to live in and will not survive if eaten by the wrong animal. Cattle and sheep can be alternated in a paddock, for example, but not sheep and goats, since they are closely enough related to be infested by the same parasites. Saving a clean pasture for calving, lambing, and kidding will minimize parasite infection when animals are at their most vulnerable.

If your young stock are failing to gain weight rapidly, or the adults look thin and their coats rough, it's time to test for worms. By looking at a manure sample under a microscope, your veterinarian can usually tell what type of worm you're dealing with, as well as how long it might survive in a pasture. Once you've identified the problem, finding an organic solution depends a lot on your individual circumstances.

Fortunately, healthy adult cattle with a balanced diet and a well-thought-out pasture rotation rarely have any serious problems with worms. But do keep an eye on the young stock — they are much less resistant, and that's where problems will show up first.

Managing around Flies

With external parasites like flies, it's helpful if you can identify the fly and look up its life cycle to find the weak spots. For example, a housefly egg is laid in manure or other moist, rotting matter and will hatch in 24 hours, take eight days to two weeks to reach adult stage, and then live for two to three and a half weeks. The adult can fly a quarter mile (0.4 km) or farther. With this knowledge in hand, it's clear that control is going to focus on getting manure and spilled feed cleaned up at least weekly, before eggs can become adults and start the cycle over again. Sticky traps and tapes are effective against adults, but you'll have to keep fresh ones up since more adults will fly in from other areas once you've cleared out the local residents.

Vaccinating

Vaccinating animals is probably the most effective and simple preventive for livestock disease that is available, and it is encouraged by the Final Rule. All vaccines are allowed under the Final Rule, except vaccines made with GMOs, which are now available but not clearly labeled, so ask your certifier about your vaccine brand. Not everyone agrees, though; some organic producers don't vaccinate, relying instead on good management, sanitation, and luck. At least ask your veterinarian which diseases are commonly vaccinated for in your area for your type of livestock, and when, and consider using them — this is cheap insurance. And note that some vaccinations may be required by state or federal regulations when you are selling or moving livestock in some areas.

Unfortunately, there are some highly contagious and devastating animal diseases that can't be vaccinated against, and if there has been an outbreak of one of these diseases in your region, you may be required to test animals in order to sell them, move them, show them, or prevent their destruction. In the summer of 2008, for example, 45 cattle owners in northwestern Minnesota signed buyout agreements with the state after an outbreak of bovine tuberculosis in the area. Their herds were destroyed in the hope that this would contain the disease.

Physical Alterations: Docking Tails, Castrating, and Dehorning

The Final Rule requires that physical alterations to animals be done only "as needed to promote the animal's welfare and in a manner that minimizes pain and stress." In addition, a producer must prove that any physical alterations were done in a way that caused the least possible pain and suffering for the animal.

This means that the fairly common conventional practice of docking cows' tails (cutting them short) is prohibited, since it is a physical alteration that is emphatically not needed to improve the animal's welfare. There is an old myth among some farmers that cutting off the tail prevents worms, but it's only a myth. The only thing docking does for sure is keep the farmer from being slapped in

Vaccinating your flock or herd is a cheap and effective way to prevent many common diseases.

Scrapie and Mad Cow Disease

Scrapie is a prion disease, the sheep equivalent of mad cow disease in cattle and Creutzfeldt-Jakob disease in humans. These diseases (from the family of transmissable spongiform encephalopathies) are all slow-moving brain afflictions that disable and eventually kill the afflicted animal or person. Scrapie was first recognized 250 years ago in Great Britain and was (accidentally) imported to the United States in 1947. Currently an estimated 0.2 percent of the U.S. flock is afflicted (one in 500 sheep), enough for many other countries to ban imports of U.S. sheep.

There is still some uncertainty about precisely how the disease is transmitted and about cross-contamination between species. Sheep in the United States are required to be identified individually in order to be sold or moved as part of the USDA's scrapie-eradication program. It's worthwhile to participate in the voluntary Scrapie Flock Certification Program, which gives sheep owners USDA certification that their flock is scrapie-free. For the certified organic producer, animal ID is required anyway, so you'll have the necessary records on hand to certify your flock. When buying animals, get them from a certified scrapie-free flock if at all possible.

In rare cases scrapie can afflict goats, but this happens usually when the goats have been exposed to infected sheep. If the goats you are considering for purchase have been kept with sheep, the sheep should be certified scrapie-free.

Unfortunately there are no inspection programs or herd certifications for mad cow disease in the United States. The good news is that it is not transmitted from cow to cow in a herd, except possibly from mother to offspring through the milk. Cows contract mad cow disease by being fed by-products of scrapie-infected sheep or other cows that had mad cow disease. Since the Final Rule absolutely prohibits the feeding of animal by-products to livestock, mad cow disease is not a concern on an organic farm. It's also been many years since the USDA allowed the feeding of by-products to any cattle in this country, which provides an additional layer of insurance.

the face by a mucky tail. Docking sheep's tails, on the other hand, which is routinely done shortly after birth, actually does promote their welfare by preventing feces and moisture accumulation near the anus, which make it a perfect breeding ground for maggots. "Fat-tailed" and hair sheep do not normally have their tails docked, nor do breeds that are naturally short-tailed. For all others, docking is done during the first week of life either by cutting or putting on a tight rubber band, and does not greatly inconvenience the lamb.

Cows are often dehorned, more for the benefit of their herd mates and the farmer than for their own health, but it's justifiable considering the intimidation and physical damage that can be done by a horned cow. Dehorning should be done within a couple months of birth, as soon as the horn buds show. There's no question that the procedure is painful for the calves: it involves applying a red-hot cauterizing iron to the horn bud to kill it. (The other common method, applying a chemical paste that kills the horn bud, is prohibited by the Final Rule.) The younger the animal, the less stressful the procedure for everyone involved, so get it done as soon as those horn buds show and consider having the veterinarian use a light anesthetic. Animals that are polled, or naturally hornless, are an excellent alternative. The Final Rule prohibits removing full-grown horns from adult animals, so if you have a mature animal with horns you have to live with it or sell it.

Likewise, castration of male animals not being kept for breeding is necessary for the safety and welfare of everyone involved, particularly with cattle. Knife castration is preferred by buyers over elastration (putting a tight rubber band around the testicles until they wither away and fall off), but if you're using the knife, using a little anesthetic

Natural Remedies for Livestock

Veterinarian Paul Dettloff has been treating livestock in southwestern Wisconsin for more than 40 years, and using approved organic remedies for his certified organic clients for more than 15 years. "All this information was known 70 to 100 years ago," he says, but it was mostly forgotten after World War II and the advent of modern drugs. "I am constantly looking at what else is out there that can be used in the organic world."

Dr. Dettloff says he uses nine basic tools to treat organic livestock:

- **Aloe.** Aloe vera, a succulent plant native to Africa, has many healing properties and is commonly used in natural medicines for humans as well.

- **Acupuncture.** "You can do anything with acupuncture — treatment, disease healing, and pain management," Dettloff says.

- **Antioxidants.** Rose hips and vitamins E and C speed healing and boost the immune system.

- **Botanicals.** Medicinal plants are used for several purposes depending on the plant. For example, St. John's wort relieves pain, and the roots of burdock, a detested weed on most farms, help cleanse the liver. Depending on the botanical and the affliction, the botanical might be fed to the animal or prepared as a tea, a poultice, or other application.

- **Homeopathy.** A school of alternative treatment that involves administering a highly diluted preparation that will cause symptoms similar to that of the illness, prompting the body's immune system to mobilize and hastening recovery.

- **Probiotics.** *Probiotics* is a catchall name for the many types of beneficial bacteria that live in animal guts and enable them to digest their food.

- **Tinctures.** Tinctures are made with alcohol, which draws the beneficial elements from a medicinal plant. A tincture of garlic, for example, is used against infection and to stimulate the immune system.

- **Micro- and macronutrients.** In other words, vitamins or nutrients needed by the body to correct vitamin deficiencies, which can contribute significantly to animal health problems, or to assist with recovery from illness by bolstering the immune system.

- **Whey.** Whey is a component of milk; whey made from colostrum can be used to help adult animals recover from mastitis and young stock from diarrhea (scours) and pneumonia.

"Conventional medicine consists of antibiotics that kill bugs, pain relievers, and hormones. Organic medicine tends to work in a more preventive mode," Dr. Dettloff says. "We work on nutrition, we work on the immune system. . . . If you have very healthy soil, then you have very healthy forage, and then you have a very healthy animal with a very healthy immune system. They just don't get very sick."

makes the procedure more pleasant for everyone involved.

Anyone attempting a physical alteration like docking, dehorning, or castration for the first time should have a good species-specific reference at hand (see the resources); even better would be to have a veterinarian demonstrate and assist. The procedures are simple, but it's important to do them correctly. With hands-on practice and advice from an experienced guide, these procedures can be done cleanly and with minimal fuss.

Deciding on Medical Treatment

Proactive management for animal health will prevent most, but certainly not all, problems. Before

Basic Birthing

Statistically a livestock producer will see the most medical emergencies when livestock are birthing. In a healthy organic herd or flock that gets plenty of outdoor exercise, birthing problems will be much less frequent than in a conventional operation, but they will still occur.

It's easy to cause more trouble than is necessary by becoming anxious and intervening too early when an animal is taking her time in labor. It's also easy to lose a baby and even a mother if things are not happening as they should and you don't act fairly quickly. It's this dilemma that makes birthing such a nervous time for livestock producers.

Timing

In a normal birth, the first thing to happen is the water breaking. The baby inside the mother is contained in a fluid-filled sack, and when this bursts the water comes pouring out (though sometimes it's more of a trickle) of the mother. Shortly afterward you should see two hooves, and as the birth progresses a nose appears between the hooves. Don't worry if the mother is still on her own hooves at this point; that's common. She will almost always lie down pretty quickly and finish the job.

A normal birth in any livestock species should take, on average, between an hour and an hour and a half from the time the mother's water breaks to the time the baby is struggling to its feet and trying to nurse. Animals giving birth for the first time may take slightly longer. Pigs, since they have so many babies at once, may take a little longer yet, but the piglets should arrive in a steady progression, followed at the end by the placenta. If there's a significant pause in the labor, it's time to check for problems.

If the baby or babies aren't on the ground within an hour or so after the water breaking, it's time to check for problems.

Positioning

When you check the mother, if you see two hooves and a nose showing but no progress is being made, most likely the baby is too big for the mother to push out on her own. If you can get a good handgrip or a soft rope (placed above the fetlock) on the hooves and pull gently when the mother is pushing, usually the baby can be eased out.

Sometimes you'll see two hooves showing but they belong to two different babies. This is most common in sheep. Fixing this traffic jam requires getting a hand inside the animal to sort things out.

If only one hoof is showing, the other leg is back. If you see two hooves but no nose, then either the baby is coming out backward or its head is bent back. If a leg or the head is back, you or the vet has to get an arm up inside the animal to push the baby back into the uterus, where you can manipulate it into the correct position.

These manipulations are not for the faint of heart, and often not for the amateur, especially if the head is back. Having a knowledgable

you have a sick or injured animal on your hands, you and your veterinarian should know what you can and can't do under the Final Rule and have some basic organic medical supplies on hand.

Many standard medications are not allowed under the Final Rule, and many others are allowed only under certain circumstances. The section of the rule governing specific medications is 205.603; it might be a good idea to give a copy of this section to your veterinarian, and also to post a copy in the barn.

Some things you can still use. Of the more common medications and remedies, aspirin, glucose, hydrogen peroxide, and iodine can be used

neighbor or a good vet on standby alert is your best bet if you're new to animal obstetrics.

Restraints

Most animals in intense labor aren't going to worry too much if you start messing around their back ends, since they're too preoccupied. For the occasional mother who needs help but objects to your presence, you'll need some way to restrain her.

A goat or sheep is fairly easy to restrain with the help of an assistant, but a cow will need to have her head caught in a stanchion or headgate if she is being difficult. (If a cow swings her head at you and makes contact, you'll wind up on the far side of the barnyard and probably have some broken ribs into the bargain, and your medical bills will probably be more than the cow is worth.)

Afterbirth

Last, the mother should pass the placenta, which looks more or less like a big raw liver, shortly after birth. Most animals will eat it, and that's natural and fine. If the placenta does not pass, consult a veterinarian. If the placenta does pass and is followed by the uterus, that's a medical emergency, and you should call the vet immediately.

These are just the most basic rules of thumb and commonest problems. The ways in which a birthing animal can get into trouble are extensive. Anything that is far out of the ordinary should prompt you to call your vet.

to treat sick animals with no restrictions. Oxytocin is allowed, but many buyers of organic milk don't allow it, since it is a hormone. Mineral oil can be used topically as a lubricant, but cannot be used internally or in feed. Others pharmaceuticals, such as atropine, tolazoline, lidocaine, and procaine, may be used only by order of a veterinarian, and

the animal must go through a "withdrawal" period afterwards, during which its meat or milk cannot be sold as organic.

The Final Rule's prohibition on antibiotics for medical treatment has been a tough rule for many producers to swallow. Antibiotics can be very effective in treating animal illnesses caused by bacteria, and it is really hard to see an animal suffer and not reach for the syringe. But the rule also states that if there is no alternative, antibiotics and other prohibited medicines *must* be used to save the animal. You are not allowed to withhold medical treatment because you don't want to risk losing an animal's organic status. But if you do use antibiotics or other prohibited drugs on an organic animal, that animal can no longer be certified as organic. Ever. With a dairy animal, that usually means selling the animal due to the inconvenience of keeping her milk separate from the rest and documenting it for a certifier. With a meat animal it's easier to keep the animal, but it can't be sold as organic.

Deciding when to use prohibited medications to treat a sick animal can be a tough call, since no farmer can look into the future and tell whether a suffering animal will survive without prohibited treatments. In an emergency situation it's especially difficult to figure out the right answer. If you use prohibited treatments, you instantly decrease the value of that animal since it's no longer certified organic, and with a dairy animal you also lose the milk production. If you don't use prohibited treatments and the animal dies, you risk losing your organic certification for not saving the animal. It's a bit of a catch-22. You and the veterinarian just have to make your best guess at the time as to what is appropriate. Just remember, a live conventional animal is worth a lot more than a dead organic one.

With most of the standard contents of the barn medicine box forbidden by the Final Rule, organic producers have turned to traditional and natural remedies, particularly in dairy operations, to treat common problems. Natural remedies are allowed by the Final Rule, but you should check with your certifier before treatment becomes necessary to verify whether a particular remedy is allowed, especially if it's under a brand name and you're not sure of all the ingredients.

Interview: Marjorie Bender

The Livestock Conservancy (formerly the American Livestock Breeds Conservancy), Pittsboro, North Carolina

It may seem like putting the cart before the horse, but Marjorie Bender, research and technical program director of The Livestock Conservancy, says that before you acquire a rare breed of livestock or poultry, you need to have a plan for marketing extra animals and for eventually getting rid of all of them.

A lot of people start with a glint in their eyes: the animals are beautiful, their stories are intriguing, and folks want to participate in protecting them from becoming extinct. But we have watched a lot of people go about gathering up breeding stock only to discover that it's really, really difficult to develop the market. Why? Because of the way chefs are trained, the quantities and cuts that people want, the producer's proximity to a marketplace and a processor, the need to maintain consistent product quality. And you have to think about the profit margin.

"Because the populations of rare breeds are so small, if people get good breeding stock their animals may be really important to the conservation of the breed," Bender says. But if the animals don't pay for themselves, too often their owners can't afford to keep them. The herd or flock is slaughtered or dispersed, and a rare breed becomes even rarer.

From a practical, economic point of view, are these breeds really worth saving? Emphatically yes, says Bender. Modern agriculture is concerned almost solely with breeding higher-production animals that give more milk, or grow faster and bigger, or lay more eggs. But this narrow focus has resulted in too many short-lived animals prone to health problems that require intensive management — not a good economic decision for the long term.

"Look at Holsteins, for example," Bender continues. "Because they're so highly inbred at this point, breeders are having difficulty getting the cows rebred, as well as problems with parturition (birthing) and mastitis. Farmers end up having to replace cows within two or three years. Is that more economically sound than raising a cow that produces less milk but actually adds to the herd through her offspring for 10 to 15 years?" she asks. "With that kind of cow there's less investment in feed, breeding, and health care."

Rare breeds are uniquely adapted to the environments in which they originated, and their unique characteristics enable you to choose a breed just right for your farm, says Bender.

It really is about putting the right animal in the right place. Look at where a breed was developed, and that is the type of place where it's best suited. . . . Breeds from Britain, where it's cool and damp, need longer hair to retain heat. The Belted Galloway, the Galloway, Kerry cattle, and the Highland are examples. Some of these will adapt, over many generations, to moving south, but if you look at the Black Angus (also a British breed and now ubiquitous in the United States), its black coat is more attractive to flies, and in the South it develops an open hair pattern, which makes it easier for flies to penetrate.

The Pineywoods and Florida Cracker cattle are much more diminutive in size. They're adapted to the scrub of the Southeast. It's not nutritious, it's highly fibrous. These cattle have been selected by the environment to survive on more browse and less graze and to still be able to bear a calf every year for about 20 years.

A whole complex of animals came over with the Spanish conquistadors — cattle, sheep, hogs, chickens, and goats. They were put out on the land. There were no fences. All these animals were selected by the environment in which they lived; some are now more adapted to drier areas, like the Navajo Churro sheep that thrive in the Southwest, and some to wetter areas, like the Gulf Coast sheep. Spanish cattle were the foundation of the Pineywood, Florida Cracker, and Texas Longhorn breeds.

The goats brought by the Spaniards founded a breed we know as the Spanish goat, which used to be pervasive throughout the Southeast and Southwest. Then Boer goats arrived in the late twentieth century and soon were being bred to the rangier Spanish does for offspring that had better muscling and a faster growth rate. Because the does are not being bred pure, we're losing the Spanish goat populations.

The short-term gain is not worth the long-term sacrifice, Bender says. "Breeders want Spanish females so they can cross them with Boer males, but one day there will be no more does to be had. And we'll lose their adaptation to the southwestern environment — they're very active browsers, they need little attention for birthing or other health reasons, they're good mothers, they're parasite-resistant, and they have resistance to some of the foot diseases. With a first-generation cross you still have a lot of those benefits, but in subsequent generations an adaptation declines dramatically, until it's extinguished. A lot of these adaptations are very heritable, but they're a gene complex, not a single gene, and you end up diluting it over time. You have to keep purebreds."

Even in a purebred herd, owners have to be careful not to lose a breed's character, Bender says. They must continue to select for those animals that thrive in their environment. "You're going to get what you tolerate. If you tolerate some level of sickness or ill

"Find a mentor. Find someone who is really good at breeding livestock."
Marjorie Bender

health, you're going to have it. Culling can be really harsh, but that's what nature is going to do, too."

"Find a mentor," urges Bender. "Find someone who is really good at breeding livestock. Find someone who knows how this animal should look now, and how it should look ten years from now, so you can continue to select for type. Because you can lose your traits through negligence just as surely as you can through extinction or intention."

The private, nonprofit The Livestock Conservancy of Pittsboro, North Carolina, serves as a clearinghouse for information on threatened breeds of domestic livestock and poultry, and on genetic diversity in general. The Livestock Conservancy website is a gold mine of information on various breeds and their status, covering cattle, horses, asses, pigs, sheep, goats, rabbits, chickens, ducks, turkeys, and geese. Members of The Livestock Conservancy receive a members' directory, which is a real help when trying to find breeding stock for your farm. (For The Livestock Conservancy contact information, see the resources.)

Profile: **Stephen McDonnell**

Applegate Farms, Bridgewater, New Jersey

Natural and organic hot dogs, sausages, cold cuts, bacon, and burger patties may not be the first thing a farmer thinks of when producing livestock, but consumers love the idea — to the tune of $100 million in annual sales for New Jersey–based Applegate Farms. The largest supplier of organic and natural processed meats in the United States, the company is committed to delivering healthy meats to consumers and to sustaining the right-sized family farm.

The only difference between the company's natural, antibiotic-free products and its certified organic products, says Stephen McDonnell, cofounder and CEO, is that the nonorganic animals are not fed certified organic grain. The company does not deal with livestock producers who raise their animals with conventional methods.

"Large industrial food producers break three laws established by Mother Nature, which in my opinion is never a good idea," McDonnell says. "First, they crowd their animals. Second, they artificially accelerate their growth rate with hormones and antibiotics. And third, they feed them cheap feed. All these things severely weaken the animals' immune system, but if the animals get sick the producers simply give them more antibiotics." This system makes it possible to have giant total-confinement livestock farms that raise large numbers of animals very quickly on a small land base, much more cheaply than raising animals in a setting and with a diet that is natural to their species.

"You cannot do organic livestock production if you are too big," McDonnell says. "And we think that's great, because we want to help resurrect the family farm."

But large-scale livestock production allows producers to capture economies of scale, keeping their costs relatively low. Because organic production requires slower growth rates, more land, higher-quality feed, and better management, production costs are higher. The tricky part in organic production, says McDonnell, "is you have to charge enough so you make money, and not so much that consumers can't afford it."

Consequently it's tough for an individual small-scale organic livestock producer to make a living. Some make it work by developing a local market for their products, but many are too busy, don't have access to good markets, or are not interested in marketing. That's where a company like Applegate Farms steps in, organizing farmers into networks to capture economies of scale. Companies such as Niman Ranch and Thousand Hills Cattle Company use a similar business model, and more farmer-network-based businesses are emerging as the demand for organic, antibiotic- and hormone-free meat rises.

As one of the largest such operations in the United States, Applegate Farms works with about 300 producers of beef, chicken, turkey, and pork, all organized into regional networks that deliver to a single slaughterhouse in their area. McDonnell explains, "We go out to the farms and set the protocols and standards and develop specifications and cuts. The slaughterhouses receive animals weekly and send the meat to 12 manufacturers, who process it to our specifications, and then we distribute it nationally."

Applegate Farms sources meat wherever enough organic or natural producers exist, a slaughtering plant is available, and production costs are reasonable. The company's chicken producers are clustered in the southeastern United States, while much of the pork comes from Canada and some of the beef from Australia and Uruguay, McDonnell says.

In return for Applegate Farms handling processing and marketing, producer networks must coordinate production to ensure a constant supply of slaughter-ready animals throughout the year, and they must use the same genetics, or breeding lines, to produce closely similar animals. A product that is consistent in size, tenderness, and taste is key to building customer loyalty, McDonnell says.

You have to start with the consumer, with what the consumer is eating and is interested in. Farmers can't be naive to market demands. You have to have a business that aligns with what the consumer wants. Then your brand becomes accepted, and that's when people are going to pay a premium for your products.

Most people start backward. Most start with an animal and then ask, 'How are we going to get rid of this meat and make it profitable?' That's an inside-out business plan. You have to start with the consumer. You have to grow what people are already eating. This country eats turkey, beef, and chicken. They eat only very small quantities of anything other than that.

Despite this, McDonnell says that Applegate Farms is currently considering adding buffalo and lamb to its product lines. "With meat, you have to offer a variety," he says, and with an established brand additional products from less common meats could expand market appeal.

Running Applegate Farms is a far cry from McDonnell's vegetarian days in college, when he quit eating meat after learning about the way livestock is conventionally raised and the additives that are normally found in processed meats. He notes, "This whole natural foods movement was started by vegetarians, and I believe their premises are correct: that meat should not be the dominant part of our diet, that animals should be humanely treated, and that there are environmental issues around factory farms."

But he missed meat. On the Applegate Farms Web site, McDonnell recounts the moment he had a change of heart: "While shopping one day, I came across some nitrate-free bacon and had an epiphany right there in the meat aisle. I could feel good about eating meat if I knew it didn't contain any of the ingredients that were bad for me. And at that point, meat and I were reunited."

McDonnell went out and purchased a maker of nitrate-free bacon. That was a couple of decades ago. Since then, the business has grown more or less steadily, and it is poised for more expansion as consumer interest in natural and organic meat continues to rise.

"We need to reform the livestock industry, not eliminate it," McDonnell says. "Life's too short not to have bacon in your spinach salad."

Pigs and Poultry

Raising Smaller Livestock for Meat and Eggs

> *To confine, whom nature has given the urge to scrap, to perch, to flap her wings, to take dust baths, in a wire cage in which she cannot do any of these things, is revoltingly cruel and I cannot bring myself to talk to anyone who does it.*
>
> **JOHN SEYMOUR**
> *Farming for Self-Sufficiency*

Pigs and poultry are different from other traditional farm species in being nonruminants. Their digestive systems are more similar to those of humans than those of grazing animals, since they have one stomach instead of four (or, if you want to be perfectly correct, they have a single-chambered stomach instead of a four-chambered stomach).

This means pigs and poultry can't process forages well enough to be able to extract their nutrient value, and they need a diet that is primarily more concentrated feed (such as a mix of grains) in order to be properly nourished. Farmers have known for hundreds of years that chickens and pigs are more tender and taste better if they have some human help with the concentrate part of their diet. For the most part this means feeding them a grain ration.

Though pigs and poultry don't process forages as well as ruminants do, greens are still an important component of a natural diet for them, providing fiber, vitamins, and minerals, just as vegetables do in the human diet. Providing pasture for pigs and poultry during the growing season allows them access to fresh greens, as well as complying with the Final Rule's requirement that animals be able to follow their natural behavior. Chickens need room to scratch and peck, and pigs need space to root and wallow. Putting them on pasture whenever possible also saves a lot of manure hauling!

In the winter or where there isn't pasture room, pigs are housed in permanent sties and chickens in permanent coops. In both cases, they must have access to an outdoor pen in order to comply with the Final Rule. You will want to be able to close the door, though, when the weather gets bad or, especially in the case of chickens, if there are predators in the area.

Chicken Management

Chickens are quite often the first type of animal raised by a new farmer, since they are inexpensive and widely available, take up little space, require only minimal fencing and housing, and are easy to handle. Turkeys, ducks, geese, guinea fowl, and other species of poultry have the same general requirements, but the details differ. We'll cover the general requirements and chicken details here. If you're interested in other types of poultry be sure to first research the specifics of their care.

Chicken Nature

Since the first requirement of organic livestock production is to accommodate the health and natural behavior of animals, it follows that the farmer needs a working knowledge of the natural behavior of the species being produced. Chickens are an entertaining study.

Chickens are descended from the red jungle fowl (*Gallus gallus*) of India and Southeast Asia (which by report still roam wild in those regions), and like their ancestors are active foragers of anything that looks edible. They use their claws to scratch at leaves, grass, and dirt to find seeds and bugs. Unlike their ancestors most breeds will fly for only short distances and at not more than a few feet above the ground. And instead of laying eggs just twice a year like the wild birds, the domesticated hen lays continuously, except for six to eight weeks off each year to molt.

Left to their own devices, chickens live in flocks of hens with a rooster or so, and they come home to the same place each evening. As youngsters they sleep on the ground, but the older birds like to be up on a roost of some sort at night. Chickens will make shallow holes in dry dirt for dust baths, and they spend their days alternating between foraging around their roosting area, resting, and dusting. Though they prefer warm weather they would rather be out no matter what the weather, except for in heavy rain or snow more than an inch or two (about 3 to 5 cm) deep. Chickens detest snow. They tolerate cold temperatures well for a tropical bird, though egg laying slows down and if it gets really cold the rooster's comb can get frostbitten and wither off.

Considering all these behaviors, plus the need to have the flock pay for itself by efficiently producing meat or eggs, the organic producer needs to provide chickens with the following:

- Room to roam outside
- A place where they can dust themselves
- Roosts for older birds
- Clean water
- Organic concentrate feed (to supplement the bugs and greens they find on their own)

Also, since every predator alive likes to eat either chickens or chicken eggs, it's essential to be able to close the flock in at night, for their own protection. Sometimes daytime fencing is necessary as well, since plenty of predators are partial to hunting while it's light. A heat lamp for extra light and warmth in the winter helps keep egg production up.

Chicken Breeds

Chicks come in a delightful, staggeringly large number of breeds of all shapes, sizes, and feather types. Hatcheries specializing in rare breeds offer dozens of choices. There is no requirement to have a certified organic source for eggs or day-old chicks, but for your chickens to be certified organic they must be on organic feed from the second day of life. The day-old chicks shipped by hatcheries have not been fed; they rely on the food they've stored from their egg's yolk. This allows the organic producer to buy any chicks and still raise them as certifiably organic.

Chickens are raised either for meat or for eggs, and different breeds and methods are used depending on their destiny. Meat chickens are bred for lots of breast meat, while egg chickens are bred for consistent egg production. Generations ago most farms used a single breed for meat and eggs, and these are called dual-purpose breeds. These old breeds are wonderfully well suited for laying flocks on small-scale organic farms, but they can be a tough sell when marketing chicken

Chickens with room to scratch and dust and peck are happy chickens, and a pleasure to watch.

Egg Color

By the way, there is no difference between a brown egg and a white egg except the color. Much of the Northeast has traditionally preferred brown eggs, while most of the rest of the country grew up on white eggs. Differently colored eggs have a certain cachet in the local foods marketplace, usually attracting rather than repelling customers if you take the time to explain about heirloom breeds. Araucana and Ameraucana chickens are often called "Easter egg chickens" since their eggs are shades of green and blue. Many producers charge a premium for those eggs.

meat because they don't carry nearly as much breast meat as modern single-purpose meat birds.

Though work is being done on breeding chickens especially for pasture production systems, the standby breed for meat production is the Cornish-Rock cross for its rapid growth and lots of meat. Cornish-Rocks don't have a lot of brains even by chicken standards, but they're as cute as any breed as chicks and grow incredibly fast. It's just 10 to 12 weeks from day-old chick to ready-for-slaughter. The heirloom breed of Jersey Giant was a standard for meat production before the Cornish-Rock appeared on the scene, but even they take longer to grow less meat than the Cornish-Rocks.

Meat bird production is a seasonal enterprise, for the simple reason that producing meat birds in winter takes quite a bit more feed and the expense of extra lighting and heat. Economically and for the extra labor involved, it's not worth it. Plus it's an inside job, as the birds can't be on pasture in the winter.

The best producing breed for eggs is the Leghorn, a white-egg layer that is scrawnier and more nervous than the old dual-purpose types. Many small farmers prefer the more sedate brown-egg laying breeds like the Rhode Island Red, New Hampshire, Barred Plymouth Rock, Buff Orpington, and many, many others. They may not produce quite as many eggs, but they are much better at taking care of themselves outside and generally are easier to have around.

Raising Chicks

Before you can expect meat or eggs from your birds, you have to raise them from chicks to feathered-out teenagers. You can order or otherwise obtain fertile eggs and hatch them yourself in an incubator. This requires excellent temperature and humidity control and regular egg turning. Most people start instead with day-old chicks ordered from a hatchery or through their local feed store. The usual minimum order is 25 birds. There are plenty of good hatcheries around the country; most can be found through poultry associations.

Chicks come either as straight-run, meaning they aren't sorted for sex and you take what you get, or sexed, which costs more per chick but allows you to choose either male or female birds. For a laying flock you'll want all female birds, since it's a waste of money to get half roosters. Quite often there's a rooster or two in the batch (sexing day-old chicks is not easy). A rooster isn't necessary; hens will lay just as many eggs with one as without. But having one in the flock mimics the birds' natural social order, and the crowing at dawn adds a traditional sound effect to the farm. (A mean rooster will chase little kids and even adults, though. If yours does this, it might be time for some chicken soup and a new rooster.)

For meat birds a straight-run order is cheaper and will give you more variety in size, since the males usually finish a little larger than the females. This can be nice since different customers will often want different size birds depending on how big their family is.

Screening Parent Flocks

One caution about accepting any chickens from other farms: Chicken diseases are fairly common, so visit the donating farm to observe the flock for health problems before buying or accepting any birds. Ordering from a hatchery gives you assurance that your new flock will be free of disease.

Setting Up

Hatchery orders come in the mail. Chicks require warmth, so you should set the delivery date for your order late enough in the year that they don't arrive in a snowstorm and die of cold before you can get them home from the post office. Ask your local post office to call you when the chicks arrive so you can pick them up. Otherwise the chicks may have to ride in the car with the mailman half the day before getting dropped off at your place.

New chicks are kept in a brooder for their first few weeks of life. The brooder should be made ready *before* the scheduled arrival date. You'll need:

- A pen, which must have access to an electrical outlet and be safe from predators
- A heat lamp to keep chicks warm
- Clean newspaper for bedding
- Waterers
- Feeders
- Certified organic starter ration, plus fine grit, such as sand

THE PEN

The pen can be any corner of the basement, garage, barn, or shed, so long as you can get a heat lamp plugged in and make the space predator-proof. In cold weather you may even want to keep new chicks in a box next to the woodstove. They are, however, amazingly dusty for such little animals and in a few days they'll need more spacious quarters that have less impact on housecleaning chores.

Chicks have been raised in everything from boxes and bins to plastic wading pools. Whatever kind of brooder you have, it should contain heat well, have a cover or sides high enough to keep the chicks from jumping out (they jump surprisingly high, and even higher as they get older), and be expandable to accommodate the increasing space needs of growing chicks. It must also be safe from predators and be kept clean and dry as the essential first step in disease prevention. With a wading pool you can insert a rim of cardboard to raise the sides higher, and add more pools for more room as the chicks grow. Another option is to use a corner of an outbuilding, completing the pen with plywood sides at least 2 feet (60 cm) high. If the pen floor is cement, put a piece of plywood or cardboard down on the floor to lessen the cold that comes up from the concrete. Cover the floor of the brooder with clean newspaper as bedding for the chicks for the first few days.

Heat lamps with shields are available at any farm store. Hang the heat lamp, with a thermometer directly under it, over the center of the brooder and leave it on for a couple days before the chicks arrive. Adjust the height of the lamp until the floor temperature is steady between 90 and 95°F (32–35°C). Never hang an unshielded lamp — if any moisture drips or splashes onto a naked bulb, the bulb may explode and start a fire.

WATERERS AND FEEDERS

Chick waterers and feeders are also available at any farm store. To start you'll need one waterer and two standard-size feeders per 25 chicks. The feeders and waterer should be filled and ready before chicks arrive. New chicks are fed a starter ration, formulated for their rapid rate of growth. Once they're feathered out, meat birds are switched to a meat bird ration, and laying hens to a layer ration.

Since birds have no teeth, they depend on grit in their craws to masticate their meals. For chicks, grit needs to be very fine. If you have fine sand in

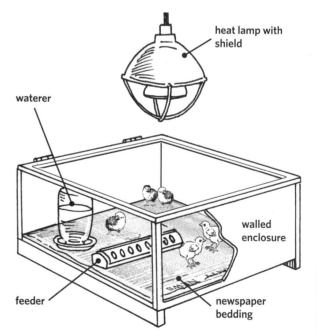

New chicks need warmth, clean bedding, water, and feed that includes grit.

the driveway scoop some up, or buy fine grit at the farm store — the Final Rule doesn't regulate where you get your grit. Sprinkle some grit over the feed every time you fill the feeder.

Caring for Brooder Chicks

When you unpack the chicks from the box, remove them one by one, dipping each one's beak into the water before you set it down. A thumb on the back of the head works. Most of them will be thirsty, and it's important to show them where the water is. Sprinkle some feed on the newspaper right around the feeder, so they find it. Count as you unpack — hatcheries often include an extra or two, but you shouldn't be short.

During the first day you should check on the chicks several times to make sure they aren't piling up in the corners because they're too hot or piling up right under the heat lamp because they're too cold. When they sit in a circle under the lamp, like people around a campfire, the temperature is just right. They will spend some parts of the day running around and eating and some time sleeping. The first time you see them all asleep you'll probably think they've all died.

Change the newspaper daily for the first several days, and fill the waterer and feeder (not forgetting to sprinkle the ration with grit) at least once a day. Both waterer and feeder should always be full or nearly so. A constant food and water supply is key to keeping chicks healthy. As the birds grow you can start using straw, leaves, dry lawn clippings, or other organic matter for bedding. You can change this natural bedding regularly or simply keep adding to it until the chicks are outside and you give the brooder its final cleaning, as long as the chicks are always on clean, dry bedding.

Adding a few big handfuls of fresh grass and clover each day seems to prevent them from pecking each other and gets them in the habit of eating greens. Many chicken farmers like to throw the occasional clod of sod and dirt into the brooder; the chicks love to scratch around in it. Any clods, grass clippings, vegetable scraps, or other treats must be certified organic if the chicks are to maintain their certification. Any bedding materials that they might ingest —and they will taste about anything — must also be certified organic. Wood shavings and sawdust are common bedding materials

for chicks and they won't feed on these, but if you use straw or old hay it should be organic.

Drop the brooder temperature about 5° Fahrenheit (about 3° Celsius) every week, continuing to use the chicks' behavior as your best guide to whether they are comfortable. Add feeders and waterers as needed as the chicks grow to ensure that all have access at all times. Enough floor space, dry bedding, clean water, plenty of feed, the right temperature, and greens will prevent health problems and pecking.

Once the chicks have grown a nearly full set of feathers (at three to five weeks, depending on the breed), you can move them from the brooder to their permanent quarters.

Pasturing Poultry

Chickens will eat nearly anything, including garden sprouts, and they will dig holes in the dirt to take dust baths in the most inconvenient places. They are also completely indiscriminate about where they leave their droppings. When there are just a few laying hens around for eggs for the family, the pleasure of having these living lawn ornaments around keeping bugs to a minimum in the yard can make up for some stray poop and dusting holes. But producing organic eggs for market usually involves too many hens and too big a mess to let them free-range around the house. For both laying hens and meat birds, a portable chicken coop that can be moved around the pasture or a stationary pen that can be subdivided for rotational foraging is a better option for giving the birds the outside access they crave.

Portable Pens

Chickens can be rotated through a pasture in portable pens or coops that are built on wheels or wooden skids and moved once or twice daily. Portable pens can be used in orchards, in pastures (following grazing animals), or in fallow areas of a garden. They offer instant fertilizer, and in a pasture the chickens will break up manure pats and eat the eggs and larvae of flies, parasites, and other insect pests.

The most famous of portable poultry housing are the "Salatin pens" designed by Virginia farmer Joel Salatin, whose 1996 book, *Pastured Poultry*

Profits, jump-started pastured poultry production in the United States. The original Salatin pens are 10 feet by 12 feet (3 x 4 m), built with a wood frame, chicken wire cover, and galvanized, plastic, or wood on the sides and top of half the pen. This means the pen is half open to and half sheltered from the weather. The pen, which will hold about 50 chickens, is moved about 12 feet (4 m) each day to fresh grass, usually by putting roller skates or a homemade set of wheels under the back, then lifting the front and rolling it. The area where the chickens just were will look trashed for a week or so, then turn green and lush, so that you can see the track of the chicken pens through the field by its deep green color.

Since Salatin came up with his pens, other farmers around the country have come up with their own pen designs suited to different conditions. These include using PVC pipe and tarps instead of wood and wire to make the pens lighter; arching the pipes so the pens are high enough to walk into for feeding, watering, and catching chickens; and dispensing with the pens altogether in favor of a portable coop on skids or wheels surrounded by portable electrified chicken fencing. Each system has advantages and drawbacks. The heavy Salatin pens discourage predators that like to dig underneath the edge to get at the chickens, but catching the chickens for slaughter is a chore since the pens are so low and the chickens can hide under the protected half. It's hard to get back in there to shoo them out where you can grab them. Farmers can

Portable chicken pens and coops make it easy to move the flock to fresh green pasture, while still giving them some protection from predators.

maneuver more easily inside pens with a higher tarp roof, but these pens can be damaged or blown over by storms and are easier for predators to get into. Portable coops with portable fencing offer a bigger area to the chickens but take much longer to move.

Figure out which system you want to try first, based on your budget, the number of chickens you plan to raise the first season, your terrain, and your inclinations. If it doesn't work as well as you like, try something else next year. That's farming.

Stationary Pens or Coops

If you don't have the room, the time, or the inclination, or you have major predator problems, you can fall back on the traditional permanent chicken pen and coop. One advantage of having the chickens always close to the house in their pen is that you don't have to walk so far every day to feed and water them, and you don't have to move pens. The drawback of having a permanent pen is that the chickens will quickly denude it of anything green, maximizing the dirt and dust. A permanent pen is also a nice spot for disease to prosper should a pathogenic virus or bacteria show up on your farm.

To overcome these drawbacks, subdivide a permanent pen into at least two and preferably more areas, so the chickens can be rotated among them. This allows the grass an opportunity to regrow and makes it tougher for disease to get a foothold. During times when there's no grass growth, throw down old hay, straw, the leaves you've raked up in the yard, cornstalks, or garden waste for the chickens to scratch through. Enriched with poultry manure, the used litter is an excellent addition to the compost pile or garden.

As a final note, if you do build permanent pens due to predator problems, consider sinking the bottom part of the chicken wire fence 6 to 8 inches (15–20 cm) into the ground, which is a lot of work, or extending it perpendicularly a foot (30 cm) or so outward at the bottom of the fence, to discourage the many predators that know how to dig under fences. Chicken wire folded at the bottom so it lays out into the lawn can be a nuisance when you're mowing or trimming weeds, so make sure it's lying flat.

Additional Considerations for Laying Hens

Chickens have the delightful habit of coming home to roost each night. That's why you can let them run free during the day. Meat birds don't live long enough to really want roosts, and they will sleep on the ground. But as laying hens approach laying age, at around five months, they develop a preference for being up off the ground at night. Their accommodations should therefore include roosts, which are usually nothing more than 1-inch-thick (2.5 cm) round or square wooden poles anywhere from 1 to 4 feet (0.3–1.2 m) off the floor. For a lot of hens, add more poles, staggered like stairs at the back of the coop.

Laying hens also need nest boxes, since they like to lay eggs in an enclosed, dark space. Nest boxes can be built out of any old scraps, as long as the wood is untreated and has no protruding nails to injure the chickens. A nest box should be a cube at least a foot (30 cm) on each side, with one side left open for the hen to get in and out. A 1-inch-high (2.5 cm) wooden board across the bottom of the box keeps the bedding (straw, sawdust, leaves, or old hay) from being pushed out, and a 3-inch (8 cm) or so lip on the outside of the box gives the hen a perch to hop onto to get into the box. The top of the box can be flat, but if you slant it steeply like a roof the hens won't sit and poop on the roof. The boxes are usually grouped together like apartments in rows of five or so, and you can stack a couple rows on each other if you're keeping a lot of hens.

With all the furniture necessary for laying hens, if you plan to rotate them through pasture it makes sense to build a floored coop rather than a floorless pen, the latter being common for meat birds. The meat birds have grass available at all times right at their feet. The hens are let out of the coop each day to do their foraging, then closed up at night to protect them from predators. Hens can be allowed to roam freely around the coop or be contained when outside by portable electric chicken fencing or permanent pens. Don't forget to walk out each evening and close up the coop. If you regularly forget to close that door at night, predators will eventually figure it out and start making regular visits.

When building a coop for layers, plan on 3 to 5 square feet (0.3–0.5 sq m) of floor space for each hen, a nest box for every five hens, as many windows as possible on the south side for winter sunlight, and a spot for the waterer where, when it leaks, it won't soak all the bedding.

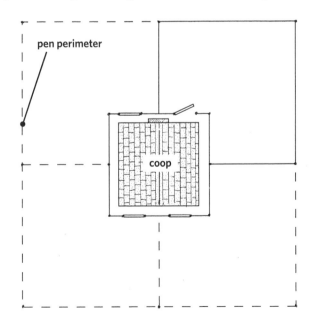

A permanent coop is a good choice when predator pressure is heavy, when you don't have much land, or as a winter home for your laying flock. This coop's pen is divided into four sections. The chickens are let out into only one section at a time, allowing the other sections time to regrow.

Chicken Feed

All chickens must be fed a grain ration in addition to the pasture. Don't believe the articles from the 1970s and 1980s promoting the concept of completely "free-range" poultry, which advocate simply turning the birds loose to find their own food. They can, but that sort of diet is sufficient only to sustain themselves. Chickens without a freely available, balanced concentrate ration will survive, but they're tough and stringy to eat, and they don't lay a lot of eggs.

Avoid Raw Soybeans

Soybeans in chicken ration should always be roasted. Compounds in raw soybeans act on the endocrine system to prohibit growth, and you can wind up with 2-pound (1 kg) cannibals instead of placid 8-pound (4 kg) broilers after 10 weeks.

A coop for a laying flock should include roosts for the chickens to sleep on, nest boxes for them to lay eggs in, feeders, a waterer placed above the bedding to keep things dry, and grit and shell feeders. South-facing windows let in much-appreciated sunlight.

The biggest obstacle in raising certified organic chickens is locating a reliable source of organic feed. In fact, you should find a source for organic feed even before ordering chicks. Poultry feeds are formulated separately for chicks, meat birds, and laying hens, or you can mix your own if you can locate organic ingredients. Commercial chicken feeds sold in farm stores should not be used because they contain prohibited additives. Commercial starter feeds, for example, contain coccidiostats to prevent coccidiosis, an often fatal disease in chicks caused by a parasite that is present in moist environments. But in a clean, spacious brooder coccidiosis is rarely a problem.

Buying a premixed ration from an organic feed supplier will cost more, but if it's available it's more convenient than trying to locate, measure, and mix all the certified organic ingredients yourself. If your organic grain dealer doesn't offer a premixed ration, you can mix your own. If you aim to be certified organic, though, be careful to comply with the Final Rule regulations. For example, the rule prohibits feeding meat scraps to poultry, though feed mix recipes in old farming books often include meat scraps. And dairy products are allowed only if they come from certified organic milk. Find a balanced ration recipe (the chicken-raising books in the resources are a good place to start) that doesn't include prohibited ingredients, and be sure to include salt and minerals in the mix.

Plan on at least 3 inches (8 cm) of feeder space for every adult bird, and just an inch (2.5 cm) for every new chick. Providing plenty of fresh water every day is critical to having healthy, growing chickens. Provide at least 3 gallons (12 L) a day for every 25 chickens, and check the level twice daily to make sure they don't run out. Water consumption can double in hot weather.

Both meat birds and laying hens also require grit, which should be housed in its own feeder. Laying hens also need a shell feeder. The shell feeder is filled with crushed oyster shell (some producers use crushed eggshell) to provide the calcium necessary for good strong eggshells.

Health Care

Chickens with access to pasture don't generally have health problems. Vaccines are available for some chicken diseases but aren't commonly recommended or needed for small flocks. Should signs of disease or infection become evident in your flock, call your vet or, if your vet has very little experience with poultry, your state animal lab (usually housed at the state land-grant university's main campus). Sometimes the modern Cornish-Rock crosses have heart attacks and die when they reach butchering weight — probably because they have been bred for extremely fast growth and a disposition to eat too much. If your Cornish-Rocks start having heart attacks, it's time to slaughter the chickens.

Processing Eggs

Laying hens will lay several eggs a week. The heritage breeds average four or five eggs per week for the first year, then quit for a month or six weeks while they molt, then lay again for another year. In the second year they produce slightly fewer but larger eggs; by the third and fourth years egg production tapers off significantly. Old hens can be butchered for soup (they are tough but exceptionally flavorful), killed and composted, or left to enjoy their retirement.

If you plan to sell eggs, collect them at least twice a day from the nest boxes and refrigerate them to preserve freshness. Eggs pick up odors, so don't store them with the onions or other strong-smelling produce. Dirty eggs should be immediately washed in cold water, with a little soap. Any eggs that are still stained after washing, or that are misshapen, cracked, or otherwise not attractive, should be kept for family use rather than give customers a bad impression.

Producers who sell eggs also usually "candle" each egg, by holding it in front of a strong light to look for dark spots in the interior. These blood spots are harmless, but consumers find them disconcerting, so keep those eggs for baking.

Butchering

Butchering chickens is not an especially difficult job, but it's a fair amount of work. It tends to be wet, since the birds are dipped in scalding water to loosen the feathers, there's a lot of washing down after gutting, and the finished chickens are chilled in cold water before being packaged. If you're very lucky, you'll have an organic poultry

processor within a few hours of your farm and you can take your chickens to it. If you're moderately lucky, you'll have a small-scale processor nearby that isn't certified organic. When you sell directly to customers, not many are likely to be very particular about whether the processor is certified or not. To sell your chickens as certified organic, however, you need certified processing.

The alternative, butchering the chickens yourself, is very doable. If you're organized and have some willing help and a good setup with a plucker, hoses, cleaning tables, and chilling tubs, it can even be a pretty good time. It's key to have customers showing up to pick up their orders just as the orders are completed, or to have enough fridge and freezer space to hold the finished chickens until the buyers show up. Many producers take their orders before ordering chicks in the spring, so they know exactly how many they need to produce and can give their customers early warning about when to show up on butchering day.

If you can't find anyone to show you how to butcher a chicken, most poultry manuals have detailed instructions and you can figure it out yourself if you take your time. But it's much easier to learn by helping someone else once or twice.

The movable "Salatin pen" popularized by innovative farmer Joel Salatin makes raising meat chickens on pasture simple and economical.

Pig Management

Pigs are generally considered the most intelligent of farm animals (goat owners may disagree), with distinct personalities, an enterprising nature, and a strong hedonistic bent. Pigs love to be comfortable and don't tolerate extreme temperatures well, so they need shade in the summer and draft-free shelter in the winter. They really do enjoy wallowing in mud — it gets them out of the bugs and heat. Though pigs have the most offensive-smelling manure of all farm animals, they are fastidious about setting aside a toilet area if you give them the space.

Pigs are omnivores (they will eat anything, including meat), and they're also terrific foragers, using their tough, muscular noses to root and dig for bugs and roots. In colonial days they were often turned into the woods each spring to fend for themselves until fall, when they were rounded up and fattened on a rich diet for a few weeks before slaughter. In fact, pigs, like cats, revert easily to a wild existence. In many rural areas of the country there are feral pigs, descended from escaped domestic pigs, and it's possible in many states to hire guides for pig hunts. But for the organic farmer, the issue is keeping pigs contained and feeding them so they grow and fatten to suit human tastes.

Obtaining Piglets

Pigs to be raised for meat are traditionally purchased as weaner pigs, at around six weeks when they're old enough to take care of themselves and have mature digestive systems (conventional producers will wean as young as 15 days old, to get the sows to breed back faster and squeeze more litters into a year). At this point they're between

Pigs on pasture will enjoy it to the utmost!

30 and 40 pounds (14-18 kg) and ready to grow without any special care. To be certifiably organic, the mother of the piglet must either be certified organic herself or have been fed only certified organic feed starting at the latest at the beginning of her third trimester of pregnancy. (If you are selling directly to customers, many won't care whether the mother was organic so long as the pig they're buying was raised organically.)

Finding any piglets at all to buy is difficult these days in most areas, and finding organic piglets even more so. Not so long ago there were always a few farmers (often the dairy farmers) in any neighborhood who kept a few sows and had piglets for sale each spring. This was extra income for the farmer in the form of piglet and pork sales, and waste disposal for the garden and milk house. Since the bottom fell out of the pork market several years ago, causing pigs to sell for less than they cost to raise, this practice has become extremely rare. Regional alternative farming networks and the Internet may be your best bet if you can't find any piglets in the neighborhood.

On the other hand, the demand for homegrown organic pork is excellent and presents a good economic opportunity for the small producer. If you develop enough of a market, perhaps you'll eventually get sows and a boar (or learn artificial insemination in pigs) and start raising your own piglets.

Pasturing Pigs

For most new farmers the best way to get started is to buy two to four weaner piglets in the spring, raise them through the summer on pasture, and butcher them in fall. Male pigs will need to be castrated, and this should be done before they are weaned. The University of Minnesota Agricultural Extension Service recommends a stocking rate of 10 to 20 pigs per acre (0.4 ha) for pigs that will weigh over 100 pounds (45 kg).

When planning where to put the pigs, it's important to realize that they don't just graze, they pretty much plow an entire area as they root and wallow. These habits are utilized by some farmers as a low-cost, low-labor method of reclaiming and fertilizing weedy wasteland or renewing a field. If you don't need any plowing done, you will have to rotate the pigs fast enough to minimize rooting, or you'll have

Breeding Pigs

A sow's gestation period is just under four months, and she will usually be ready to breed again a month after having piglets, so a producer can figure on two litters per year. Litter size varies by breed, age, and nutritional level. The average for U.S. pork producers is ten piglets per litter.

Conventionally raised sows are confined when farrowing (birthing) to pens so narrow they can't turn around, with low guards at the sides under which the piglets can escape so their mother doesn't roll on them or eat them. Wild pigs and organic sows with individual sheds for shelter and enough pasture space will take excellent care of their piglets and defend them against all comers, though there are bad mothers in any species, including pigs.

Keeping pigs for breeding increases housing costs, since in cold weather pigs should be inside, especially if you plan for the sows to farrow during that time. Any sort of building will do for shelter, as long as it is draft-free and holds enough heat and clean bedding to keep the pigs warm. For example, plastic-covered hoop houses with deep straw get a lot of solar gain during the day, and the straw keeps the animals warm at night. Low-ceilinged sheds, again with deep bedding, are more traditional, but they're hard to clean if you're tall. Whatever arrangement you set up for your pigs, be sure it is large enough for the pigs to have a distinct toilet area.

to plan on smoothing the pig pasture with a disc or drag after the grazing season is over. Young pigs don't do as much damage as adult ones, but pigs grow really fast, so it's only a temporary reprieve.

When pigs were still commonly kept outside, farmers used to put rings in their noses to stop them from rooting. The organic farmer who keeps pigs outside, where they naturally should be, and who needs to keep them safely confined and to discourage their destructive behavior, may wish to do the same. The Final Rule allows this.

To limit their disruption of the soil to designated areas, pigs have to be confined. Pigs prefer to go under rather than over obstacles but aren't afraid to try climbing, so perimeter fences need to be both high and low, and they must be strong. Electric fence is excellent for subdividing pastures for rotational grazing, if you first train the pigs to it by running a wire in front of a solid fence. After they've touched the electric wire a few times with their noses, they'll be wary of touching it again, even if it isn't backed up by a solid fence. At least one wire has to be at nose level (which is quite low) or they'll just go under it. In a pen or pasture, check the wire daily, since the pigs may bury it (shorting it out) while rooting next to the fence.

Barbwire fence *may* work for a large area where the pigs aren't as anxious to see what's on the other side, if the wires are close together and the bottom wire is at or in the ground. Woven wire is preferred, with a board or strand of barbwire at dirt level to prevent the pigs from digging underneath and escaping.

Several metal gates wired together as a movable pen will hold a couple pigs, and this arrangement allows you to move the animals regularly, so that a sizable area can be pig-plowed in a season without having to invest in a fence. Small pigs aren't strong enough to lift the gates up to escape underneath them, a lesson that they usually continue to believe as adults.

As is the case for chickens, rotational grazing for pigs will minimize parasite problems and maximize available forage.

Pigs in a portable pen of wired-together gates can be moved to where they're useful in rooting up overgrown, brushy areas.

Supplementing the Forage

Compared to grazing animals, pigs take up relatively little space and have the additional advantage of progressing from birth to slaughter in a single growing season. A newborn pig weighs just a few pounds (1–2 kg) and the finished product six months later will top 200 pounds (90 kg), gaining on average a pound (0.5 kg) per day. That takes a lot of feed. How much, exactly, depends on the type of feed. Feeding nothing but a premixed balanced pig ration takes less total feed but costs more, while letting the pig forage some of its own food from the pasture, and consume as much waste and leftovers from garden, orchard, and dairy as you care to feed it, requires more feed overall but a much lower cost for purchased feed. Pigs love table scraps, too, but unless yours are certified organic or your buyers don't mind, these are off-limits for pig feed.

Balancing pig rations is quite a science. To start it's probably best to purchase a certified organic premixed ration from an organic supplier. Salt, vitamins, and minerals will be part of a purchased complete feed. If you decide to mix your own, you can find instructions in pig-specific reference books (see Resources) or, usually, on the Web sites of land-grant universities with swine programs. You'll need to purchase a vitamin and mineral mix formulated specifically for swine, and include that in your ration at the level recommended by the manufacturer. Remember, only FDA-approved vitamins and trace minerals can be used, and all components of the mix must be natural in order to be used in organic production. No artificial color, flavor, dust suppressants, or flowage agents are allowed by the Final Rule. For example, anticaking agents like yellow prussiate of soda are common ingredients and are prohibited by the Final Rule. As with chickens, any soybeans in the ration must be cooked or they will inhibit growth.

Like all animals, pigs should have clean fresh water available at all times. The trick is to have a waterer that they can't tip over or destroy. If you're rotating pastures or have the pigs in movable pens, the waterer has to be light enough to move easily — the concrete trough isn't going to work. Try attaching a tub to a metal fence post or two. Posts are both sturdy and quick to pull and pound if you have the right tools, which are available at farm stores. Feeders also need to be nontippable, and they should be under cover if you live where it's very rainy.

Pigs are truly pigs about eating — it's a frenzy, not a meal. Don't get between them and the trough when you're filling it with feed, because you'll get run over.

Pasture Shelter

Sheds for pastured pigs can be built from scrap lumber, but they have to be solid. Pigs push, tip, dig under, and crush things, not because they're malicious, but because they're pigs. It's what pigs do. Lean-tos and A-frames on skids are traditional. Metal or plastic hoop houses are also used. A plastic hoop house should have board sides at least 4 feet (1.2 m) high to keep the pigs from destroying the plastic. There should be enough room inside for all the pigs to lie down without crowding.

In a temperate climate pigs may need no shelter at all in the summer, if they have tree shade available for hot days. Pigs will uproot small trees and damage larger ones, though, so it might be best to start with the pigs on one side of the fence and the trees on the other, or with some brush that you'd like to have rooted up anyway.

Handling Pigs

Handling pigs when it's time to send them off to the plant is completely different from handling cattle, sheep, or goats. They don't herd, they don't scare easily, and you can't push them around once they weigh 200 pounds (90 kg). If pigs knew how to play football, all the teams would use them. They're powerful, solid, and low to the ground, perfect for tackles. For these reasons it's important to design handling and loading facilities carefully. Temple Grandin, who was mentioned in chapter 9, has some excellent suggestions in her book and on her Web site (see resources).

You may read about the old standby of putting a bucket over the pig's head and backing it into where you want it to go. This doesn't really work so well — a pig with a bucket on its head will back up, but rarely in the direction you would like. A better idea might be to park the truck or trailer in the pigs' pasture for a few days and feed them inside of it. They will learn to load themselves.

If you intend to sell pork as certified organic, the processing plant must itself be certified organic, as is the case for poultry. It is possible to butcher a pig for home use, but you'll probably need help. On-farm butchering on a commercial scale is not feasible (and is illegal in many states).

"I say, start with an assortment. See which breeds best suit your needs and your environment, and then the next year you'll know the one that's best."

GLENN DROWNS
Sand Hill Preservation Center, Calamus, Iowa

Profile: **Glenn Drowns**

Sand Hill Preservation Center, Calamus, Iowa

Genetic diversity is a foreign concept in most of rural Iowa, the beating heart of the Corn Belt, where hybrid and GMO crops cover the ground as far as the eye can see. But preserving the genetic diversity of domestic plants and farm animals is exactly what the certified organic Sand Hill Preservation Center near Calamus is all about. Here owners Glenn and Linda Drowns meticulously care for one of the largest private collections of heirloom vegetable seeds in the world, and what is probably the largest collection of heirloom poultry breeds in the world.

Glenn, a science teacher in the local school district, works long days:

> *I get up at quarter to five and work on chores until 7:00, then I'm off to work. In the winter it takes 78 5-gallon (20 L) buckets to water all the critters in the morning. I'm at a high trot. Home at 4:00 for chores. We have 230-plus breeds of poultry, and all the eggs have to be collected and coded. I'm not milking the cow anymore; that used to tie us down even more.*
>
> *We grow 90 percent of our seed offerings on the farm, and we work with friends to produce the others. We have a total of about 2,000 varieties of seed. It's a bookkeeping nightmare to keep track of all of them. Isolation distances, hand-pollinating, making sure a variety is grown out often enough, making sure there are no bugs in the seed jars — it's a commitment.*

Glenn bought the 40 acres (16 ha) in 1988, when prices were low. "It's a quarter section that nobody else wanted, because a sand hill divides it east to west. Almost half of it is not tillable, and my sheep and pigs and cow are pastured there. We have about 10 acres (4 ha) in seed production and evaluation, another 10 or so in hay and orchard, and the rest is part pasture, part unused to be left natural. The poultry buildings take up 1 to 2 acres (0.4–0.8 ha). All the young poultry have the run of the orchard, and in summer they run everywhere until I gather them up. I have to keep the breeds separate, so when they reach breeding age we put them in pens."

The Drownses offer vegetable seeds and day-old chicks for sale on their Web site. Glenn says, "The seed part of the business makes money, but the poultry part does not, so it sort of comes out as a wash. To rescue a rare flock I sometimes spend a huge amount that I may never see back. But it's something I feel I have to do, because whenever a crisis occurs — like the feed price crisis in 2008 — people abandon things right and left. I used to send my wife on chicken rescue expeditions. She'd drive 500 miles (800 km) to pick up a chicken flock that someone couldn't take care of anymore."

Glenn says his interest in genetic diversity began early:

It goes back to when I was eight or so years old, growing up in Idaho. I started reading in third or fourth grade about the dodo bird and all those species that had gone extinct, and I was fascinated. I had had my own garden since I was five, and by the time I was eight I was ordering my own seeds and no one else had anything to do with it. We had impossible soil — a pH of 8 or 9, a last frost around the tenth of June and first frost by Labor Day, and no water. I spent a lot of time trying to find things that would grow there.

The poultry thing came around age eight also, when I persuaded my parents to let me have some poultry. By the time I was eleven or twelve I'd joined the Society for the Preservation of Poultry Antiquities. I had ten to fifteen poultry varieties by the time I went to college.

By high school Glenn had become involved with the Seed Savers Exchange in Decorah, Iowa, a nonprofit organization devoted to preserving heirloom vegetables and now apples and farm animal breeds as well. "I made many pen-pal friends in the Midwest, and it convinced me that Iowa would be the best place. I moved out here sight unseen after I graduated from college and never regretted it. It really is, to me, the garden of Eden."

Once he was established in Iowa, his seed and bird collections kept expanding, Drowns says. "About the mid-1990s the world had reached a very low point in terms of genetic diversity in poultry. The commercial hatcheries had dropped many varieties because they weren't selling. If a hatchery was getting rid of a flock that wasn't common, I would take over. Especially for turkeys, there wasn't anyone else. I used to freak out if a hawk flew by — what if the last of a breed died here?"

By the late 1990s the American Livestock Breeds Conservancy had begun doing a lot of breed preservation work. "So now," Glenn says, "I can relax."

Why is genetic diversity important? "It's the best insurance policy you can have," Glenn says. "Everyone now is too young to remember the corn blight that took so much of the crop in the 1970s, when we were first trapped in the hybrid narrow-line stuff. The narrower your genetic base, the less likely you'll be able to survive a plague, a virus, a change in climate. This summer (of 2008) was a good example. All the people who grew hybrid tomatoes around here didn't have tomatoes this summer — they blighted and died. I was picking bushels of tomatoes until the first hard frost. In 1987 the same thing happened, and we'd have these little old ladies stopping and asking, 'Why aren't your tomatoes dead?' It was a teachable moment."

The mission of the Sand Hill Preservation Center is education and preservation, Glenn says, "for the long-term sustainability of the planet." New farmers can help, but they should proceed with caution, he advises. "People will pick up an article that says you need to raise, say, Delaware chickens, and they run off and get chickens, never having

"**The narrower your genetic base, the less likely you'll be able to survive a plague, a virus, a change in climate.**"
Glenn Drowns

raised poultry before. I say, start with an assortment. See which breeds best suit your needs and your environment, and then the next year you'll know the one that's best."

And don't think that if you've raised modern breeds — Broad-Breasted White turkeys or Cornish-Rock chickens — that you understand heirlooms, Glenn says. "The old breeds of turkey are completely different. They roost, and they become attached to people, especially the young hens. At 10 to 12 weeks they love to see how high they can fly and what they can roost on — they're free spirits when they're teenagers. Ours are off foraging in the hayfield and eating apples off the trees right now. I have about 150 young turkeys that refuse to go in the shed at night now. It looks like I've decorated for Christmas with roosting turkeys. Next week I'll start pruning feathers and putting them in breeding pens, but some I won't get until we get a snow."

Ruminants for Meat

Raising Goats, Sheep, and Cattle on Pasture

> *A skill is just like speaking a foreign language or learning to swim. One must do it badly before one can do it well. With grazing, this learning curve is at least three years long.*
>
> **ALLAN NATION**
> *Grassfarmers*

Cattle, sheep, goats, and other ruminants, with their multichambered stomachs, are designed to extract a complete diet from low-nutrition forages such as grasses, clovers, and other plants. Ruminants swallow before they chew, gulping down bites of grass and clover that are stored in the first chamber of the stomach (the rumen) as they graze.

After they're done grazing, they take a break to begin burping up, mouthful by mouthful, the recently consumed greenstuff. They chew each cud thoroughly before swallowing it again, this time passing it on to the second and subsequent stomach chambers for full digestion.

Ruminant guts are packed with specialized microorganisms whose job it is to digest the cellulose in plants and extract nutritional value that can't be used by any other animals. That's how ruminants turn grass that is of no dietary use to humans into meat and milk. This specialized digestion is also why ruminants need a diet of primarily forages to remain healthy. Their systems aren't set up to handle a high proportion of grains and other concentrates, any more than human digestion is adapted to an all-sugar diet. It may taste great, but in the long run it will make you fat and sick and then kill you. (The beef feedlot industry's no-forage feeding practices, which essentially race to make cattle fat before the diet makes them sick, were detailed by journalist Michael Pollan in "Power Steer," published in the *New York Times*

Magazine on March 31, 2002. Pollan quoted one feedlot vet as saying, "[Heck], if you gave them lots of grass and space, I wouldn't have a job.")

Feeding Livestock

For ruminants intended for slaughter, the farmer's goal is to put on fat and muscle as rapidly as possible. Young, lush forage is terrific for this purpose. Using pasture for fattening usually involves leaving animals no longer than three days on one paddock, to keep them on the very richest grass. For breeding animals such as ewes and cows, taller, lower-feed-value forage and a slower rotation may work better, because you don't want them as fat.

Hay and Silage

Except in those lucky areas where pasture is available year-round, organic farmers have to buy or grow additional forages to store as hay or silage for those times when grass isn't growing. Silage, which is fed primarily to dairy cows, will be discussed further in chapter 12. Hay, which is fed to both dairy and meat animals, is the grazing-livestock producer's constant concern and delight. The smell of fresh-cut hay is right up there with the smell of baking bread and wood smoke on an autumn evening as one of the great olfactory pleasures of rural life.

If you hope to make any profit on your meat livestock enterprise, or at least to break even, you should avoid buying hay. The exception to this rule is if you have little land, just a few animals, a small species, or a high-value product. Then buying hay can make sense. For example, it's quite common for small dairy goat operations to buy all their hay, since a few goats just don't eat that much. But it's

Equine Grazing

Horses and other equines should be rotated very slowly or not at all, since they quickly become too fat in a management-intensive grazing (MIG) system. In a very slow rotation, the difficulty lies in transitioning a horse from a closely grazed paddock to a lush one. This can be a recipe for colic, a dangerous upset of the digestive system that can cause death or ruined hooves. The transition to new pasture must be gradual, starting with an hour a day on the new area and gradually adding time until the transition is complete. An alternative would be to strip-graze a pasture, moving a portable electric fence each day to open up a small new area of fresh grass.

rare for cattle operations. Though there will always be drought, exceptionally long winters, or other occasional emergencies when it's necessary to purchase hay, it is too costly to make sense for most meat-animal operations with more than a couple head. And certified organic hay is almost always more expensive than conventional hay, if you can even find it within a reasonable distance.

Making your own hay requires land and equipment, as discussed in chapter 7. If you have those, it's not too difficult. If you have land but not equipment, it's usually feasible to find someone in the neighborhood who will make the hay for you. Traditionally the person who does the work takes half the hay and the landowner gets the other half. If the neighbor does not have organic certification, make sure the equipment is purged (see chapter 3) before it processes your half. The landowner traditionally is responsible for maintaining soil fertility. If the hayfields are permanent (not being used for any other crops) this takes some planning ahead. Soil amendments that aren't incorporated by tillage take much longer to reach the root zone, from a season to three years depending on the grind and the type of amendment.

Balancing the Ruminant Diet

Ruminant diets are easier to balance than those of nonruminants, because forage grown on good soil will supply most of their nutritional needs. The only other really necessary nutrients are salt and minerals; these should always be available in free-choice feeders.

Forage can be stored for the winter as large or small hay bales, as silage, or in ungrazed paddocks that are "stockpiled" for use when the growing season ends. (Legumes, especially alfalfa, don't stockpile nearly as well as grasses.)

silo for storing silage

hayloft for small square bales

stockpiled paddock

freestanding round bales

One Hundred Percent Grass-Fed Meat

It is possible to give ruminants just their natural diet of forages, without feeding them any grain. One hundred percent grass-fed organic beef, goat (cabrito), and lamb are steadily gaining appreciation from knowledgeable consumers for their flavor and nutritional benefits, especially their higher levels of conjugated linoleic acids, which research indicates are important for good health and preventing heart disease. However, raising a good-tasting, tender meat animal solely on grass is not for amateurs; it takes a sound knowledge of good genetics, constant monitoring of animal performance, and superb pasture management. In the past couple of decades, enough consumers have bought all-grass-fed meat that was gamey-tasting and tough that initial consumer enthusiasm for grass-fed meat suffered a significant setback. Fortunately some experienced producers now are offering superb 100 percent grass-fed meat and repairing the damage.

Certified organic grass-fed meat offers the highest potential return for the producer if marketing and production are excellent. It certainly is something to consider once you have the hang of livestock production, or if you can find a mentor to guide you. The bible of the grass-fed meat sector is the *Stockman Grass Farmer*, published monthly by Allan Nation and his staff.

Salt and Minerals

All animals, including humans, need salt and essential minerals in their diets. However, many salt and mineral mixes contain prohibited ingredients or additives that a producer simply doesn't want to feed to livestock, such as artificial colors, flavors, dust suppressants, and flowage agents. Even "pure white salt," for example, often contains the anticaking agent yellow prussiate of soda, which is prohibited under the Final Rule. (Dairy white salt, on the other hand, is usually unadulterated, but you should check it all the same.) If your feed store doesn't have a mineral mix that will meet Final Rule requirements, you will have to order one from an organic supplier. The Web site of the Organic Materials Review Institute (OMRI) lists regional suppliers (see the resources).

Minerals must be mixed differently by region; for example, iodine is notoriously short in Midwestern soils ("the goiter belt") and selenium is overabundant in western soils. Feeding the wrong amounts in the wrong places causes illness, particularly for birthing and lactating livestock, whose nutritional demands are highest. Mineral mixes appropriate for each region will be sold in that region.

Salt and mineral feeders can be purchased or built. The only specifications are that they be protected from rain and sturdy enough not to be tipped over by a scratching animal. Instead of a mineral mix, some producers feed kelp and other natural minerals in a buffet-style feeder where each mineral is segregated in its own little box. This allows livestock to pick and choose, so they can balance the minerals themselves.

Grain

Grain isn't necessary for a ruminant, but it usually is fed to even out growth rates and ensure tenderness in meat animals. Grain is also a terrific training tool; if livestock are used to receiving their grain ration in the handling area each day, it's simple to get them in when you want them.

Management of Beef Cattle

Beef cattle are tough. They can handle more weather extremes than other livestock, and except for newborn calves, they aren't bothered much by predators. Beef cattle will walk farther than sheep or goats to find feed, though they are a bit fussier about what plants they will eat. On decent pasture, they will graze about eight hours a day and spend the rest of their time chewing their cud and lying around. Cattle usually will run rather than fight, but cows will defend their calves, and bulls are unpredictable.

All domestic cattle are thought to be descendants of the massive, now extinct wild cattle of Europe and Asia (also known as aurochs, or *Bos primigenius*). Like their ancestors, cattle are herd animals. They maintain a hierarchy, usually with a tough old cow in charge and the bull somewhere in the middle. They deal better with cold than with heat, though there are a few breeds, such as Brahman, that are well adapted to hot climates and others, such as the long-coated Scottish Highland, that prefer really cold areas. Because they're so self-sufficient, beef cattle take the least amount of daily chore time of any livestock.

Getting Started

Cattle take the longest of all traditional domestic meat animals to mature, at least a year for "baby beef" (pretty bland stuff) and 16 months to 2 years for a well-marbled, "finished" product, depending on the breed and how much grain is fed. Before corn finishing became the industry rule in the mid-twentieth century, it was common to let a steer grow 4 years or more before slaughter.

The beef production cycle begins with breeding a cow, then delivering the calf nine and a half months later. Traditionally breeding is done in the summer for spring calves, but some producers

The Final Rule on:
Meat Animals

For meat animals, the brood animal (the mother) does not need to be certified organic to produce certified organic offspring. If she is raised under certified organic management from her third trimester of pregnancy and her offspring are under certified organic management for their entire lives, those offspring will be certified organic. The organic status of the father is unimportant; he does not need to be certified organic for his offspring to be certified.

time the season for fall calves, so they can sell fresh beef a year and a half later in the spring, when it's rarer and therefore commands a higher price. Conventional beef operations commonly use synthetic hormones to bring cows into heat, but the Final Rule prohibits this practice.

Calves usually are weaned at four to six months, and often they are sold at that point. If you are interested in trying beef cattle, a good way to start is to pick up a couple of these "feeder calves" in the fall and raise them to slaughter weight. Many producers start weaned calves on a grain ration

Grazing cattle are a soothing sight, the epitome of a farmstead landscape.

immediately, though it's also common to "rough" them through the winter on hay alone and to delay feeding expensive grain until 90 days, or even just 60 days, before slaughter. Starting a daily grain ration early may result in more tender meat, and a steer accustomed to coming into a pen every day for grain certainly will be easier to pen and handle at shipping time, which saves a lot of stress for both cattle and people.

Pasturing Cattle

Beef cattle do not need indoor housing except in areas where icy rain is common all winter or where there is no wind protection. They can stay on pasture all year. In summer they can be grazed rotationally on good pasture. Through the winter, they can be fed hay on poor spots in the fields so that they add manure and organic matter (uneaten hay) to the soil.

Winter Feeding

For winter feeding, you can deliver hay each day to round bale or other feeders in the pasture, or you can set round bales out in rows in the fall — 15 to 20 feet (4.5–6.1 m) apart is a good spacing — and surround them with an electric fence. As the cattle finish a row of bales, you move the fence back a row, then roll the feeders over and flip them onto the bales in the new row. This is an excellent method for putting a lot of manure and organic matter on a poor spot in a field or pasture, and it saves a lot of manure hauling in the spring.

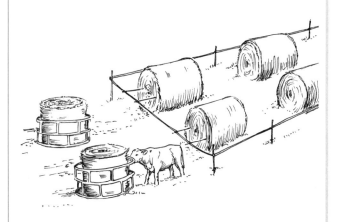

Round bales can be set out in fall for winter feeding, using a portable electric fence to ration out bales through the cold season.

For convenience, many producers prefer to feed their cattle in or near the barn during the winter. When everything thaws in the spring, the piled-up manure and uneaten hay should be cleaned up before warm weather or there will be major fly problems. The easiest way to get this job done is with a manure spreader and a skid steer or a tractor with a front-end loader.

Cattle should have fluid water available at all times, despite any stories you've heard about how they can get by on eating snow. They probably could, but they won't hold their weight as well and they will be more stressed, and therefore more susceptible to health problems. In cold areas, this means watering twice daily or plugging in a water-tank heater, so the water tank has to be near an electrical outlet, usually in the barn. If the cows are in the pasture during winter, they will have to walk to the barn each day for a drink, which provides good exercise for them and a good chance for you to check them for lameness and other problems.

If you're feeding a grain ration, you can do so when the cows come for water. If you feed at the same time every day, they will show up at the same time every day. The grain can be fed in metal bunkers, homemade feeders, or along a fence line; just make sure there is room for everyone. If one of the cows is a bully and won't let the more timid ones feed, try setting up two feeding areas. Or ship the cow.

Summer Grazing

During the summer, it's important to maintain constant weight gain on fattening cattle to ensure a tender carcass. This is best accomplished with rotational grazing, which gives the cattle more consistently palatable forage through the summer.

On some farms and ranches it may pay to run waterlines with hookups along the paddocks and move a water tank with the cattle, but on most small operations the cattle are never so far from the barn that they can't walk back to get a drink.

Fencing

Cattle can be contained with a barbwire fence; the standard is 4 feet (1.2 m) high with three or four strands of wire. A single electric wire will suffice to subdivide pastures for rotational grazing, but it should be checked frequently for shorts. Cows are quick to sense when the fence isn't working — they will put their noses quite close to check — and will walk through it if the grass looks greener on the other side.

Breeding

Because beef cattle need so little day-to-day maintenance, they are a common choice for a livestock enterprise on a small farm, especially if the owners have off-farm jobs, as is usual. Many people who start with some feeder calves decide to expand to a cow-calf operation, in which they breed and raise their own calves. This allows the certified organic producer to raise beef for sale without the yearly stress of finding certified organic beef calves to buy.

A beef cow that is well cared for commonly will produce 10 or more calves, one per year, through her lifetime. (And then she makes great hamburger!) If you have a small butchering facility nearby and you have worked with your cattle so they aren't unduly stressed by handling, you can give an animal a wonderful life in the sun and wind, followed by a very quick and painless death. That should be the goal of all livestock producers.

The one catch with the small cow-calf operation is getting the cows bred. This can be done either by artificial insemination (AI), which is allowed under the Final Rule, or by buying a bull and turning him out with the cows about nine and a half months before you want to see the first calves. Both methods have drawbacks. Artificial insemination means the owner has to observe the cows two or three times a day to see which ones are in heat, then get those cows into a headgate so an AI professional can get them bred. This involves sticking an arm deep into a cow to deposit a strawful of frozen bull semen. Often the cow doesn't settle (become pregnant) on the first attempt and you have to do it again in three weeks, when she is back in heat. Quite a few producers take AI classes and buy a liquid nitrogen tank for semen storage so they can do the job themselves, which saves a lot of waiting for the cows and the owners.

Keeping a bull is a much easier way to breed cows; he will detect cows in heat and breed them all on his own. But good bulls are expensive, and

Be Wary of Bulls

Never, ever trust a bull. If you're in a pasture with a bull, don't put yourself in a position where you can't quickly get under a fence or behind or under some machinery. A full-grown bull can weigh a ton or more, so if he charged and made contact, it would be similar to getting hit by a pickup truck.

you should buy a good one, since he represents half the genetics in your herd. Prices vary some with the year and the region, but $1,200 to $1,500 is a reasonable price to pay for a good bull as of this writing. A bull is useful for two seasons; then you have to sell either him or his daughters to avoid inbreeding.

All bulls are potentially dangerous. Many producers pen bulls by themselves when they're not needed for breeding, but this too often results in a permanently disgruntled bull. Others leave the bull with the cows year-round, which makes the cattle happy but is a little trickier for the producer when it's time to move the herd to a new paddock or transfer hay feeders to new locations. If you have to cross a pasture that contains a bull, don't walk; take a truck or ATV.

As with so many farming decisions, each farmer or rancher has to decide what will work best for his or her own operation. One way that producers can have the ease of breeding provided by a bull with less danger is to purchase a yearling bull in the spring and have him delivered in midsummer for a late spring calving the next year. At that age, bulls are still very young, and most of them haven't really figured out yet that they are, indeed, bulls. (But you still should never turn your back on the bull.) The bull stays with the herd through the next summer and breeds the cows again, with any luck in 30 to 90 days after calving. Then the bull is sold before he can breed with his daughters and hopefully before he gets old enough to have, as farmers say, "an attitude." The money from the sale of the old bull in early fall is put in the bank to buy a new bull the next spring.

Culling Is Essential

If you establish a breeding operation, each year you will have extra animals and you will have to choose which to keep and which to sell. A breeder always keeps the best: those animals that grow quickly, have good feet, legs, and body form, and stay healthy. Because cattle are so large and potentially dangerous, it's also extremely important to cull animals with poor dispositions. Any cow or calf that is overly aggressive or nervous and flighty should be sent off on a truck, preferably for slaughter, since you shouldn't get rid of your problems by selling them to someone else. Bulls and cows injure and kill farmers every year. Don't be one of the victims.

Handling

Cattle need to be handled occasionally for weaning, vaccination, castration, and shipping, and you can't just wade in and push them around as you can with sheep and goats. A good cattle handling facility is important and should include, at a minimum:

- A chute for channeling the cattle in the right direction
- A headgate at the far end of the chute for catching an animal's head so you can work on it
- A crowding pen or tub with solid, strong fencing at least 5½ feet (1.7 m) high and a swinging gate to push animals into the chute
- A loading site where a trailer can be backed in to load cattle, either through the chute or from a pen

The layout of these facilities should be designed around cattle's natural tendencies, such as their inclination to move toward light and around curves so long as they can't see a dead end. Any livestock owner, and especially any cattle owner, who is interested in minimizing animal stress and maximizing human safety should review Temple Grandin's handling rules and facility designs when setting up an operation, as discussed in her book, *Humane Livestock Handling*, and on her Web site (see resources).

Because cattle are potentially dangerous when excited, the most important step a small producer can take (after building a solid and well-laid-out handling facility) for low stress and safe handling is training the cattle. Making them walk through the handling facilities every day (except when the cows are too pregnant to fit in the chute) after they've had their water or grain makes a world of difference when you're actually working with them. They will be used to you and used to the chute and much less likely to get panicky or upset. This practice also puts you in compliance with the Final Rule's requirement to minimize pain and stress for the animals.

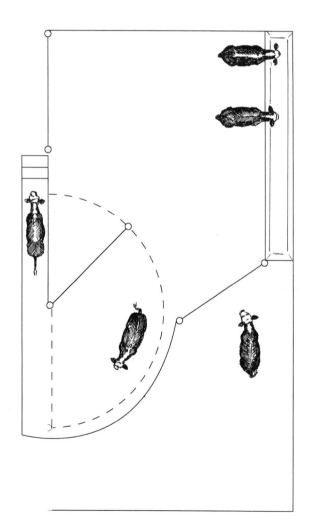

If you train your cattle to walk through your pens, chute, and headgate several times a week for a month or two before vaccination and sorting time, cattle working days will go much more smoothly and calmly.

Sheep Management

Though sheep may not be the brightest of farm animals, they have many endearing qualities: They are fairly cheap to buy. They are easy to handle because of their small size, docility, and strong instinct to stay in the herd. They will eat a wider variety of plants and browse than cattle will eat. And they will do less damage on and make better use of steep terrain than cattle. For the small producer, the biggest problem in keeping sheep is finding necessary support services nearby: a shearer, a plant that will process sheep, and other producers in the area who can supply rams and lambs.

Spring lambs traditionally are butchered in the late fall. Meat from sheep that are not butchered until they are a year old or more is called mutton. It has a much stronger flavor that is loved by some, detested by others. There usually is a ready local market for lamb, while mutton can be a harder sell.

Sheep Nature

Descended from wild sheep of Asia Minor and adjacent regions, domestic sheep (like their ancestors) are best adapted for dry upland regions, though they are nearly as widespread as cattle around the world. As prey animals, sheep have an instinct to flee quickly in response to any unusual human, animal, noise, or activity. Getting your flock accustomed to a routine, to you, and to the handling facilities makes raising sheep much less stressful for people and for sheep.

Sheep are raised for milk, meat, or wool, with each different breed having one of these primary purposes. Some breeds have fleecy coats, while a few have hair coats. Meat producers often choose a hair-coat breed to save the trouble of shearing.

Shelter

Sheep are very tolerant of cold weather, but they don't deal well with wet conditions. In dry areas, or even areas where there's snow but little cold rain, they can get by with no shelter at all if they can get out of the wind behind a bank or in some woods. A south-facing, open-sided shed works well, too. In rainy areas where sheep regularly can get soaked, inside shelter is important to preserve the quality

A llama that has been raised with sheep makes an excellent guardian against coyotes and other predators.

Sheep as Weed Control

Interestingly for organic producers, sheep are being used in some areas as an alternative to herbicides to keep power-line rights-of-way clear of brush and weeds and to suppress grass and forb competition in tree plantings. Cattle would eat or trample the young trees; sheep will graze around them. Flock owners who have a truck and trailer might be able to develop another income stream by renting out the flock's grazing services.

of the wool and to protect the flock's health. Sheep are more prone than cattle to parasites and disease; wet, shivering sheep are much more likely to get pneumonia, colds, and other problems. This is especially pertinent for the organic producer, who can't rely on the battery of dips, drenches, and shots that are standard in conventional sheep production.

Pasturing

With their little mouths and teeth, sheep can graze a pasture right down to the dirt. Unless you have far fewer sheep than the carrying capacity of your pastures, MIG or rotational grazing is the best method to maintain pasture quality and production. If you

Hair breeds of sheep are raised for meat and don't need shearing, a substantial advantage in areas where there's no good wool market, or where shearers are hard to find.

stockpile paddocks (see page 239) at the end of the growing season, rotational grazing will considerably reduce the need for hay as well. Having sheep follow cattle through paddocks allows an owner to make fuller use of pastures, since the sheep will eat what the cattle leave.

Because sheep graze plants lower than cattle do, they pick up more parasites; there are more larvae closer to the ground. In cases of heavy parasite infestation, certain precautions should be taken:

- The sheep should be kept off the worst pastures for a year.
- Wet areas should be fenced off.
- Producers should select parasite-resistant breeds or animals.
- Producers should consider running cattle or another species (not goats, because the relationship is too close) ahead of the sheep on the pasture to break up the parasite cycle.

Note that parasite loads are lower in western states, so sheep brought from out West to eastern states, where loads usually are high, will take a year or more to develop resistance.

Fencing

Sheep fencing has to be tight. Woven wire is the traditional choice, but it's expensive and hard to install. Barbwire will work if the strands are close together and taut. Many producers use three strands of electric wire as a perimeter fence, with internal subdivisions of a single strand of wire. Lambs (like calves) will trot under a single wire, but they will come back as long as their mothers stay in place. Portable, electrified woven wire is handy and relatively inexpensive compared to nonelectric permanent fencing, but it's a lot of work to move.

As with all livestock, sheep should be trained to the electric wire by having the wire in front of a sturdy nonelectric fence so that when they touch the wire with their noses, the shock will send them backward instead of through it. A fleece does not transmit electricity, so it's important that sheep make nose contact when they first meet an electric wire.

Breeding and Lambing

Most sheep producers keep a ram for breeding, though some producers practice artificial insemination. As with cattle, the ram must be kept away from ewes when they are fertile but it's too early for them to breed. Fortunately, rams are easier to contain than bulls. Put a wether (a castrated male sheep) in with the ram for company when he can't be with the ewes.

Sheep have a gestation period of about 145 days, or four and a half months, but they naturally lamb only once a year, traditionally in the spring. By age-old custom, lambing ewes are put in separate small pens in a shed or barn so that a shepherd can watch them. The small quarters also help ewes bond well with their lambs. Since sheep quite frequently have twins and triplets, birthing complications are more common than in cattle. When two or three babies are trying to be born at once and getting mixed up in the process, the shepherd or a veterinarian might have to help sort them out. If you can learn to do the sorting yourself and not call your veterinarian a lot in the middle of the night, you and the vet will be on better terms. Some producers lamb on pasture, which usually reduces disease problems, but a sudden cold rain or snowstorm can result in a lot of cold lambs in need of attention, not to mention a really cold shepherd.

Handling

Herding and handling sheep is similar to handling cattle, though sheep are more flighty and can be physically pushed around. They also prefer to be packed more tightly than cattle. Put too many cows in a crowding pen and they'll push each other around, but you can pack sheep like sardines and they'll be happy. Handling facilities for sheep are built along the same lines as for cattle, just in miniature. Again, Temple Grandin's Web site and book (see the resources) are the best starting points for researching the principles of handling and facility design. There's no better source for learning about humane, low-stress handling.

Shearing

Fleece-coated sheep traditionally are sheared once a year in the spring, when temperatures are

Scrapie

Many small processing plants (the kind that will deal with small-scale livestock owners) are reluctant to process sheep because of the fear of scrapie contamination and the difficulty of disposing of the offal (by-products) from processing. To put your processor at ease, it's worthwhile to participate in the voluntary Scrapie Flock Certification Program, which gives sheep owners USDA certification that their flock is scrapie-free.

pleasant but there is time before it gets hot for them to regrow enough wool to insulate their delicate skin from summer sun. Some producers shear before lambing, but this can be hard on heavily pregnant sheep, especially if it's still pretty cold outside. An alternative is to only clip the area under the tail before lambing, to keep things tidy.

Shearing sheep is a skill, and it requires both training and a strong back. Sheep are tipped on their butts to make them hold still and restrained with one hand, while the other hand holds the shears.

Raising Livestock for Fiber

Wool, which is what most people think of first when they think of sheep products, is the most difficult sheep product to market profitably. There are niche markets for hobby and artisan spinners and weavers. Organically produced and processed wool for chemically sensitive customers also has potential as a niche market. Keeping fleeces reasonably clean and free of burrs and debris is a must in that market; the best way to do that is to keep pastures free of burr-bearing weeds and to be sure that bedding areas are clean and dry. If you're interested in raising sheep for fiber, see the resources for information on organizations and publications devoted to this topic.

Or you can do the usual thing and lamb first, then shear.

Finding someone to shear your sheep can be difficult. Shearing is plain old hard on the back and does not pay especially well as a profession. In some regions, shearers may have to travel long distances to numerous small, scattered flocks to make a living. If you find a good shearer, treat him or her well. Or you could learn to do it yourself; many state extension services offer short courses in shearing. If you take a liking to it, you can make extra money by shearing for other flock owners.

Goat Management

Goats are as curious and smart as pigs, except that while pigs go down and under, goats climb up and over. Blame it on their mountain-climbing ancestors. They are better adapted for warm climates than for cold ones, though they are successfully kept throughout the United States. Like sheep, goats have fewer health and production problems in dry areas.

Aside from the Angora and cashmere breeds, which are raised for fiber (see the sidebar at left), goats are raised either for meat or for dairy, with different breeds better suited to each purpose.

Healthy livestock on green grass warms the heart of any farmer.

Shelter

In colder areas, goats should have protection from rain, snow, and wind. As with sheep, a shed with the south side open is adequate in most circumstances, though in the northern tier of states an enclosed, draft-free barn is necessary. It'll also save on veterinarian bills and feed costs, since the animals won't have to work as hard to stay warm. Putting a shed on skids allows a producer to move it with the flock through a pasture rotation.

Pasturing

Goats generally prefer brush and forbs to grass, and they are excellent for clearing brushy pastures or getting rid of noxious weeds that cattle won't eat. As with sheep, some goat producers make money by renting out their herds to clear weeds organically, particularly on terrain where using a sprayer is difficult.

Fencing

Goats are tough to fence. If they can't climb it, they'll squeeze under it or wiggle through it. Barbwire fences with more strands and more closely spaced posts will hold them; running a strand or two of electric wire on the inside of the barbwire is better yet. Portable electric fencing, either woven wire or several strands, is effective if the goats are trained to it.

Breeding and Kidding

As is the case for sheep, most goat producers keep a buck, who must be separated from the does when they are not ready for breeding. Goats naturally kid just once a year, in the spring. The gestation period for goats is about 150 days, and spring-born kids are ready for slaughter that fall, at 50 to 70 pounds (23–32 kg). The greatest demand for kids is at holiday times, including Easter, the Fourth of July, and Christmas, as well as ethnic holidays. The producer who has finished kids ready just before those dates has a real marketing advantage. Meat goats are in demand among many ethnic groups that hail from warmer countries where goat meat has been a mainstay for centuries: the Mediterranean, parts of Africa, and Mexico and other Latin American countries.

Handling

Handling facilities are similar to those of sheep. Some provision needs to be made to prevent the goats from trying to leap or scramble out of chutes and crowding pens, such as building walls or fences higher, making sure they don't offer hoofholds for climbing, or putting bars over the tops of chutes and pens.

"Beginning organic grass-fed livestock producers should make a careful assessment of what they are willing to do, what their resources are, and what markets are available."

RICHARD PARRY
Fox Fire Farms, Ignacio, Colorado

Profile: **Richard and Linda Parry**

Fox Fire Farms, Ignacio, Colorado

Third-generation farmer Richard Parry and his wife, Linda, produce and direct-market certified organic, 100 percent grass-fed lamb, goat, and beef on their mountain ranch in southwest Colorado. Fox Fire Farms also produces and sells organic eggs and pork from pastured poultry and pigs, and has recently introduced organic wine and apples.

Well-known in the grass-fed community, Richard is a sustainable agriculture consultant and frequent speaker at grass farming conferences, as well as a regular contributor to the *Stockman Grass Farmer*. Following is his description of Fox Fire Farms' production system.

OUR OPERATION HAD BEEN A TRADITIONAL western range sheep operation. We ran 1,500 ewes year-round, spending the summers in the mountains and then migrating to the desert to spend the winter. In 2002, and for several years afterward, we experienced a severe drought. As a result, we had to sell all but 200 of our ewes.

We consider ourselves to be health conscious and we were becoming more aware of the health benefits of grass-fed meats. In 1998 we had started a small direct-marketing business for some of our grass-fed lambs. When the drought of 2002 forced us to liquidate most of our sheep, we decided to make lemonade out of lemons, and we rebuilt our ranching operation to direct-market all our livestock. Since then, our direct-marketing meat business has doubled in volume every year.

We have considered ourselves "grass farmers" since attending Allan Savory's Holistic Management and Stan Parson's Ranching for Profit schools in the 1980s. It is not a very big step from grass farming to certified organic production. We became certified organic to add an additional layer of value for our customers, and to codify our own organic values. Only the certified organic producer has "put his money where his mouth is." Some are saying that "local" is the new organic; we believe that we need to be both local and certified organic.

We have 560 acres (227 ha) of irrigated pasture and approximately 500 acres (202 ha) of native pasture. We have divided the irrigated land into 10-acre (4 ha) semi-permanent paddocks. We have planted productive cool-season grasses such as meadow bromes, orchard grass, garrison creeping foxtail, and perennial ryegrass. Once the irrigated grasses are established we overseed Dutch white clover and red clover for the legume component. Because we are located in the mountains, the cool-season grasses with legumes are very productive and of "finishing quality" all season long.

Our dryland native pastures are much less intensively managed. They are grazed only once or twice a year. We are experimenting with ultra-high-density stock grazing and are seeing a good response from the native grasses. We do not like to put finishing animals on the dryland native pastures; we use them mostly for dry stock or our goat herd.

We have divided the irrigated paddocks into two grazing cells: one for the sheep herd and one for the cattle herd. After each species of animals has completed a rotation, the two herds swap grazing cells. Sheep now rotate to where the cattle were, and cattle are now where the sheep were. The rotation has two purposes: First, the large cow-calf herd conditions the grasses for the sheep by taking down the coarse grasses, which regrow in a leafy, vegetative state. Sheep do not like to graze on coarse grasses. Second, each species is a dead-end host for the internal parasites of the other. This is an important component of certified organic livestock production.

We have a management problem with internal parasites because of irrigation. We have an eight-point program for internal parasite control:

1. Expose parasite larvae to sunlight by grazing the pastures fairly short.
2. Do not graze the last 2 inches (5 cm). Infective parasite larvae seldom climb higher than that.
3. Give pastures 30-day or longer rest periods. The majority of infective parasite larvae perish after 21 days.
4. Rotate dead-end host species: cattle following sheep following cattle.
5. Cull carrier animals. We all have chronic carrier animals in our herds.
6. Give copper supplements. Research has shown that increased supplementation of copper plays a role in reducing internal parasite pressure. Be careful with sheep and goats, as they are susceptible to copper poisoning.
7. Encourage genetic resistance to worms. Certain breeds have resistance, and you can also select your replacement breeding stock based on individual resistance to worms.
8. If you are not certified organic, you should have a strategic program for the sheep and goats such as drenching before winter and again before the "spring rise" in parasite populations.

We graze finishing animals first through the irrigated paddocks. They "top graze" and get to select the very best diet possible. We follow the finishing animals with the large ewe-lamb herd or the large cow-calf herd. We especially use the clover-dominated paddocks for the finishing animals.

We keep the goats separate. Our goats have specific pastures that are dominated by brush, willows, and other woody species. The goats are not grazed with the sheep or after sheep, as goats are very susceptible to sheep internal parasites.

We have portable hen houses for our laying hens. Each house holds 300 laying hens, and we move the houses almost daily by ATVs. Each hen house is guarded from

"Our livestock are trained to shift paddocks at the sound of a whistle. All we do is open the gate to the next paddock, blow the whistle, and get out of the way."
Richard Parry

predators by two Great Pyrenees dogs. The hens are the last component of the grazing rotation. They serve as a sanitizing cleanup crew. . . . The hens work the livestock manure piles, control insects, and leave their own very beneficial manure on the pastures.

We have recently added pastured hogs to our livestock species mix. We chose heritage grazing breeds: Large Black and Red Wattle. The hogs follow the large cattle or sheep herds, before the laying hens.

The size of the paddocks and the size of the herds must be adjusted so the animals never stay longer than three days in a paddock. The finishing animals that are top grazing have a very short paddock graze. They take only 10 to 15 percent of the pasture before being moved on. We want the large herds of cattle and sheep to take the rest of the pasture down to a one-third residual, following the "take two-thirds and leave one-third" rule of thumb.

Our livestock are trained to shift paddocks at the sound of a whistle. All we do is open the gate to the next paddock, blow the whistle, and get out of the way.

We are blessed with several unfair advantages: We have a bountiful supply of irrigation water that comes from the nearby mountains and is renewed every winter by the snows. And our high elevation and cool growing season temperatures allow us to grow a very high-quality finishing pasture at all times, and our livestock do not suffer from summer heat and humidity.

Beginning organic grass-fed livestock producers should make a careful assessment of what they are willing to do, what their resources are, and what markets are available. Certified organic production requires a whole new set of skills, a willingness to do paperwork, and a willingness to learn new ideas. Grass-finished production is not as challenging as certified organic production; however, a high-quality grass-finished meat product is rare.

If you want to direct-market your animals you will become a salesman, a marketer, a distribution agent, a bill collector, and a customer service agent. Are you willing to let your life get that complicated, and are you willing to work the long hours it takes to get a new business off the ground?

Do your climate and land resources allow you to produce grass-finished animals, or are they better suited to producing animals for someone else to finish? Do you have the financial resources to get a new business off the ground? Direct marketing requires more time and money than being simply a livestock producer. Assess your proximity to markets. If you are in a disadvantaged location, it may be better to concentrate on grass-fed or organic livestock production instead of direct marketing.

Dairying

Raising Cows, Sheep, and Goats for Milk

> *Many farmers have found that a transition to organic dairy production has allowed them to keep the family farm profitable. Strong milk premiums and expanding markets support a larger milk check, especially when combined with the lower input costs of rotational grazing.*
>
> **JODY PADGHAM**
> *Organic Dairy Farming: A Resource for Farmers*

Dairying produces the highest-value product of any traditional farm enterprise; it also has the highest start-up and operating costs and demands the highest level of management. In other words, the gross income may look good, but the net income is what really counts, and that can be tough to control.

A small-scale, organic, grass-based cow dairy requires far less in capital investment than a conventional total-confinement dairy, but the money needed to get started is still significant. Sheep dairies and goat dairies require considerably less in start-up costs, since milking equipment and facilities are much smaller, but some cash is still necessary.

The good news is that certified organic cow's milk has received a premium price in the marketplace for several decades, and demand continues to grow. In fact, demand has often outstripped supply in recent years, so organic milk processors and marketers have begun actively recruiting dairy farms. The price paid for organic milk also has been fairly steady from month to month, while big price swings have become common in the conventional marketplace. Milk from organic sheep and goats is a different story; there are few processors or wholesale markets, so producers often process and retail the milk themselves. As a result, the price they receive usually depends on their marketing ability.

On the other hand, there's no guarantee that the increasing demand and premium price for organic cow's milk will continue, and if your farm isn't in the right place at the right time, you might have difficulty finding a processor. At least producers can sell organic cow's milk on the regular milk market while waiting for a spot on the organic milk truck, or if the organic processor closes. Producers of milk from sheep and goats usually don't have another processor to go to, if they were even able to find one in the first place.

The Dairying Life

Dairying requires great physical and mental stamina, since it's a twice-a-day, every-day-of-the-week enterprise. You can't take a day off from dairying unless you can find someone else to do the milking or you milk seasonally (as is common with sheep and goats, but not with cows), so that all the animals are dry during part of the year.

In traditional dairying regions, community activities used to be organized around the relentless dairying schedule. Meetings and events were held later in the evening or at midday to avoid milking times. This has changed over the past 20 years as the number of dairy farmers has dwindled dramatically. Now, many producers find themselves unable to attend most of their kids' sporting events or to be part of community groups, since everything seems to happen around milking times, when it's more convenient for everyone else. Don't underestimate the importance of this problem if you have children and want to see their games and school shows. Without a doubt, the relentless schedule is one of the biggest drawbacks to a dairy operation. But for some it's also the biggest attraction: a daily routine that creates a close relationship with the animals and continues a tradition that stretches back thousands of years.

If you're interested only in a single cow or a couple of goats to produce organic milk for your family, dairying isn't as hard to get started and it isn't as tough to get a half day off. You don't need automated milking equipment and a precisely balanced feed ration, since you aren't pushing for production or dealing with a state inspector. And if there's an event you'd like to attend in the evening, you can milk early or just put the calf or kid in with its mother and let it do the milking. *The Family Cow* by Dirk van Loon is a thorough and inviting introduction to this topic; though published in 1976, it hasn't lost any relevance.

Commercial Organic Dairying

Selling organic milk as part of a livelihood requires a much higher level of management than keeping a family cow, and there are some significant differences between conventional dairying and certified organic dairying. These differences include:

- Which animals are eligible to produce certified organic milk
- How animals are housed and fed
- Which disinfectants and cleaners are used to keep facilities and equipment clean and sanitary
- Where the milk is processed and sold
- How the young animals are raised
- How health problems are handled

We'll discuss these differences in the context of how dairying works, but this is not by any means a complete set of instructions on how to set up an organic dairy. You can find further resources for beginning dairy operations in this book's resources section. Or as dairyman Tony Azevedo urges in the accompanying profile, get yourself a job on an organic dairy. There's no better way to learn the business.

Processing and Marketing

In the marketplace, cow's milk is the easiest to sell, hands down. Goat's milk and sheep's milk are specialty items, but cow's milk is in every grocery store and is the standard for making cheese, yogurt, ice cream, and other dairy products. In other words, cow's milk is a commodity item, while goat's milk and sheep's milk are niche markets.

Organic cow dairies are concentrated in the Northeast, the upper Midwest (particularly Wisconsin), and the West Coast but can be found in other parts of the country, along with the necessary certified organic processing plants. Processors of goat's milk and sheep's milk are much rarer, and producers are thin on the ground. In areas where there are some producers but not enough for a processor to establish a daily or even three-times-a-week truck route to pick up milk, the solution has been for producers to freeze milk between pickups. This is feasible because the volume of milk is relatively small, but it requires freezer space and strong, sterile plastic bags.

If no processing is available, the only option is to process the milk on the farm. Processing entails an additional level of equipment and expertise, plus the job of developing a market to sell the milk. A totally on-farm operation is really three separate businesses — production, processing, and marketing — and it requires more available labor and careful planning for long-term success.

Traditional hand milking is very nearly a thing of the past, since it takes too long. But if you're keeping just one or two milking animals, it's still the most economical and efficient method.

How Milk Happens

Ubiquitous as dairy products are in our lives, their production is a mystery to many people. Everything starts with a pregnant animal. For the purposes of this discussion, that means a cow, a doe, or a ewe. (Humans also milk mares, donkeys, camels, water buffalo, and who knows what else, but the only viable markets in the United States are for milk from sheep, goats, and cows.) We'll talk about cows and calves here to simplify the grammar, but you can substitute sheep (ewes and lambs) or goats (does and kids). The basics are the same.

A cow, like all mammals, begins producing milk when she gives birth. Dairy breeds of sheep, goats, and cows have been bred for their propensity to produce far more milk than is needed to feed their offspring. For the first few days, the milk is especially rich and loaded with antibodies and other immunity-boosting factors. This first milk is called colostrum. Though humans generally don't consume it, colostrum is essential for the new baby to drink shortly after birth (preferably within 8 hours, certainly no later than 24 hours) for the extra immunity boost. Many dairies separate calves from cows immediately after birth and feed the colostrum by hand. Others leave the calf

Most conventional and organic dairy farms sell their milk to a processor rather than processing it on the farm. Bulk trucks collect the milk from farms and deliver it to the processing plant for pasteurization and bottling.

with its mother for a few days to a week, to make sure the calf gets both the colostrum and some good mothering. Because colostrum is so important for calf health, many certified organic farmers keep a supply of frozen colostrum on hand for emergencies, since the Final Rule prohibits synthetic alternatives.

When the cow stops producing colostrum and starts producing regular milk, the organic calf has to share the milk with the dairy owner.

Feeding the Calf

Though the farmer has begun milking, the calf still has to be fed. How calves are fed is a major difference between organic and conventional dairies. In a conventional dairy, it's common practice to feed calves with a purchased milk replacer, and some producers who have goats and sheep may do so as well, as a precaution against some species-specific diseases that are transmitted through milk and to have more milk to put on the milk truck. But the Final Rule prohibits the use of milk replacer. Organic young must be fed organic milk from organic cows, ewes, or goats. If they are not, those babies can't ever be certified as organic (there is one exception, which occurs when transitioning an entire herd to organic, that we'll discuss later in this chapter; see page 315).

The one glitch with this program is that some diseases can be transmitted from mother to young in the milk. For this reason, if there is a possibility that any of your cows have Johne's disease or that some of your sheep may carry the virus for caprine arthritis and encephalitis, you should pasteurize the milk before feeding it to the young stock.

Separating the Calf and Cow

The mechanics of sharing the cow's milk between humans and calves vary. Generations ago it was common to keep the cow and calf separated half the time. The cow was milked once a day, at the end of the separation period, when her udder was full, and the calf took care of the second milking by sipping away all the time it was with its mother. Very young calves appreciate an extra feeding during the 8 to 12 hours in which they are separated from their mothers, and this can be done with a bottle or a bucket. They should have water available to them at all times, though it may be quite a few days before calves begin to drink it. Calves also should have some fine hay or green grass to nibble and, if you wish, some ground grain, though it will be four to six weeks before their digestive systems gear up enough to need or get the full benefit from solid foods.

This method is still practiced in some places in the world, and on the occasional small farm in the United States. Though it decreases the amount of milk available for the farmer, this method allows a farmer to be flexible about milking times, and the calf is available to keep the milk coming when the farmer isn't there.

But on most conventional dairy farms, calves are housed and fed separately from their mothers, and farmers milk twice a day (though once-a-day, three-times-a-day, and three-times-in-two-days systems exist). Many organic producers also separate cows from calves at birth, but many don't. Other practices vary from leaving the calf with the cow for the first day or two to leaving the calf with its mother for a week to several months. Some producers use a "nurse cow," generally an easygoing, lower-producing older cow, to raise a few calves at a time while the real mothers rejoin the milking string. This can be a lovely way for the calves to get all the milk they want for as long as they want, without greatly affecting the amount of milk available to be sold.

There are pluses and minuses to each system. Complete separation reduces any chances of disease transmission (Johne's disease is the prime motivator in this case). Keeping a calf with the cow for a couple of days ensures that the calf gets both some mothering and the colostrum without the hassle of feeding the calf yourself, at least for that short period. Separating them is more stressful, though, after they've bonded. Leaving the calves with their mothers for several months is uncommon but seems to grow excellent calves. This practice significantly reduces milk production for sale, but it also eliminates the labor and cost involved in feeding calves and seems to pretty much eliminate problems with calf health and thriftiness as well. A nurse cow system takes one or more cows out of the milking string (depending on the number of calves on hand), but the benefits in saved

chore time and healthy calves may far outweigh the disadvantages.

The Final Rule is silent on the issue of when mothers and babies can be separated, so how you get the milk from the mother to the baby is up to you: bottle, bucket, barrel with nipples, or let the babies and mothers handle it themselves.

Bottle and Bucket Feeding

Calf feeding can be done with buckets or bottles. With either, calves have to be taught to drink. Bottles are by far the easiest; the first time or two you might have to back the calf into a corner and push the nipple in, but then the calf catches on. Calves take a little longer to get the hang of buckets, but buckets are much easier to wash than bottles. Bottles, on the other hand, help satisfy the calf's strong natural instinct to suck. (Appropriately sized bottles and nipples are available from farm stores and dairy equipment suppliers.)

Weaning

Conventionally raised calves generally are weaned by eight weeks of age; many organic farms also wean this early, but many don't. The Final Rule has no requirements on this issue. If the calf has been separated from the cow at birth, weaning is not traumatic. It consists of a gradual transition from milk to solid food, which is done by decreasing the milk ration and increasing the pasture, hay,

A barrel on a stand fitted with a row of nipples is a convenient and portable method of feeding young dairy calves when they're out on pasture.

or grain ration over a period of about two to four weeks. For this reason, it's common to start providing a little hay, grain, or grass in the calf pen each day by the time the calf is about a week old (the timing differs from farm to farm) so it will begin nibbling out of curiosity and thus teach itself to eat solid food. The cow will continue to give milk, with production peaking about two months into the lactation and then tapering off until she is "dried off" about 10 months after giving birth. Ideally the cow is bred within 60 to 90 days after calving, so that after she has her 6- to 10-week vacation from milking, she will deliver another bouncing baby at about the same time as she did last year, and begin the cycle again.

If the calf is allowed to nurse from the cow, weaning is done at any time after eight weeks or so of age, using the same method of gradually reducing the calf's access to the milk and increasing its solid food ration.

Differences between Cows, Sheep, and Goats

Size is the most obvious difference between cows, sheep, and goats, and there are significant differences in their milk, the markets for it, and the yearly cycles. All are important factors in deciding what type of organic enterprise is best suited to a new farmer. And because of the complexity of any type of dairying, getting some hands-on experience through an internship, short course, or other learning opportunity is even more important than it is with any other farm enterprise.

Milk

Milk from goats, sheep, and cows differs in taste and structure, and the markets for each variety also differ. Goat's milk and sheep's milk have smaller fat globules that humans digest more easily. Sheep's milk has the reputation of being the mildest-tasting, but it is so rich in fat and other

Homogenization

In cow's milk, the fat globules are big enough that most of them rise to the top. That's the cream. There's fat in goat's milk and sheep's milk too, but it's in much smaller globules and doesn't rise. So milk from sheep and goats has cream (or fat), but the cream doesn't separate on its own as it does in cow's milk. Homogenization is the process that breaks fat globules into smaller bits in cow's milk so that the cream stays mixed in and doesn't rise to the top. Sheep's milk and goat's milk are naturally homogenized, in a manner of speaking. If you buy homogenized whole milk at the grocery store, the cream never rises to the top even though it's present in the milk (that's why it's called "whole milk"). If you buy milk raw from a dairy farm and let it sit for half a day, you'll find a lovely top layer of cream that can be scooped off for coffee or strawberries or to make your own butter.

solids that it usually is not drunk. Most sheep's milk winds up in cheeses. Goat's milk is similar to cow's milk in butterfat and protein levels, but some people prefer it to cow's milk for drinking because of its different taste and because many people digest it more easily than cow's milk.

Size

On the farm, the differences between sheep, goat, and cow dairying are due to the size of the animals. A milk cow weighs about 1,000 pounds (455 kg) or more, while sheep and goats are more in the range of 100 to 300 pounds (45–136 kg). Sheep and goats are small enough to be physically moved by a human (or two), but cows can't be pushed around as easily. This is a decisive factor for many new dairy producers when picking a species.

Everything in a cow dairy, from feeding and housing to the size of the milking equipment, is on a much bigger scale than in a sheep or goat dairy. That means that up-front costs are considerably higher for everything from the animals to the milking-system components. Feed costs are higher as well. A cow eats a lot compared to a sheep or a goat, and organic feed is considerably more expensive than conventional feed. Nevertheless, while five sheep or five goats can graze a pasture that will feed one cow, a single cow will produce much more milk than those five sheep or goats.

Cow Milking Equipment

Cows traditionally are housed and milked in stanchions (open stalls), using a wheeled cart to move the milking units and teat dip from cow to cow. In most stanchion barns, the units attach to a pipeline that runs the length of the barn and are moved from coupler to coupler as the milker moves down the line of cows. This system takes a toll on human knees and hips over the years, from all the bending and crouching to attach the milking units. Most new operations (and many old ones) use milking parlors instead, either built new in a separate building or retrofitted into an old barn. In a parlor system, the milking equipment and people doing the milking are stationary. Cows are admitted to the parlor and milked as a group, then let out to be replaced by the next group. Overall, milking is quicker and less physically strenuous.

Sheep and Goat Milking Equipment

Sheep and goats are small enough that it's too hard to bend down and attach milkers, so they traditionally have been milked on a stand, which is essentially an elevated one-animal milking parlor. So it's

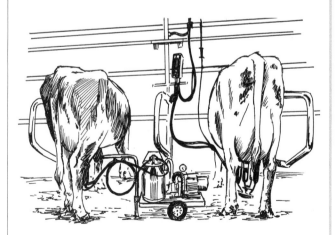

A traditional stanchion barn provides a stall for each cow. A wheeled milk cart is used to hold the milking units, udder wipes, and other equipment as the farmer moves from cow to cow during milking.

a relatively small step to a bigger parlor, where 6, 8, 12, or however many you can afford equipment and labor for are milked together. You can build a ramp to an elevated milking station or dig a pit so that you can get low enough that the udders are at about shoulder level (cow parlors are either pits or flat; pits are easier on the human body). Whether you have a stand or a parlor depends on your budget and the number of animals being milked. In either case, though the equipment may be harder to find, it should cost less than cow-size components.

Housing

A milking parlor (or milking stand for sheep and goats) eliminates the need to confine animals in individual stalls. If you have a stanchion barn in good condition, it's fine to continue to house and milk cows there, as has been done for generations and still is done on many organic dairy farms.

If the barn isn't suitable, or you have a milking parlor and don't need the old stanchions, or you have sheep or goats, you can get creative with your animal housing as long as you meet the requirements of the Final Rule. These are:

- Appropriate temperature level and ventilation
- Room for natural maintenance, comfort behaviors, and exercise (that is, the animal is able to bend and turn to scratch and lick itself, has room to stretch out when reclining or shift position when standing, and to walk and run)
- Appropriate clean, dry bedding

The goal for dairy animal housing is a well-lit, well-ventilated but not drafty, easy-to-clean, well-bedded barn or shed, with plenty of room for the animals to walk around and lie down, and with access to the outside.

A cow dairy farm usually will have at least three distinct groups of animals to house separately and to rotate separately through paddocks: the milking cows, the dry cows and heifers (teenagers), and the young calves. Sheep and goats generally are all in milk or dry together, so there's no need to run two groups of adults.

A pit parlor brings the cows to the farmer, speeds up milking, and is much easier on the knees and hips than the traditional stanchion barn.

Free-stall barns give animals freedom to move about, lie down, drink, and eat as they like.

Housing Adult Animals

Since cows are prone to bullying each other a little, it helps to have open stalls that allow a cow to lie down without too much threat of another cow higher in the herd hierarchy coming along and stepping on her udder or making her give up her warm spot. These free-stall barns, with rows of open stalls along the back and feed bunkers down the middle or front, interspersed with waterers, are now standard on big, conventional dairy farms. Like a spanking new milking parlor, a new free-stall barn usually is beyond the financial reach of beginning small-dairy farmers. But again, creativity can substitute for money. Some barns can be retrofitted for free stalls fairly easily, or sheds can be converted for the same purpose.

Though free stalls save space and some animal stress, they aren't necessary, especially for sheep, goats, or a small herd of cows. In fact, some producers successfully winter their cows or sheep entirely outside or in an open shed on a deep bedding pack, where they are dry and out of the wind. But goats need warm, draft-free, inside housing in cold weather.

Housing Young Animals

With young calves, the first decision is whether to house them in individual pens, called hutches, or in group pens. Hutches prevent disease transmission and keep the calves from sucking on each

Low-Cost Milking Parlors

Building a milking parlor is kind of like buying a car. The price range between a fancy new car and an old used one that needs some work is amazing, yet both are quite capable of getting you where you want to go.

A big, brand-new milking parlor in a new building can cost several hundred thousand dollars, a price well beyond the means of the beginning, small-scale organic dairy farmer. But you can retrofit an existing building, buy used equipment, do a lot of the work yourself, and get by in many cases for less than $10,000. Milking parlors for sheep and goats can be even cheaper.

Since building a low-cost parlor is a creative process of combining the facilities you already have with what you can find that's cheap, each one is going to be a little different. That means there is a lot of room for error. Doing your research and talking to other farmers who have built parlors can prevent many mistakes, such as having a too-shallow or too-deep pit, poor drainage of wash water, and an inconvenient gate system for moving the animals through. (See the resources section for information on plans for building low-cost milking parlors.)

other, as they're prone to do if there's no mother's udder handy. Pens, on the other hand, allow the young animals to be in their natural social group. Or you can use a combination: start the calves in hutches until they're old enough to have developed a good immune system and to have lost some of the desire to suck everything they can get their mouths on, then move them into group pens. Lambs and kids almost always are kept in group pens.

During the growing season (a term producers use to describe the period in which plants are actively growing), the young stock must have access to a pasture. In other words, if the grass is growing, the young stock should have access to it.

Housing calves together in an airy, sunny pen with an open-sided shed for weather protection is pleasant and natural for the calves, and efficient for the farmer.

Housing calves in individual hutches prevents the spread of infectious disease and ensures that each calf gets enough to eat.

Health Issues

The other major difference between cows, goats, and sheep is the type of health problems most likely to occur. The biggest issue with dairy sheep and goats is internal parasites. In an organic system, pasture management is the primary means of controlling these parasites, as discussed in chapter 9. If parasites are a problem, producers should not allow pastures to be grazed shorter than 5 inches (13 cm) or so, since the parasite larvae usually don't climb higher than that on the grass stems. This is easier said than done in many situations; the better solution is to move the animals to pasture that hasn't been grazed for at least a year, which likely will not be infested. In some cases, renting additional grazing land may be necessary, if you can find certified organic land or land that is certifiable (see page 242).

Cows are more resistant than sheep and goats to parasites. In a well-managed organic system, cows tend to have far fewer health problems than in a conventional system, in which digestive and birthing problems are commonplace.

Birthing

Because they frequently have twins or even triplets, sheep and goats often need assistance with lambing and kidding to get the babies sorted out so they are delivered one by one (instead of all trying to emerge at the same time and getting stuck as a result). Cows, which usually deliver just one calf at a time, don't have this problem, though of course human assistance may be necessary at any birth. All animals that are close to giving birth should be checked on frequently, since problems need to be recognized and dealt with quickly (see chapter 9). If you are not putting animals in stalls or pens for birthing, then saving a clean pasture for calving, lambing, and kidding will minimize parasite infection when animals are at their most vulnerable. Birthing on pasture is the most natural way to get the job done, though not the most convenient for the farmer, as it's quicker and easier to check on a penned animal than to check on one who could be anywhere in the pasture. Especially if it's dark and raining.

Scrapie and OPP

Sheep have more health problems in humid areas, but contagious ovine progressive pneumonia (OPP) and scrapie (see page 256) are major concerns in any area. Sheep should be purchased only from flocks that are certified to be free of these diseases.

Johne's and Bang's Diseases

Two diseases are of special concern in dairy cows because they are incurable and can be transmitted to humans in milk: Johne's disease and Bang's disease.

Johne's disease, also called paratuberculosis, is a contagious bacterial ailment of the intestinal tract that is most common in ruminants, particularly dairy cattle. The illness can spread from an infected animal through a herd without an owner being aware of it, because symptoms may not show up for months or years. Young animals are especially susceptible to the disease, which they can acquire from their mother's milk or from nibbling on manure-contaminated bedding or feed. Though it hasn't been confirmed, many people believe there may be a connection between Johne's disease in cows and the very similar Crohn's disease in humans.

Infection can be hard to detect. Though an infected cow continues to eat well and doesn't show a temperature, she begins to "waste away." Diarrhea and rapid weight loss are the primary signs of infection, but these symptoms also occur with other diseases. In the advanced stage, the cow becomes weak. There is no cure.

The only way to prevent Johne's disease is to keep young animals completely separated from adults and feed them only pasteurized milk. Then test the entire herd, and any purchased animals, for Johne's and cull them if they are infected.

Bang's disease, also known as brucellosis or undulant fever, causes abortion in cows and severe illness in humans. It can be transmitted to humans by contact with infected, unpasteurized milk. Prevention is by a onetime vaccination, but to be effective, the vaccine must be given while the animal is still young. In fact, this cheap, one-time vaccination can save so many problems later on that beef cattle producers should consider it as well.

Profitability

Once you have an idea what species you'd like and how you're going to house and milk them, the next question is how many animals you'll need to make a living. That's a tough question, and there are no hard- and fast- answers. Will this be a full-time living for a couple, or will one person work off the farm? How much income do you need? Will you milk year-round or seasonally? How volatile is the milk price in your area for the type of milk you plan to produce? Do you have the management skills to maintain milk quality and components so you can depend on premiums? How much will you pay for organic feed, or can you grow your own? How much money do you owe on land and equipment? What level of production can you count on, given your herd's genetics and your management abilities? All these variables affect your profit margin.

With the caveat that this wide range of possibilities makes any estimate extremely general, you can use 200 to 300 sheep or 80 to 100 cows as a starting point for making a full-time living. Since goats produce more milk than sheep do and for a longer period of the year, a livelihood may be possible with as few as 100 milking goats, if you have a strong market. Processing the milk on-farm to capture more of its retail value reduces the number of animals needed, but increases the amount of facilities and labor.

Only you can answer the question of how many animals you need to make your dairy operation profitable and it's best to do so before you start, because of the substantial investment and high risk of failure involved. Planning is no guarantee against disaster, but it sure helps. And remember, if you can't sell the milk, it doesn't matter how many animals you have; you still won't make a profit. Drawing up a detailed business plan that includes up-front costs as well as operating expenses and projected income will help you decide.

Make a Family Business Plan

Aspiring dairy farmers also should ask themselves, "What do a significant up-front financial investment, a demanding work schedule, the need for technical expertise in several areas, and still having enough energy to do the paperwork add up to?" The answer is an enterprise that needs commitment from all family members and that generates income quickly enough to cover operating expenses. Small dairies fail when a spouse isn't interested and gets resentful of all the time his or her partner spends in the barn. They also fail because they don't generate enough money for the level of work involved. The best way to minimize these sorts of problems is to hold a lengthy, thorough family discussion and to make detailed business plans that (again) involve the whole family *before* starting a farm. A business plan is an excellent idea for any farm enterprise; with a dairy operation it is nearly essential. There are a number of resources available (some are listed at the end of this book) to help you write a business plan. This should be the first step in setting up a dairy.

Sell before Producing

The other essential step before beginning to milk is selling the milk. Because milk is highly perishable, you need to know who is going to buy it before you start producing it. Milk piles up so fast that you can't store it while you look for a market, and you can't stop production just because no buyer has turned up. Unless you have an on-farm processing plant (we'll get to that in a minute), the milk has to be picked up regularly and taken to a processor to be pasteurized and put into bottles or made into dairy products such as yogurt, ice cream, and cheese.

To get the milk picked up, you have to be on a milk truck route. And for a route to exist, there must be enough producers in your area to make it economically feasible for a processing plant to function — there has to be enough milk that is not so widely scattered that it takes more fuel and time than it's worth to pick it all up. This has been a real issue with producers of sheep's milk and goat's milk, because processing plants are few and far between in the United States.

Just because your neighbor is on an organic milk route doesn't mean you'll get instant access. Milk truck and processing plant capacities have to be balanced with production and demand.

Sometimes that means a new operation has to wait before getting a spot on the truck. Once you've found a processor or buyer, stay in touch as you move toward beginning your dairy to ensure that when you have milk, there will be space on the truck.

In many cases, a milk buyer will sign a contract with a dairy at the beginning of its last year of transitioning to organic certification. The buyer may even pay a premium for the milk during that year, though this practice may depend on current demand for organic milk and competition among buyers.

Milking Equipment and Procedures

Unless you're milking just one or two animals, doing the job with hands, bucket, and stool probably is not going to be cost-efficient. Mechanical milking systems are used on cow dairies almost universally and on many commercial sheep and goat dairies. They include three basic parts:

- Electric mechanical milking units, to extract milk from the animal
- A bulk tank, to store milk
- A pipeline, to carry milk to the bulk tank

The milking unit consists of rubber-lined, stainless-steel cylinders (four to a bundle for cows; two for sheep and goats) that are open at one end for the teat and have a tube at the other end to feed the milk into a pipeline (or sometimes just a covered bucket). The rubber lining, or inflator, delivers a pulsed vacuum to suck the milk out of the teat and down the tube to the pipeline, which delivers the milk straight to the bulk tank in the milk house, or to a portable tank or bucket that is wheeled to the milk house by hand and dumped into the bulk tank. The insulated, stainless-steel bulk tank is electrically cooled and contains an agitator to make sure that an even temperature is maintained from top to bottom and that the cream in cow's milk doesn't all float to the top.

All this equipment must meet government regulations for materials, configuration, temperature control, cleanliness, and sanitization procedures or you won't be allowed to sell the milk. That's why it gets so expensive. You can build the pipeline yourself if you have the skills, but if you go this route, be sure to talk with your state inspector beforehand to review regulations and possible pitfalls. In fact, talk to your state inspector before you start work regardless of who is doing the job. This could save a lot of trouble, and you'll start off on the right foot with the inspector, who will be a regular visitor as the years go by.

Most producers buy equipment, and many pay to have someone else install it. Used equipment is much cheaper than new equipment, though it may take some searching to find all the components you need and some tinkering to make them all fit together. The Internet and farm auctions are good places to start when looking for used milking equipment.

Sanitation

The Final Rule has some specific clauses concerning dairy sanitation. Federal and state regulations call for stringent cleaning procedures for all dairy equipment, so one of the first things a new dairy operator should do is obtain a copy of the rules from the local milk inspector.

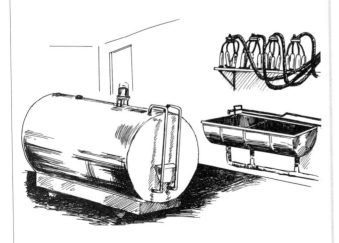

Milk houses must be kept scrupulously clean and must conform to state sanitary regulations in order for their operator to be able to legally sell milk.

All sanitizers and cleansers, including chlorine and iodine, that are used in conventional dairying also are specifically allowed in organic dairies as long as there is no residual in the milk. The exception is quaternary ammonia, which has high residual qualities. The Final Rule prohibits quaternary ammonia for use in sanitizing equipment unless you can prove that it leaves no residual in your milk. It is also allowed as a boot wash for visitors and dairy workers. The Final Rule also restricts phosphorus-based sanitizers that are used in pipelines to particular uses rather than regular use, and the first milk through the line after this type of sanitization must be discarded. Organic producers use purchased test strips to document that there are no residual chemicals in the milk.

Milk Quantity

A dairy farm's gross income depends on how much milk each animal produces each day, and for how many months. The amount of milk an animal produces depends to a large extent on her genetic heritage. Her milk production also depends on how quickly her youngster is weaned, how energy-rich her feed ration is, and how many days you continue to milk her. But it's extremely important to remember that the net income, not the gross, is what counts. The dollars of profit that a cow makes for a farmer during her lifetime are more important than the pounds of milk she produces. Feeding cows an unnaturally rich diet to boost production has a downside in increased health and breeding problems and shorter life spans. A cow on a high-forage diet may produce fewer pounds of milk, but she will have far fewer veterinary bills, and she will produce more offspring and spend many more years in your milking string. All these factors contribute significantly to the bottom line.

These variables added together create a very wide range of production averages, which are measured in pounds of milk per animal. (A gallon of milk weighs approximately 8 pounds, or 3.6 kg.) In general:

- Sheep are milked three to eight months a year and may produce anywhere from 100 to 1,100 pounds (45–500 kg) of milk in a season, depending mostly on the breed of sheep.

- Goats average 10 months of lactation and 1,500 pounds (680 kg) of milk per year, but again the range is wide, from as little as 800 pounds (360 kg) to as much as 5,000 pounds (2,268 kg) per year.

- Cows generally are milked for 10 months and produce between 17,000 and 18,000 pounds (7,711–8,165 kg) of milk, though seasonal, grass-based cow dairies are often more in the range of 12,000 to 14,000 pounds (5,443–6,350 kg), and some total-confinement operations feeding very high-energy rations will hit as much as 25,000 pounds (11,340 kg) or more with individual cows.

Animal Well-Being Trumps Production

On a conventional dairy, it's common to find all other considerations sacrificed in the never-ending pursuit of higher production from each cow. But an organic producer must take into account the whole cow and the whole farm system, being careful not to sacrifice a cow's well-being to the gods of high production. In practical terms, this means the animal selection, feed ration, and weaning program have to be geared for animal health and happiness as well as for higher production.

This approach makes sense on economic, ethical, and environmental terms. A cow (or ewe or doe) that is allowed to exercise, graze, and eat an overall diet that is reasonably adapted to her digestive system rather than to forcing additional milk production generally lasts many more years in the milking string. As a result, most organic dairy farms raise more young cows than they have room for in the milking string. The sale of these excess heifers provides substantial additional income for the farmer. Lower veterinary bills and higher forage rations also can translate into a higher net profit, even though production may be somewhat lower than it is on a conventional confinement operation.

The Milking String

The milking string is the group of animals that are currently producing milk. An animal enters the milking string immediately after the birth of her first offspring and stays there until she is no longer able to become pregnant and produce milk.

A cow produces more milk early in her lactation, peaking at about two months. Then her production gradually tapers off until she is "dried off" about 10 months after "freshening," or giving birth. To ensure next year's production, she is bred between one and three months after having her calf, so that she will freshen again at around the same time the next year, after about two months off with the dry cows. Given a balanced diet, plenty of time to graze and laze outside, and good treatment, a cow might stay in the milking string for eight or more years. (In the United States, the average cow is shipped for slaughter after two or fewer lactations because of digestive ills, feet and leg problems, or other disabling factors. Of course the average U.S. cow spends her life inside on concrete and is fed an unnaturally high-energy ration, so it's no wonder she gets sick easily.)

Goats also generally are milked for about 10 months per year, while sheep have shorter lactations, ranging from 3 to 8 months. Sheep and goats are seasonal breeders, meaning they naturally go into heat only once a year, so they all will freshen around the same time and dry up around the same time, giving the farmer a nice long vacation. But such a concentrated lambing or kidding season is intense. Producers don't get a lot of sleep during those weeks because they often get up at night to check on the flock and assist any animals that are having trouble.

Some cow dairies use this seasonal approach as well. They breed the cows so they all freshen within about eight weeks of each other and they can be dried up at the same time, with no milking for a month or two. On the other hand, some dairies with sheep or goats use artificial lights to mimic longer days and so induce a second breeding season six months after the normal one in order to have ewes or does milking year-round. The Final Rule allows this physical manipulation of breeding cycles (at least, it's not specifically prohibited), but other methods for artificially manipulating when an animal comes into heat, such as heat-synchronizing hormones, are prohibited.

Prolonging a cow's lactation or artificially inducing a second season in sheep and goats increases overall milk production and therefore gross income, but the costs of doing so may or may not make the effort worthwhile.

Milk Quality

Processors of both conventional and organic milk commonly pay bonuses, or premiums, for high-quality milk, and the standard measure of quality is the somatic cell count. These cells, also known as white cells, are the dairy animal's response to infection and disease.

A high somatic cell count is related to lower production and lower percentages of fat, lactose, and casein in the milk, as well as a shorter shelf life and more off flavors. Udder infection is the primary cause of high somatic cell counts. Environmental stresses such as heat, sore hooves, and other sources of discomfort also make an animal more susceptible to infection.

The Final Rule's requirements for animal housing and comfort are common-sense methods to reduce somatic cell counts. Lower counts help the producer get an additional premium for the milk and improve the farm's bottom line.

Processing Milk

Though it's common for dairy farmers to drink milk just as it comes from the cow, milk generally is processed in some way before it is sold to retail customers. Usually this is a job for a separate processing business that owns or hires trucks to pick up milk from dairy farms and deliver it to a processing plant. There, it is pasteurized and bottled, and it may be homogenized or have its fat, and maybe other components, separated to make butter, cheese, ice cream, or yogurt. To capture more of the milk's retail value and to deliver a unique product to consumers, some dairies process their milk on the farm. This allows them to create gourmet items such as raw-milk aged cheeses, sheep's milk yogurt and ice cream, milk soap, and other wonderful items not commonly available elsewhere. But on-farm processing also can focus on basic items such as cream-line (nonhomogenized) certified organic milk that is so fresh and tasty, potential customers who try it won't even think of getting their milk at a grocery store again.

On-Farm Processing

For some organic dairies, on-farm processing and sale of milk and milk products has been a delight and the road to economic viability. It's also been the straw that broke the camel's back for plenty of operations. The added work and costs can turn out to be more than a producer can handle and the market may not develop fast enough. Sometimes there just aren't enough hours in the day, or enough support from the family. Again, your best bet for success is careful planning that involves the entire family.

On-farm processing is as closely regulated and inspected as milk production. Processing facilities and equipment must all meet government standards. A state inspector will check regularly to make sure you are adhering to regulations regarding procedures and sanitation. Contact your area inspector to get information and advice before investing in equipment.

Making a small dairy with on-farm processing work for the long term requires reliable help, either hired or from the family. Also necessary are management talent and unrelenting attention to detail. Lastly, someone on the farm has to have marketing ability. Put these together, simmer for a few years while developing the market, and success can result. Or not. It's fairly common for a small dairy operation with on-farm processing to go great guns for a few years and then close because the owners burned out from overwork. Having help is important if you want to do this long term, and in a young business, it's hard to find the cash to pay good help. Your business plan should address this dilemma.

Even with good planning and extra help, it's highly unlikely that you will process and sell *all* the milk you produce right out of the starting gate. Most new dairy operations get on a milk route, even if the eventual plan is to process all the milk themselves. Unless you're willing to dump a good portion of your product down the drain while ramping up on-farm processing and sales, you still need to line up a milk trucker before you start milking.

As with milking facilities and equipment, you can buy a complete and pricey new processing plant or construct one yourself from used components for much less money. To be certified organic, it's also necessary to use only food additives and equipment cleaning products that are approved by the Final Rule and by your certifying agency, and not to add to your products any stabilizers, emulsifiers, or similar agents that are not approved for organic operations.

The Raw Milk Debate

Pasteurization is simply heating milk to a required temperature for a required length of time to kill disease-causing pathogens that may be present in the milk. It is illegal in many states to sell "raw" or unpasteurized milk, though there is a significant national movement to change those laws and there are methods to set up your business structure to make it legal.

Retail milk, by law, is pasteurized to kill the many disease-causing organisms that can flourish in milk. Some consumers and researchers believe that pasteurization greatly impairs the nutritional value of milk for humans. These

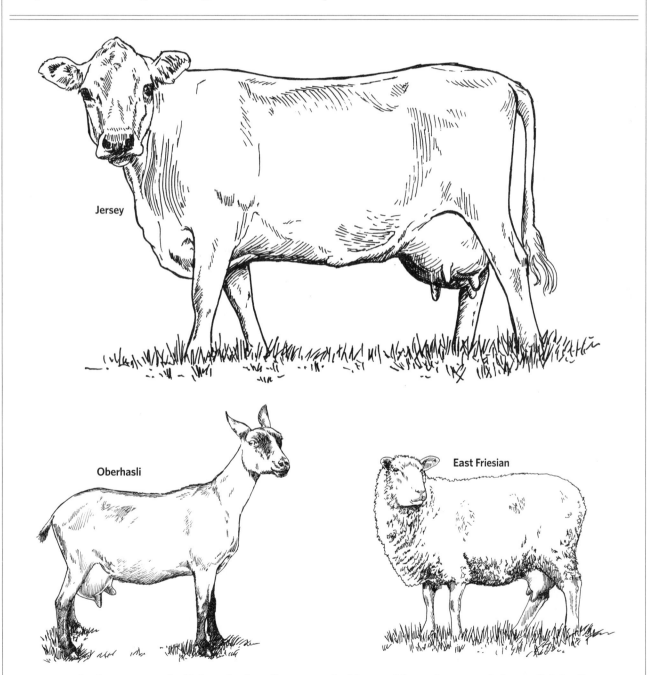

Dairy animals of any species ideally have high, well-supported udders with longish teats and when well fed will put the calories into the milk bucket instead of onto their backs as fat.

health-conscious customers also tend to want organic milk, so the growing market for raw milk is concentrated almost entirely on organic producers.

Dairy producers who wish to supply this market have come up with some innovative ways of making it possible for consumers who want raw milk to obtain it. These methods center on the fact that while it may be illegal in most states to sell raw milk to retail consumers, it is not illegal to drink raw milk from your own cow. So if a farm incorporates and sells shares in either the cows or the farm business, the shareholders legally own the cows and legally can drink their milk. To sell shares, a company or other legal entity has to be created and registered. Shareholders then are entitled to pick up milk from their cows, and the fee they pay per gallon is recorded as a bottling or boarding charge to cover costs. Some states tolerate this legal loophole; others are waging a battle to shut down raw milk sellers. If you are interested in the raw milk market, take the time to learn about your state's regulations and how vigorously they are enforced. The Weston A. Price Foundation (see the resources) serves as a clearinghouse of information and offers some assistance to farmers interested in this market.

Buying Dairy Animals

After you commit to an organic dairy enterprise and put together the land, facilities, and marketing plan, you'll start looking for your first livestock. What should you look for in an organic dairy animal? Buying with high production as the primary objective is common with conventional dairy operations, but that doesn't fit an organic, holistic approach. Good production is important, but finding animals with good feet and legs, and udders that won't break down and drag on the ground after a few years, is just as important, along with the animals' parasite resistance and ease of breeding. This approach entails looking at the animal and also the animal's mother and grandmother, or

their records. How long did they last in the milking string? Did the vet have to treat them often? Did they calve about the same time each year? Is this cow accustomed to grazing? (Animals used to being completely confined don't know how to graze.)

Still, a low-producing animal will eat almost as much as a high-producing animal, and feed is expensive. Production potential definitely should be high on the list of considerations when buying animals, but as part of the whole package and not as the only criterion.

Buy docile animals, and consider older animals rather than heifers or first-calf heifers if you're new to dairying. It's easier to learn on a cow that is used to being milked than if you and the cow are learning at the same time. She doesn't have to be show quality, but she does need to be healthy. In most years, certified organic dairies will produce more replacement animals than they need. Those farms are an excellent place to start looking.

If no organic cows can be found in your area, or you can't afford them, consider nonorganic animals. Check local farmers' networks, food directories, and the Internet for other dairies. Farm publications carry ads and auction notices; check the ads and ignore the notices. Unless you know the farm and the owner very well, buying your first dairy animals at auction is an excellent way to acquire a lot of problems in a hurry.

The ideal dairy animal will have:

- A gentle disposition
- An excellent appetite
- A high udder with longish teats
- A skinny but not emaciated frame (an animal that is turning food into fat instead of milk is not going to be cost-efficient)

Any cow prospects should have been vaccinated for brucellosis (Bang's disease) before the age of 11 months and should have been tested for Johne's disease. Sheep should come only from certified scrapie-free herds, and goats should come from flocks that have not been in contact with sheep unless those sheep were certified scrapie-free.

Breeding

Cows, ewes, and does must be bred each year to produce the calves, lambs, or kids that prompt milk production. Male animals or artificial insemination (AI) can be used; the Final Rule doesn't regulate this issue. If a male is used, he does not have to be certified organic to produce certifiably organic offspring. Only the mother must be certified, or she must be managed organically from the third trimester of pregnancy. This allows a farmer to choose from a much wider range of potential males, and often to find one close to home at a reasonable price. A single male (bull, ram, or buck) can breed from 25 to 50 females in a season, and more if he has all year.

With artificial insemination, a single bull can breed thousands of cows. Most cow dairy farms don't even keep a bull around, because the farmers prefer to use artificial insemination to breed cows. Since cows are being observed twice a day during milking anyway, it's much easier to detect when they are in heat than it is with meat animals.

Artificial insemination can be used on sheep and goats, too, but it's often easier to obtain a ram or billy goat and let him take care of things. That way, you don't have to worry about detecting when the animals are in heat, finding an AI technician

The Final Rule on:
Dairy Animals

The Final Rule has detailed requirements covering the origin of your milking herd, and they are a little different from the requirements for meat animals. Meat animals must be under certified organic management from the last third of the mother's gestation, but the dairy animal rule is not as stringent if you're just getting started. For a start-up dairy operation, a dairy animal must be under organic management for a year before her milk is sold as certified organic. This means a beginning farmer can buy a nonorganic herd and transition it to organic production along with the farm, though the cows will never be able to be sold as organic meat. The final year of the three-year transition for the farm can coincide with the one-year transition requirement for dairy animals. During that time, the herd can be fed hay and grain that was grown on the farm and harvested between 24 and 36 months after the last prohibited substance was applied. Transitional hay or grain may not be purchased for use on a dairy during the final year of transition; it must be grown on the farm.

After a farm is certified organic, any purchased replacement heifers must have been managed organically from the last third of gestation to qualify as certified organic. You may not buy nonorganic animals and transition them into your organic herd after your farm is certified organic. Your own replacement heifers, of course, will be managed organically from conception.

When an animal is going to be retired from the milking string, it can't be sold as certified organic meat unless it was under organic management from the last third of its mother's gestation. This usually isn't a big problem because in most areas, there is no shortage of people eager to buy homegrown burger regardless of who its mother was.

You also can't move a dairy animal in and out of organic certification. Aside from the exception that allows beginning certified organic producers to convert a herd to organic milk production after a year, a cow is either organic or not. There's no switching. If you treat the animal in some way that doesn't conform to the Final Rule, such as administering antibiotics or feeding nonorganic grain, the animal will be nonorganic forevermore.

Keeping a Bull

Years ago, many dairy farmers would keep a Hereford (or an Angus) bull for the heifers. Herefords have a reputation for gentleness and are easy to come by; they also have smaller calves than Holsteins do. A smaller calf means a smaller likelihood of calving problems. That was seen as a good tradeoff for the drawback of having calves that weren't of any use in the milking string, though they made decent beef animals. Using smaller, easy-calving bulls for heifers is an option for organic farmers interested in minimizing their heifers' calving problems.

Though AI is common and fairly convenient with dairy cows, many pasture-based organic producers question whether they might be better off breeding their own bulls than relying on semen from bulls that have been bred for production rather than for soundness. To breed dairy cows that have good grazing instincts and are adapted to the local climate, parasites, and forages, it's becoming more common for organic producers to work with other nearby farmers to select and raise bulls. Several farms working together and keeping good records will have enough unrelated cows to avoid inbreeding, and young bulls can be housed as a group on one of the farms. This is a natural living situation that tends to result in less temperamental bulls. (Since they compete with each other, they are less likely to see a human as a threat or competitor.) Also, because bulls are not required to be organically certified to breed with organic cows, you can work with neighbors who have nonorganic farms as well as those with organic ones on this sort of project.

who handles sheep and goats and who can get there while your animals are still fertile, or taking the class and finding semen for doing the artificial insemination yourself. Rams and billy goats are easier to have around than bulls simply because they don't present nearly the danger or handling problems that bulls do (though they still can put you in the hospital if they catch you when you aren't being careful).

Female Offspring

Heifer calves, doe kids, and ewe lambs — in other words, the girls — are raised to replace their mothers in the milking string. Heifers are bred to have their first calf at about two years old, sheep and goats at about one year.

Male Offspring

Male animals are essentially a waste product of the dairy industry; they have no place in the dairy. But male calves, lambs, and kids can be castrated and raised as meat animals. If you have the pasture and feed available, you can turn surplus males into a profitable side business in organic meat.

In fact, quite a few dairy sheep and goat producers start with meat animals and add dairy later by introducing dairy-type genetics into their flocks. That's a good way to learn how to manage the animals before taking on the steep learning curve of dairying. Dairy-type animals of any species don't produce as much meat per animal as meat breeds do, but if they are given a good ration, the meat will be just as tender and tasty.

ORGANIC FARMING AT WORK

Organic dairy farms typically have far fewer veterinary bills than conventional dairy farms.

Organic Dairy Nutrition

Because dairy cows, ewes, and does are performing at such a high level — it's a lot of work for a body to make all that milk — they are more sensitive to a poorly balanced diet or a change in ration than animals raised for meat. Human athletes adjust their diet for more energy; the same is done for dairy animals.

The organic dairy farmer needs to feed a ration that balances the desire for higher milk production with the desire for a healthy cow that has no digestive problems and will spend many years in the milking string. It really isn't that hard. As with all farm animal diets, there are three components to consider in the dairy animal diet: forages, concentrates, and supplements (salt, vitamins, and minerals).

Supplements: Salt, Vitamins, and Minerals

Salt, vitamins, and minerals can be fed separately in free-choice feeders in the barnyard or pasture, or they can be mixed into the ration. Salt is necessary in all animal diets, and the Final Rule does not regulate the source of pure salt. But check the source anyway, because if the salt has any synthetic additives to prevent caking (such as the prohibited and commonly used yellow prussiate of soda) or to make it flow better, it will not be allowed.

The Final Rule allows the use of any vitamins and trace minerals that are approved by the U.S. Food and Drug Administration, but many organic producers prefer kelp and other more natural sources to the commercial mineral mixes available at the feed mill. It makes no difference to the organic certifier which you choose, as long as your animals have salt, vitamins, and minerals that are free of prohibited synthetic additives. Your local or regional organic feed dealers or farm suppliers will have a mineral mix suitable for your area.

Natural additives, such as molasses for palatability, are allowed if they are certified organic.

Concentrates

Some producers feed only forages, but most include some grain in the ration to make it easier to balance nutritional components through the seasons and to boost milk production and make it more consistent.

Balancing the dairy ration can be so detailed and complicated that you end up hiring a consultant to do it for you, or it can be as simple as watching your animals' behavior, weight, and health and adjusting a grain ration up or down accordingly, and making sure they have good pasture and hay. You can feed straight corn if you live in the corn belt, but most often a mix of grains is fed, for better nutritional balance and to save on feed costs when possible. Deciding what to use in a ration and in what proportions requires some research on various feeds' nutritional qualities. You may want to consult with a veterinarian or feed specialist before committing to this course. Your best bet is probably to talk with some organic farmers you admire and find out what they feed their cows and how they mix and adjust the ration for the time of year and stage of lactation. The Final Rule's only requirement is that all feed be certified organic.

Forages

The Final Rule requires that forage (pasture, hay, or silage) be included in a dairy animal's ration, but it does not specify what proportion of the ration must be forage. But since cows, sheep, and goats have digestive systems that are built for processing forage, organic producers generally should aim for a dairy ration that is 60 percent or more forage as a health measure, according to veterinarian Hubert J. Karreman, author of *Treating Dairy Cows Naturally*.

The All-Forage Option

A basic decision for every pasture-based organic dairy producer is whether to feed any grain at all. As with just about every farming question, there are arguments for and against grain use. In favor of an all-forage diet is the fact that cows, sheep, and goats are ruminants, with digestive systems designed for 100 percent forage, and so a diet of only forage should be easier on their digestive systems, which in turn improves health. Second, grass-feeding proponents say that milk from their cows at the height of grazing season (May and June in most places) not only has a distinctive rich flavor that results in gourmand-quality butter and cheeses but is higher in conjugated linoleic acids that protect against cancer and heart disease. Third, an all-forage diet fits well with a seasonal dairy operation; many of these producers dry off the whole herd at the same time and take a vacation for a couple of months each winter. There's no milk check during that period, but there's no grain bill, either. Fourth, if you're not feeding grain, you're not buying or growing grain, so even though your milk production is probably lower, your feed costs are, too. Last, if you're not feeding grain, you can keep your land in permanent pastures and hayfields and dispense with a lot of machinery and labor.

On the other hand, maintaining a balanced ration on an all-forage diet can be difficult, as pasture growth explodes in the spring and then dwindles during the summer. Feeding added grain (concentrates) will bump up milk production and make it simpler to keep protein, fats, and carbohydrates balanced. Growing your own organic grain puts your hayfields back into the crop rotation, which allows you to till in soil amendments and aerate the soil during crop years. That may fit perfectly with your soil program. Feeding grain makes it easier to maintain production levels year-round, so there's less variation in the monthly milk check. (Sheep producers will have the winter off from milking whether they want to or not. Similarly, goat producers will have time off, though it will be a shorter vacation, and goat producers can choose to have part of the flock kid in the fall and milk year-round.)

Grazing animals are happiest and healthiest when doing what nature designed them to do.

Dairy Grazing Management

Regardless of whether you decide to feed grain or not, don't forget the Final Rule requirement that all livestock have access to pasture. Chapters 7 and 9 discuss hay making and pasture management, but there are some additional considerations when dealing with dairy animals, since the temptation is always to push for a little more milk production. This often is done by managing the pasture rotation so that the milking string is always on fresh, lush, relatively short (under 12 inches, or 30 cm) grasses and legumes. This type of diet produces the most milk but may be hardest on the animals, since it doesn't have enough fiber to linger long in their digestive systems.

Watch any herd that's been on spring grass for a day or two, and you'll see their manure is almost liquid, like cake batter. In recent years, some widely respected organic producers, including Gary Zimmer and Art Thicke, have shifted to grazing their dairy cows on longer, more mature grasses. Longer rotations and older pastures seem to make cows calmer and healthier and the pastures more productive. (See chapter 3 for profiles of both Art Thicke's and Gary Zimmer's operations. Though they agree on forage lengths, their methods are otherwise very different and in both cases successful.)

Silage

Animals in most of the country can't graze year-round and at some point must eat stored feed. In addition to hay, many dairy producers feed corn silage or hay silage (haylage). Silage is made by cutting the crop and gathering it while still quite wet, ideally at moisture levels between 50 and 70 percent (in contrast, hay should be below 15 percent moisture, and less than 10 percent is better). The wet plant material is blown into the traditional upright silo or into a long plastic bag on the ground. Larger producers also will use concrete bunkers or simply make big, flat piles covered with plastic that is held down with old tires. Smaller producers may make small, individual haylage bags, which look remarkably like giant marshmallows.

Whichever method is used, the goal is an anaerobic (oxygen-free) environment that causes fermentation rather than curing. This gives silage its distinctive sour odor. The cows love it.

Silage is an efficient way to store high-quality forage for the winter, but it requires a lot of equipment. Silage bags save the expense and hassle of maintaining a permanent silo, but they create large quantities of nonrecyclable plastic.

BENEFITS AND DRAWBACKS OF SILAGE

The biggest advantage of silage is you can cut it one day and put it in the silo the next. This dramatically cuts the chances of rain ruining the crop. Other advantages are that silage preserves a higher proportion of the plant's nutritive value than hay does, and if you have the right equipment and setup, it's easier to feed.

Disadvantages begin with the need to have quite a bit of equipment: a flail chopper or forage harvester for cutting, some chopper boxes (wagons) for transporting the crop from the field, a blower to put the silage in the silo or bag, and an unloader or tractor with a bucket to take it out. In addition, if you use the bag system, there's a lot of plastic to dispose of, and if the bag is punctured (deer hooves are frequent culprits), the silage spoils around the hole. Also, silage in any system can leach moisture that is very high in nitrates. If this is excessive and not contained, it can cause water pollution problems and could result in loss of certification.

Storing silage in an upright silo has its own dangers. If the top of the silage freezes or the unloader breaks down, the farmer has to climb up the silo, open the doors along the side to find the silage level, and try to fix the problem. But fermenting silage gives off nitrogen dioxide, a deadly gas that is heavier than air, so a layer of it will sit on top of the silage. You can't see it and you can't smell it, and if it doesn't drain off when you open the door, it will kill you in minutes.

INOCULANTS AND PRESERVATIVES

Inoculants commonly are added to silage to jump-start the fermentation process and prevent the wrong bacteria from getting a foothold,

Disposing of Plastic

Agricultural plastic, such as the kind used to wrap silage, is a disposal problem. Many farms burn it, but this releases a lot of pollutants into the air. Though organic regulations currently don't cover this issue, it's certainly discussed a lot and probably will be regulated at some point. In the meantime, the choice is to burn the plastic or to send it to a landfill, and the landfill is probably the lesser of two evils.

while preservatives also are commonly used to ensure that the silage keeps through the year. Preservatives are prohibited by the Final Rule. Inoculants, on the other hand, are live microorganisms that speed up the fermentation process, and they are allowed in organic production as long as they are natural. Be on the safe side and check with your certifier before using any inoculant to make sure it is approved.

Forage Quality

Whether your forage is going into silage or hay, it needs to be high quality for good milk production. Quality first of all refers to proper harvest and storage. Hay should be green, not brown or moldy, and silage should be unspoiled by mold or exposure to air. Quality also refers to the maturity of the forage that is harvested. Here again there is some disagreement over whether young, lush forage that boosts milk production is good for animals in the long run, or whether they should have some hay from more mature plants. But haylage made from overly mature plants won't pack tightly, which makes it difficult to prevent spoilage from oxygen exposure.

There's also plenty of discussion about whether it's best to feed straight alfalfa hay (again, this route maximizes milk production) or mixed hay with plenty of grasses and maybe other types of legumes and some forbs, which means less production, but more balance. All organic farmers have to decide these questions for themselves, taking into consideration what is best for their cows, their soil program, and their personal preferences.

Dairy Health Care

Health care for organic dairy animals is a constant topic of discussion among farmers. Maybe that's because dairy animals seem to have more health problems than any other kind of farm animal, or because this is where organic livestock care differs most radically from conventional practices, both in what you can do and what you can use. There's no question, though, that most longtime organic dairy farmers have developed good systems over time and have few health problems, if any. There are quite a few stories about veterinarians thinking a local dairy farm has gone out of business because the farmer never calls anymore; what really happened is that the farmer went organic and doesn't need to call anymore.

Chapter 9 noted the Final Rule's requirement that livestock owners maintain preventive health care practices that include:

- Selection of genetically suitable livestock
- An adequate and balanced ration
- Appropriate and clean housing
- Access to pasture
- Appropriate vaccinations and physical alterations

But despite all preventive and proactive measures, medical problems arise, and on an organic dairy operation, or any livestock operation, it's important to be prepared.

The first step is to explain to your veterinarian that your livestock are organic and that only certain medications are allowed. You might give the vet a copy of the Final Rule section that lists allowed and prohibited substances for use on organic animals. Talk about what alternatives might be used, including herbal and homeopathic remedies.

Maintain Comfortable Living Conditions

An animal that is in good condition and has a clean, comfortable environment will deal with health problems much more easily than one that

Natural Pasture Medicines

An excellent discussion of the types and uses of natural medicines, and of organic veterinary medicine in general, is found in *Treating Dairy Cows Naturally*. Author Hubert Karreman notes that a mixed pasture will provide a variety of beneficial forbs that cows will seek out on their own, such as dandelions, wild garlic, purple coneflower, chicory, and celandine.

is stressed by an unbalanced diet, dirty conditions, or too much confinement. When there is a problem, an observant farmer who catches it before the infection can really take hold gives the animal a much better chance of recovering quickly and completely.

Good fly control in the barn and farmyard contributes significantly to cow comfort. Follow the guidelines in chapter 9 to create a healthy environment (starting on page 252). Physical controls that will help keep flies away from cows also include:

- **Sticky tapes.** Place these around the barn to trap adult flies.
- **Vegetable oil sprays.** Little research has been done on how these work, but farmers report good control from spraying oil on barn walls. The oil probably works as a repellent and may also suffocate eggs and larvae. Be sure to use only certified organic vegetable oil or formulated sprays approved by the Final Rule.
- **Lime dust.** As with oil sprays, lime dust appears to act as a repellent, and its dessicating effects may also inhibit fly reproduction. Only natural lime that has not been hydrated, slaked, or otherwise chemically altered or added to may be used.
- **Barn fans.** In hot weather, when drafts are not a concern, fans can be used to blow flies out of the barn.

Practice Good Nursing

Good nursing goes a long way toward maintaining animal comfort. If an animal is sick, put her somewhere clean and dry, with plenty of water and hay. She should be able to see her herd mates, because

Plastic booties for visitors from other farms are an elementary precaution against the spread of disease.

a cow with no company is a stressed cow, and that stress will slow recovery. Know what natural remedies are available (see page 257) and how to administer them.

Enforce Biosecurity

An important preventive measure now common on dairy farms is biosecurity, that is, making sure that people and other animals that come onto your farm from other farms either don't have contact with your herd and facilities or are clean and disinfected. Biosecurity prevents disease transmission from farm to farm. Most veterinarians are well aware of disease transmission issues and are scrupulous about disinfecting between farms. On the other hand, the feed delivery person and the milk trucker don't need to go into the barn.

Foot traffic is a primary vehicle for disease transmission. Farms may keep a supply of plastic booties on hand for visitors to wear or have them step through a foot wash.

If you purchase an animal from another farm, verify that it has had all appropriate vaccinations, and that the other farm is free of species-specific diseases such as scrapie and Johne's. If you don't own the trailer or pickup that transports the animal, it's also a good idea to make sure that the vehicle is thoroughly cleaned beforehand.

"The main point is to make the system simple, and the simple thing to do is to capture the milk that's made from fresh grass."

DONN HEWES
Northland Sheep Dairy, Marathon, New York

Profile: Tony Azevedo

Stevinson, California

Organic dairy farmer Tony Azevedo has some pithy advice for new farmers interested in starting an organic dairy operation: "Get a job on an organic farm, even if it's for less than minimum wage. A lot of people out there are looking for somebody to pass this on to. There are opportunities out there."

For those who are looking for land, he recommends searching out "the oldest living farmer in the area. You can usually find him sitting in a coffee shop or on a tractor. These people are gold mines of information. They can tell you what that farm was good at growing, or who was successful on that farm, or who wasn't successful on that farm. They can tell you the flaws of that farm. Does it have stray voltage, does it have some kind of parasite so that after six months your calves would be dying? I'm always amazed when people buy a farm and don't talk to the neighbors."

Azevedo knows what he's talking about. He's a second-generation dairyman, his California farm has been certified organic since 1996, and he has a deep respect for the knowledge of past generations. "My father was an organic farmer without knowing it. Back then they passed the knowledge on from generation to generation, and they knew what they were doing," Azevedo says. But when he took over the farm, he began using the chemicals and animal drugs that were available. "I milked 160 cows for 18 years, and it got to where I was not happy with what I was doing." In addition, he was on the verge of losing the farm because of the rising costs of land, water, and every other input.

Fortunately for Azevedo and for his community, he was able to keep the farm afloat because he had built a museum for the antique agricultural tools and implements he collects. He began offering tours and renting the space for weddings, reunions, and parties. "If it hadn't been for the entertainment business, I would have lost the dairy," he says.

"One day I decided I would mortgage what little I have left and double the herd to 300 cows," Azevedo continues. He figured that the additional income would make the dairy farm a viable proposition. He went to visit a neighbor who was milking 300 cows, but the neighbor said he wasn't making it with 300 and planned to expand to 600. So the two went to visit another farmer who was milking 600 cows, and he told them he wasn't making it with 600, and he planned to expand to 1,000. The three farmers went to visit a new dairy facility that had just opened and had 1,000 cows. "And that farmer said he could see right now 1,000 cows wasn't going to make it," Azevedo says. "This is a true story. I came home just as discouraged as when I'd left, and I knew I wasn't going to beat the problem by expanding. So I decided to sell my dairy products at our entertainment facility and put the cows back on pasture."

At this point, fate intervened in the form of a field man for Organic Valley dairy cooperative in Wisconsin, the largest farmer-owned organic dairy cooperative in the country. Organic Valley was looking for more farmers to satisfy the accelerating national demand for organic dairy products (and it is still looking) and happened to knock on Azevedo's door. The timing was perfect. Azevedo already had decided to return to his father's way of farming, and Organic Valley offered the support and the milk price to make it work. The upshot was that Tony Azevedo's farm became the first organic dairy in California's San Joaquin Valley, and now it's a going proposition. In the past 12 years, he's recruited more than 40 other farmers to organics. Even more important for his family, the farm is profitable enough and fun enough that his son, Adam, saw a future there and returned home to take over day-to-day management.

There aren't many places in the United States where this would have been harder to accomplish. California is home to the biggest dairies and some of the highest farmland prices in the country. And water is not free. "It probably costs $70 per acre per summer to irrigate," Azevedo says. "My land is marginal, and it's worth $15,000 per acre (0.4 ha). What do you think that does for your taxes? Right now things are kind of tight, due to rising costs, which started with fuel prices."

His neighbors have responded to rising costs by expanding or exiting the dairy business. "I see these dairymen building these huge dairies, and because of the smell they move to town and their help lives at the farm. They don't get to do what they wanted to do," Azevedo says. "They're in awe that I was able to put the brakes on and enjoy what I do."

These days, the Azevedo dairy milks about 325 Holstein cows (a small operation by California standards) and has a total of 800 head of cattle, many of which are young stock. "We're very good at raising animals. California's cull rate is 32 percent; ours is about 17 percent," Azevedo says. His dairy owns and rents a total of 600 acres (240 ha), and the main crop is pasture. Corn silage and Sudan grass supplement the pasture for feed. Pastures are grazed for five or six years before the legumes start to disappear; at that point they're "freshened up" with new seeding, usually the "Azevedo mix" developed for these acres by Tony's father.

"You have to farm according to your pluses and minuses. You have to look at your soil," Azevedo says. His family's farm started with considerably more minuses than pluses. The land was poorly drained, subject to flooding, and alkaline to the point that he and his sister used to play in the salty dust as children, pretending it was snow. "My father realized he needed to control his own destiny. I remember him saying, 'No one is going to fix up a farm for me to move into,'" Azevedo says on his farm Web site. "He began to revitalize this wasteland. I remember as a young boy walking through pastures

"In the West, especially, we're so intent on expansion that we forget about quality."
Tony Azevedo

324

"When I got into organics, I had just about managed to breed all the grazing ability out of my cows, and in the last 15 years we've been trying to get it back into them."

Tony Azevedo

laden with clover and other lush grasses. His passion for grazing was phenomenal. He was renowned for his ability to know what to plant."

The first step was to drain the land, so the salt could leach away.

If you don't have good drainage you're not going to get anywhere. Our area was originally very fertile, but when they introduced irrigation around the turn of the century, with no drainage, that brought the salts up. My dad had it figured out. When he moved here in the 1940s, he started putting in large ditches, in some places as deep as 8 feet (2.5 m). We drained that water out of here. It took less than a generation to turn it around. Over the years, we've eliminated the ditches and gone with underground tile, and in some areas we were able to eliminate the drainage altogether.

Pumps bring the salty groundwater to the surface. That water is mixed with the surface irrigation water to dilute the salt, which reduces the amount of water the farm needs to buy and keeps the salt deep in the soil from rising to the surface. With the salt washing away, the next step was to develop pastures. "We have a grass out here that most people think is a nuisance, Bermuda grass," Azevedo says. "It's one of the grasses that will establish first, and it lays the groundwork for your clovers and trefoils. It builds that top 3 to 4 inches (8–10 cm) of soil, and it's a good starter grass."

Getting Holsteins to be aggressive grazers has been a challenge for many farmers, including Azevedo. "When I got into organics, I had just about managed to breed all the grazing ability out of my cows, and in the last 15 years we've been trying to get it back into them. Over the last 35 years in this country we've bred a cow to stand in one spot and eat, and now we're scrambling around trying to get that grazing ability back into them. Fortunately I had some of my dad's cows. . . . The disappointing thing is that my father died before he saw me return to his philosophy of farming."

Another innovation at the Azevedo farm is letting calves stay with their mothers for 24 to 48 hours after birth, to get colostrum and mothering. "We think that's important," he says.

Like any farm, this one has dealt with its share of problems, but the dairy uses only organic remedies, many of them rediscovered by looking back to generations past. Pinkeye has been cleared up by running boar (billy) goats with the herd, just as Azevedo's father did. He doesn't know why it works, but it does. A debilitating parasite problem was eliminated with a more scientifically understood method, by shifting the cattle to different pastures.

The farm's agri-tainment business, built on the concept of honoring past generations of farmers, also is prospering. The Double T Acres Ag Museum features antique farm equipment and implements, horse-drawn carriages, and a real centerpiece: a very rare Union Pacific 737 steam locomotive. A railroad dining car is used as a "dinner train," where groups can enjoy a meal and a documentary film. A remodeled barn at the site has room for dancing and socializing, and there's an aviary and garden as well. In addition to providing income, the museum helps people understand where their food comes from. In a June 27, 2007, interview with the California Farm Bureau, Azevedo said, "One of the philosophies I have is to get people who don't have the opportunity to go to a farm out to a farm. We have to connect these people so that they see what we're up against in order to produce food."

Organic is important, says Azevedo, "not only because it's good for you and the environment, but also because it supports an entire community of family farms."

"You could call them small farmers but I refuse to call them that. I call them hands-on farmers. In the West, especially, we're so intent on expansion that we forget about quality."

Profile: Donn Hewes and Maryrose Livingston

Northland Sheep Dairy, Marathon, New York

By any standards, Northland Sheep Dairy is tiny. Donn Hewes and Maryrose Livingston milk just 45 ewes and make and sell just 1,800 pounds (816 kg) of gourmet-quality cheese a year to a loyal customer base at the farmers' market in Ithaca, New York. It's enough. The farm income and Hewes's off-farm, part-time job as a firefighter combine to make a satisfying living for the couple.

"We feel we have a very high standard of living," says Livingston, "even though our income may not look like it. We raise almost all our own food, and Donn built our straw-bale house, and it's beautiful. We live a wonderful life. We're really lucky."

Their farm and firefighter income is supplemented with sales of sheepskins and lamb. Hewes also offers short courses in working with horses and mules. But the real secret of their success has several sources: the foundation laid by Jane and Karl North, the couple who started the dairy; the experience and attention to detail that Hewes and Livingston bring to daily farm operations; and an underlying philosophy of simplicity, working with natural systems, and living lightly on the earth.

Jane and Karl North had homesteaded in France before returning to the United States to begin a sheep dairy using the principles of holistic management. Starting from scratch, they built the farm and the market into a beautifully integrated, organic, and holistic way of life. Nearing retirement age, in 2002 they offered Hewes and Livingston the opportunity to work on the farm and gradually transition into managing and finally purchasing the operation in 2007.

"One of the things Jane and Karl did was to make everything as simple as they could, as simple as the regulations would allow," Hewes says. "That's what allows you to afford to do what you want to do. Karl built open sheds that house the sheep, and open sheds for the hay, and simple milking facilities."

Livingston and Hewes carry on that philosophy of minimizing facilities and inputs, and the necessary corollary of maximizing management and product quality. Every detail of production and processing is carefully thought out, monitored, and modified whenever necessary to meet changing conditions or improve the operation.

Hewes notes, "We are 100 percent grass-fed, and that requires a lot of attention to detail. It's more important not to let anything slip, because you don't have the fudge factor that you get with grain. The hay quality has to be high all the time; the pasture quality has to be high all the time."

Because sheep give such small amounts of milk, Hewes and Livingston select them for increased production, as the Norths did. "But with such a small number of animals, genetic selection is a very slow process," Hewes says. "You have to be realistic about the limitations of a small group."

Lambs are weaned after one month, and ewes and lambs are grazed rotationally in two separate groups. "The milking flock is moved after every milking, so during the times of year when we're milking twice a day, about four months, they're getting fresh grass twice a day," Livingston says.

"For a couple of months after that they're moving once a day, when they're being milked once a day," Hewes adds. "The lambs are moved once a day, so we're feeding them like dairy animals. We have high-tensile perimeter fences and electric netting for interior paddock divisions. We vary the size of the paddocks from week to week depending on the state of the grass. We try to keep plant height between 6 and 12 inches (15–30 cm). In May and June we're struggling with grass that's too tall, and later in the summer we're getting shorter and shorter and trying to keep from overgrazing."

Their secret grazing weapon for parasite control and forage quality, the couple says, is the seven-head horse and mule herd. "We graze them at night, and we think they're interrupting the parasite cycles. They eat the overripe grasses the sheep don't want," Livingston says.

As on any sheep operation, parasites are a primary concern. Besides using equines to short-circuit parasite reproduction, the couple also selects animals for resistance and keeps the flock well fed and healthy. "At different times since we joined the farm we've had parasite issues, and we've dealt with them in the same way," Hewes says. "Rather than looking for a single solution we've looked for a handful of solutions. Animal health is critical to giving them resistance and heartiness."

Using the horses and mules to do most of the farmwork minimizes energy inputs, and compost replaces purchased soil amendments. "The only thing we use the tractor and bucket for is to turn the compost and load the spreader. I've gradually replaced each tractor job with horses. The last couple of years we've made 100 percent of our hay with horses, and we do all of our pasture clipping, firewood hauling, and compost spreading with them. Over a period of two to three years we cover the whole farm with compost, 60 to 80 loads a year," Hewes says.

In keeping with their philosophy of keeping things simple and working with natural cycles, the couple has no intention of ever being anything but a seasonal dairy. "Sheep are very seasonal, but some sheep dairies are trying to go year-round," Livingston says. "I love milking sheep, but I love not milking sheep when it's zero degrees in my milking parlor." Hewes adds, "The main point is to make the system simple, and the simple thing to do is to capture the milk that's made from fresh grass."

"We do tend to take vacations in the winter," Livingston says. "I like having a break from making cheese. There's still plenty to do, there's still animal feeding." Asked about the cheesemaking operation, she says:

"When we think about changing the farm, we don't think about making it any bigger, but we do think about making it more diverse."
Donn Hewes

I do as much farmwork as Donn. I move animals and fences, make hay, shear sheep, and so on. It's important to me to be seen as a farmer first and foremost. I started cheesemaking in my kitchen in Olympia (in Washington state), with books, basically. I took a three-day course at Washington University, and that gave me the science background that I was lacking. We went to Europe for four months and spent two months with Mary Holbrook in England. She's a very talented cheesemaker, and her style was very appealing to me. She has a very simple approach, but she makes very refined cheeses.

Especially with cheesemaking, it's really easy to be led down the path of buying a lot of expensive equipment. You can spend tens of thousands of dollars up front and not even know if you're going to make a cheese that's worth a darn. Keep your equipment really, really simple. You can always upgrade.

Take the time to develop the ability and the desire to make a really good cheese. That's how you can make 1,800 pounds (816 kg) of cheese a year and make a living from it.

Though the dairy was one of the first to be certified organic in New York, the Norths eventually dropped certification. Hewes and Livingston adhere to the Final Rule, but they have not recertified. "It doesn't really help us sell or market cheese," Hewes says. "Years ago we would have seen certification as a political statement, but today it's not. For Maryrose and I to market our farm as being local is more important to me than being certified."

Expansion is another thing the couple isn't going to undertake. "When we think about changing the farm, we don't think about making it any bigger, but we do think about making it more diverse," Hewes says. "The scale is a very valuable lesson that we've learned. . . . Both of us are in our mid-40s. That's a double-edged sword. There are days when you wake up with a little bit of a backache, but it's helped in that we were old enough and mature enough not to rush in headlong. That's worked to our advantage," Hewes says.

Though neither of them grew up on a farm, both had acquired farm experience before arriving at Northland Sheep Dairy. For Hewes, "I was basically a gardener who was overtaken by horses and mules, and they started making my garden bigger." For Livingston, "I decided as a young kid that when I grew up I wanted to be a dairy farmer. I went to college to milk cows, and I dropped out of college to be a herdsman on a dairy."

Both believe there is room in sheep dairying for more farmers, as long as they're prepared to learn fast and work hard. On-farm processing is a good idea, too, they say. Livingston says, "I think there's growth in small sheep dairies, and that people who are going to do their own processing are finding it very successful. We just did a road trip and met a lot of sheep dairy people that are doing very well."

Hewes adds, "Start small, and have a vision."

Biodiversity

Managing Habitat for a Thriving Farm Ecosystem

> *The landscape of any farm is the owner's portrait of himself.*
> **ALDO LEOPOLD,** *For the Health of the Land: Previously Unpublished Essays and Other Writings*

Birds, bugs, brush, wildlife, wildflowers, woods, and water: that's what makes the countryside a place where we want to live. Fortunately, organic farmers are required by the Final Rule to preserve their farm's natural resources and landscape in all its natural diversity. What a pleasant job, though not simple. Ecosystems are complex, and they are almost universally degraded in farm country, so that thought and effort are required to comply with this part of the rule.

Preserving the wild corners of your land has added benefits in the form of potential additional harvests. For most people the most obvious is firewood from a woodlot, but there's a long list of others, from gathering Christmas greens to collecting native plant seed to harvesting wild berries and nuts, willow for baskets, native medicinal plants, and finally, wild meat for the larder in the form of deer, rabbits, and game birds.

Whether your plan is to harvest or simply enjoy, it's first necessary to preserve the native species that are on your farm and to work to reestablish the ones that are no longer there, if there's room in the habitat. Or, in the words of the Final Rule, the organic farmer must promote ecological balance and conserve biodiversity.

Normally at this point in the discussion there'd be a diatribe on the sad shape of both local and global ecosystems. But let's skip all the usual hand-wringing over the ongoing human-caused extinction of species and disruption of natural processes and cut right to the chase by first defining ecological balance and biodiversity and why they are important, and then talk about all the really cool things you can do on your farm to restore and maintain the incredible array of species that would like to share it with you.

Or, in other words, though there is plenty to discuss about the uncertain future of our planet, unsustainable lifestyles, misguided government environmental policies, and who are the good and bad actors in the world's unfolding ecological drama, this is the wrong place. You are already aware of these problems, or you wouldn't be reading this book. This discussion will be confined to the practical aspects of improving your farm's ecosystem, once we've pinned down why it's worthwhile for a farmer to spend time and thought on this aspect of the operation.

Why Biodiversity Matters

The term *biodiversity* can be broken down into *bio*, which is Latin for life, and *diversity*, or variety. So *biodiversity* means a variety of life, and that's what the Final Rule requires farmers to maintain. This begs the question of why biodiversity matters.

In their discussions of soil, crops, and livestock, the preceding chapters should have made crystal clear that biodiversity plays a critical role in the efficient cycling of nutrients and in making farm production resilient in the face of disease, pests, and bad weather. Now broaden the perspective to include the entire farm — all the unused edges, ground and surface waters, woodlots, and rock piles, as well as the fields, pastures, and orchards — and see it as a single ecological system, because that's what it is. The livestock and crops are systems

The Final Rule on:
Wild Crops

The Final Rule has two things to say about harvesting foods and other products from the wild, whether from your land or public land:

- Such products may be sold as certified organic only if they have been harvested from a designated area that has had no prohibited substances applied for the previous three years, and you can document this fact.

- Such products "must be harvested in such a manner that ensures that such harvesting or gathering will not be destructive to the environment and will sustain the growth and production of the wild crop." In other words, leave no trace that you were there, and always leave enough plants or animals for them to reproduce and sustain the population.

within this larger system, and they depend on the larger system just as surely as we individual humans depend on our larger society for our functioning and quality of life. Coffee, books, music, roads, medical care, stainless steel pocket knives — there's an endless list of things society provides that we can't as individuals. Being part of that larger system makes our lives richer, healthier, and more likely to last for the entire natural life span. So it

is with an ecosystem. The more species occupying an area and going about their daily business, the richer life is for all of them and the better chance they all have of surviving as individuals and as species. There are more nutrients being utilized and recycled, cleaner water sources, more pollinators and species to pollinate, more choices of food and shelter, and more potential mating partners.

There's also more hope for the ecosystem as a whole to last through time. Biodiversity equals resilience in an ecosystem. It's the elastic that holds everything together. Every species is another strand in the elastic, making it stronger and allowing it to stretch more easily to absorb stresses like oak wilt or a tornado or a wildfire, and then snap back to a functioning system.

Harvard-based entomologist Edward O. Wilson, who has won two Pulitzer prizes for his lucid science writing and has been a lifelong champion of biodiversity, says it best in his book *The Diversity of Life*: "Biological diversity is the key to the maintenance of the world as we know it. . . . But diversity, the property that makes resilience possible, is vulnerable to blows that are greater than natural perturbations. It can be eroded away fragment by fragment, and irreversibly so if the abnormal stress is unrelieved. . . . Eliminate one species, and another increases in number to take its place. Eliminate a great many species, and the local ecosystem starts to decay visibly. Productivity drops as the channels of the nutrient cycles are clogged."

Windbreaks have recreational benefits, too. Hunting pheasant on a crisp autumn day with a good dog is an energizing break from farmwork, and if you're lucky, a nice dinner as well.

A brushy edge of wildflowers, native grasses, and shrubs provides food and cover for birds, small mammals, and insects.

Hunting

Human hunting is needed in many areas to restore the ecological balance that was upset when large predators were extirpated by humans. A given area can support only so many members of a single species before food and shelter run short, and if predators aren't present to keep numbers in check, then disease and starvation set in. This support limit is called the carrying capacity of the land, and it applies to deer and mice as surely as it does to cattle and chickens. For example, in many areas of the northeastern United States where there are few wolves or cougars left to keep the deer population under control, those excess deer that aren't killed by cars (taking some people along with them) will starve to death in hard winters.

The rest of the ecosystem suffers as well. Deer are sometimes called "the rats of the forest," since they eat so many plants, and large numbers of deer can slow or stop regeneration of many plant species. Old-growth cedar swamps in particular suffer, since cedars are so slow-growing. Where the deer herd is large, the numbers of white cedar sprouts are vanishingly small, since it's a favorite deer food. As cedar swamps are lost and not regenerated by new growth, so are the many orchids and other rare species found in this special niche, and biodiversity spirals downward.

Hunters and hunting organizations have also been leaders in habitat preservation for many decades. Though they focus primarily on game habitat (wetlands for waterfowl, woods for deer, and brush for game birds), all wildlife has benefited from the activities of Pheasants Forever, Ducks Unlimited, the Ruffed Grouse Society, Whitetails Unlimited, and the innumerable local hunting clubs that work to conserve habitat and increase wildlife populations. Since many hunters are farmers as well, they also understand the issues involved in combining agricultural and wild areas.

Good hunters and trappers are excellent resources for comprehending and planning wildlife conservation and habitat management, simply because they spend so much time in obscure corners of the woods, watching what happens. If you have enough land to hunt, consider inviting one of the many ethical and careful hunters who don't have land of their own. Both of you will benefit. In fact, having hunters you trust out on your land during game seasons will go a long way toward keeping "slob" hunters from trespassing on your property, which is a perennial problem in many areas.

Or consider learning to hunt yourself. As Steven Rinella, author of *The Scavenger's Guide to Haute Cuisine*, wrote in a December 14, 2007, op-ed piece in the *New York Times*, game meat can just as well be called "free-range, grass-fed, organic, locally produced, locally harvested, low-impact, humanely slaughtered meat." If you're on a tight budget and the deer keep eating your

Modern agriculture is of course an abnormal stress on an ecosystem, and the biggest user of land in the world. In the United States, crop and grazing land occupies about two-thirds of the entire country. Every time a monoculture of corn or an overgrazed pasture replaces the original woods or prairie, biodiversity is lost. (But note that a well-managed pasture can actually enhance biodiversity!) On monoculture farmland, the variety of plant species is reduced, the soil life suffers, and as a result the ecological niches available for birds, bugs, and animals get thin on the ground. With fewer places to live and eat, fewer forms of life can make a home there. Enough local losses can add up to regional and then ultimate losses. If the farm is still productive in terms of human food, does that really matter?

It does. Besides the joy of living in a rich natural environment, there are immediate practical reasons why we need to preserve biodiversity.

newly planted trees, venison in the freezer will help in more ways than one.

Knowing how to handle a gun can be handy in other situations as well, such as when you catch a coyote sneaking up on your lambs or an obviously sick fox in the barnyard at midday. To learn, contact your local agricultural extension office to find out when the next hunters' safety course is being held. This course is required in most states to be able to purchase a hunting license of any type (unless you are age-exempt) and is an excellent way to learn the basics of handling and shooting a gun. Generally it's held one or two nights a week for six to eight weeks, and though it's geared for 12-year-olds, a significant number of adults enroll. Additional shooting practice can be had at your local shooting range; just ask your hunting safety instructor where to find it.

Hunting is as old as the human race, and a strong cultural tradition in most rural areas. Besides being an essential tool of game management, hunting still supplies meat for many rural families, and more important than anything else, inculcates a love of the outdoors and wild areas in hunters. What people know and love, they will work to preserve.

Though there are slob hunters out there, most hunters are ethical and careful. As Rinella wrote, "Hunters need to push a new public image based on deeper traditions: we are stewards of the land, hunting on ground that we know and love, collecting indigenous, environmentally sustainable food for ourselves and our families."

Wilson points out that loss of species equals the loss of "vast potential biological wealth . . . still undeveloped medicines, crops, timber, fibers, pulp, soil-restoring vegetation, petroleum substitutes, and other products and amenities will never come to light. It is fashionable in some quarters to wave aside the small and obscure, the bugs and weeds, forgetting that an obscure moth from Latin America saved Australia's pastureland from overgrowth by cactus, that the rosy periwinkle provided the cure for Hodgkin's disease and childhood lymphocytic leukemia, that the bark of the Pacific yew offers hope for victims of ovarian and breast cancer."

But the value of potential discoveries pales in comparison with what all these other organisms are already doing for us. Biodiversity creates the atmosphere we breathe, cleans the water we drink, and builds the soil that grows the food that feeds us, Wilson explains, with most of the work being done by the small, obscure organisms that humans tend to ignore. Plants are the lungs of the planet, chemically balancing the atmosphere with their respirations to exactly suit human needs. Healthy woodlands, wetlands, and grasslands filter ground and surface waters, providing billions of dollars worth of purification services. Insects, birds, and animals provide pollination and seed-dispersal services, keep many plant populations in balance with each other through their feeding and disease-carrying activities, and help to maximize nutrient cycles with their droppings and, finally, dead bodies. Most of the work is done by the smallest organisms, simply because there are so many more of them.

These "weeds and bugs," Wilson continues, "support the world with such efficiency because they are so diverse, allowing them to divide labor and swarm over every square meter of the earth's surface. They run the world precisely as we would wish it to be run, because humanity evolved within living communities and our bodily functions are finely adjusted to the idiosyncratic environment already created. Mother Earth, lately called Gaia, is no more than the commonality of organisms and the physical environment they maintain with each passing moment, an environment that will destabilize and turn lethal if the organisms are disturbed too much."

That's why biodiversity matters to all of us, and especially to farmers. Because it directly impacts the most acreage, agriculture bears much of the responsibility for the loss of habitat, which is the essence of biodiversity.

Habitat Controls Biodiversity

Biodiversity, in practical terms, equates to habitat diversity. The more diverse the habitat, the more diverse the plant and animal life it contains. And habitats don't have to be expansive to be diverse. Bears and eagles need a lot of space, but many bugs and plants need hardly any space at all. And, after all, it's the bugs and small plants that form the basis of the food chain for nearly every other life-form. So even if you don't have the acreage to provide homes for the large birds and animals that need a lot of space, you certainly have odd corners and field edges to provide homes for the "weeds and bugs." There is room for natural habitat on every farm.

Habitat is the physical and the biological features of an area. Physical features include available light, how hot and cold it gets, how much water is available throughout the year, how windy and how humid it gets, the soil type, the topography, how much cover is available and of what type, and how often fires happen. Biological features include predators, competitors, parasites, disease organisms, and the availability of food.

Every organism requires different arrangements of physical and biological features in order to exist and reproduce. For example, ground wasps need undisturbed, soft ground to establish their hives, and rabbits do best in a combination of brush for cover and grassland for food.

Different habitat features have diverse functions. A brush pile, for example, yields rotting vegetation for insects, shelter for small birds and mammals, and maybe a place for young trees to sprout where the deer can't get at them. Tall vegetation around a farm pond creates cover for small critters to drink out of sight of predators, a constantly moist area for frogs, and shady, cool water for minnows and tadpoles on hot days. Native wildflowers provide much-needed nectar sources for bees and other pollinators during those times when the apples, alfalfa, and vegetable crops are not in bloom, shelter for predatory bugs, and a food source for bug- and seed-eating birds.

Taken together, an organism's requirements and the role it serves are its ecological niche, or its unique place in the ecological system. Fewer types of habitat equate to fewer ecological niches and less diversity. Monocultures, whether of corn or pine, can be ecological deserts. On the other hand, a greater diversity of habitat draws increasing numbers of species, which in turn create more ecological niches, so that an upward spiral of diversity is activated.

The goal of the organic farmer is the setting in motion of nature's natural tendency toward diversity. The farmer does this by fostering quality habitat — diverse in plants and physical features — on

A field-scale monoculture, as is common in the production of soybeans, is more of a biological desert than most actual deserts. A single type of plant and plenty of chemical sprays ensure that this field won't be home to many other forms of life.

A diverse agricultural landscape, with fields, pastures, woods, and brush, is deeply pleasing to the eye and resilient to disease, pests, and changing market conditions.

the farm, setting aside areas for natural (undisturbed) habitat in order to encourage that upward spiral of diversity. This returns us to a basic tenet of organic farming, that of fitting the farm into the natural system, rather than seeking to conquer it. Fortunately, organic farmers have found that diversified organic farming can fit hand in glove with a healthy native ecosystem. For many, this is both the most fun work of organic farming and its very essence.

Defining Habitat Types

Ecologists categorize habitats in various types, such as grasslands, woodlands, deserts, high mountains, tundra, and rain forest. Though all are fascinating ecosystems, throughout the history of agriculture farms have been made in two of these types: grasslands and woodlands. Surface waters (ponds, lakes, streams, or wetlands) are part of both woods and prairie, so that a farmer will be concerned with at least two and often all three of the habitat types found in farm country.

Woodlands and grasslands come in a mind-boggling number of subtypes, or characteristic combinations of plants and animals, depending on the specifics of the local climate, surface and groundwater, soil type, and available species in the bioregion. In the 1959 classic *The Vegetation of Wisconsin*, author John E. Curtis describes two bioregions and 19 distinctive types of plant communities found in just this one state, each in turn with its own subtypes. Expand that sort of habitat diversity through all 50 states, and the potential variety of distinctive local ecosystems or habitats in this country is awe-inspiring.

Though most plant species are found across wide ranges, Curtis writes, any particular small location "is likely to have a very different assortment of species than another location…they may differ from one another in the kinds of species they contain, in the relative amounts of the same species, or in both ways." It follows that the species and numbers of insects and animals will vary considerably by location also, depending on the type of ecological niches available. For example, if you live near a swamp you're going to have a lot more mosquitoes than your neighbor up the hill.

Grasslands

Grasslands, or prairies, are native to regions with between 10 and 30 inches (25–76 cm) of precipitation per year and a well-defined dry season. In presettlement North America, prairie covered California's Central Valley, the Palouse region of Washington and Oregon and extending into Idaho and Canada, and the Great Plains from the Mississippi River to the Rocky Mountains, and from southern Texas well into Canada.

Prairie plant communities vary depending on the amount of precipitation. Short-grass prairie is characteristic of the far West, with the three dominant native species being buffalo grass, blue grama grass, and galleta, while mixed prairies with

Even a small landscape element like a rotting log provides important habitat for small creatures ranging from salamanders to centipedes, wood lice, fungi, and other organisms important for the decomposition and the recycling of nutrients.

Pollution's Effect on Biodiversity

After physical loss of space, the second biggest factor that decreases biodiversity is pollution. Agriculture is a primary contributor here also, through the use of pesticides and soil-destroying chemical fertilizers, manure runoff, and monocultures. By farming organically, you are already helping to reverse habitat loss and pollution by not using synthetic chemicals and by rotating crops, utilizing a variety of genetic resources for seed and livestock, and returning nutrients to the soil. Thank you!

The compass plant, native to tallgrass prairie, is adapted to the soil, weather, and insects of its home.

species like side oats grama and western wheatgrass are found in middle areas like Kansas. The famous tallgrass prairies are furthest east, characterized by plants like big bluestem and compass plants growing 4 to 6 feet (1.2–1.8 m) high. Native prairie species, like all grasses, have growing buds at or below the ground, enabling them to survive fire and grazing.

Woodlands

Woodlands are found wherever there is enough rain to support them. Trees fall into two basic categories: deciduous and coniferous. There are forests consisting of all conifers, all deciduous, and mixes of all sorts. Which type is present in your locale depends on whether the site is dry or wet, how much rain falls each year, how cold it gets in the winter, its stage of succession, and the natural and human disturbances that have occurred in the past. Each woodland type has characteristic understory plants such as bloodroot, mayapple, or Solomon's seal.

Wetlands

Wetlands are found in all bioregions and provide essential habitat for a broad range of species, from waterfowl and wading birds to beavers and muskrats, and many types of insects. Wetlands are possibly more important to more species than any other type of habitat. There are several different types:

- **Swamps** are characterized by woody vegetation such as willows, cedar, alder, and ash.
- **Marshes** lack brush and trees, having instead cattails, sedges, rushes, reeds, and other wetland grasses.
- **Bogs** are dominated by sphagnum moss and characteristically contain orchids like the lady's slipper and carnivorous plants like the pitcher plant.

Determining Your Farm's Natural Habitat

Your farm is a piece of a bioregion that was once (or still is, if you're lucky) home to a local ecosystem built of subspecies of plants, animals, and insects uniquely adapted to the farm's microclimates, soils, waters, and all the other little variables that make it different from anywhere else. It follows that if you can reinvigorate or reestablish at least some part of the original ecosystem, you'll have the constellation of life-forms that best suits your farm.

Determining which habitats your land once sustained can be very helpful in shaping your plans to develop appropriate biodiversity on your farm. A little research can reveal what your land was like before settlement, with quite a bit of detail if you're lucky. The best source for this information is the original land survey records, especially if the survey was done before the land was settled, as is true of most of the Midwest and West. (Unless, as was occasionally the case, the surveyor just made up a description rather than go to all the work of brush-busting to mark the survey's section corners.) These surveys noted individual marker trees and general vegetation type, wetlands, and other major features of the landscape. Unfortunately, they're generally handwritten and can be hard to decipher and interpret. You may be able to find a copy at the county courthouse.

Fortunately, others have been there before you in most areas and compiled the survey information, along with early pioneer accounts, into more accessible resources. Because you are looking for very local information, the trick is to find the local resources. Naturalists at state and county parks are usually excellent resources, as are local foresters, biology and ecology instructors, many

Habitats and Indicator Species

Ecology is essentially the study of adaptations. Adaptations are the characteristics and behaviors of an organism that allow it to be successful under the physical and biological conditions in its environment. Adaptations that work under one set of conditions often won't work under another set, which is why most species of plants and animals are found in only some habitats and not others. Or, in other words, different types of plants, bugs, birds, and animals need different types of habitat.

Groups of plants and animals that are adapted to similar physical conditions are found together in communities, where they have adapted to living with each other. Any particular farm's location and topography will be suited for some types of plant and animal communities and not others.

Being able to identify a few basic plant species allows you to identify the type of habitat, or ecosystem, you're looking at. For example, plants with deep roots are adapted to dry areas, while those with shallow, spreading roots grow best in wet areas. Most pines (not all!), sunflowers, and vetches have taproots and grow best in dry areas, while sedges, willows, and white cedar have shallow roots and prefer wet soils. But red pine and willows won't tolerate growing in shade, while white cedar almost demands shade and white pine tolerates it pretty well. In fact, white pine will wait in the shade for many years until a bigger tree falls and gives it enough light to grow again. Most grasses need a lot of sun and aren't common in mature woods, except for the nonnative, more invasive types, while many forbs are adapted to the low-light conditions of forest floors and won't tolerate full sun.

Being able to identify a few basic animals, birds, and bugs can help you identify ecosystem types in your region, though not necessarily on your farm, given how far these creatures travel. For example, animals that feed on acorns and pinecones, like turkeys and red squirrels, are common in dry woodlands of oak and pine. Mosquitoes are thickest in wetland areas because they need the stagnant water to lay eggs in, while many types of biting flies prefer sunny, drier open areas or grasslands. Redwing blackbirds nest in open marshes and feast on the abundant insects, while grouse will eat the buds of wetland trees like alders in winter but prefer aspen buds and nesting in drier uplands in spring.

Every organism has a preferred habitat that is made up of the light, moisture, and temperature conditions it's adapted to and that offers enough food and shelter without too much competition from other organisms so that it can successfully reproduce. If you know what plant and animal communities are suited to your land, and whether there are any ecological niches in the farm environment that are not being filled, then you have the basis of your biodiversity plan.

state department of natural resources specialists, and local and regional organizations devoted to conservation and preservation. The local historical society may have descriptions of the landscape made by the first explorers or pioneers in the area, while libraries and bookstores are good resources for books on an area's native flora and fauna. There's also information on the Internet, where many states have posted maps of original vegetation patterns.

Knowing your farm's natural vegetation pattern — which native plants grew on your farm, whether there were wetlands that are now drained, whether your woods have changed dramatically in species composition, and whether your woods were once grasslands or your grassland once woods — gives you a solid foundation for planning how to restore, maintain, or enhance an appropriate ecosystem on your farm. Re-creating an exact replica of the original may not be possible or appropriate, but you will at least know what has worked in the past.

"People who live in rural areas or urban residents who drive through them may not know that they are seeing a biologically impoverished landscape, because they have no knowledge of its diversity before modern agriculture," writes Dana Jackson in *The Farm as Natural Habitat*.

Finding out what was there before the farm lets us envision the sort of biological richness that is still possible on an organic farm.

Taking Stock of Existing Habitat

Once you know what your farm once had, poke around and see if any of it is still there. Maybe some of the original prairie and wetland species are still around in odd corners, and an organic farmer can stabilize them by giving them a little more room or by removing invasive species that threaten to overwhelm the original inhabitants. Woodlots now are shadows of the original forests that once covered most of eastern North America and parts of the mountain and coastal West, but the small second-growth trees are often the same species as some of the originals. Sometimes just leaving some land alone for a while will give long-dormant native seeds the opportunity to sprout.

Making a Biodiversity Plan

Making plans to enhance habitat on your farm ranks right up there with looking through garden catalogs as one of the great winter pleasures of the organic farmer. Building habitat is usually the sort of fun job the whole family can do, with no pressure to get it done by a deadline, and the results are wonderfully gratifying.

Providing Year-Round Diversity

When thinking about creating habitat for bugs, birds, and animals, it's important to think year-round. Much food and shelter that is available to wildlife in the summer disappears in the winter. If there aren't alternatives available when crops aren't blooming or in cold weather, even overwintering species will disappear from the area. Providing good year-round cover and food sources for bugs,

birds, and animals is an important consideration in habitat management. Select plantings with an eye to when they produce flowers and seeds and whether they will provide cover through winter.

For example, grasslands that provide cover for many small mammals and game birds (rabbits, quail) during the growing season can be flattened by winter winds and snow. Bushes, brush piles, tangles of wild vines, or young evergreens with ground-protecting low branches are needed if those species are to survive in that area during the winter. Since there are no bugs or greens to eat, weed seeds and spilled grain, along with berries that stay on the plant all winter are needed. Plants that hold on to their seeds and stay upright (a distinction including many weed species, in fact) can be critical to overwintering birds and small mammals.

Evaluating Succession

Habitats change over time through the process of succession, or change in the composition of a habitat as its member organisms, especially plants, respond to and modify the habitat. For example, as a woodland matures and gets shadier, ground-level vegetation diminishes, since fewer small plants are adapted to shade than to sun. This in turn reduces food for animals that depend on those plants, such as ground-feeding birds, deer, and squirrels, and many of those animals will leave for greener pastures. But a mature woodland also increases habitat for increasingly rare birds such as vireos and warblers.

Winter is the hardest season for wildlife, when food and cover are in shortest supply. Brush piles, rock piles, conifers, and plants that hold seed or berries through the winter are all important for wildlife survival.

Each stage of succession in a woodland is characterized by a different group of plants and animals. In woods, succession begins with fast-growing, sun-loving trees and understory plants. These are eventually crowded and shaded out by slower-growing species better adapted to more competitive conditions.

In native grasslands, succession begins with annual weeds, which are gradually replaced by slower-growing but more tenacious species. The diversity of plant life continues to increase the longer the area is left undisturbed, but only up to a certain point.

When left alone, both the plants and the animals in an ecosystem change over time until the ecosystem reaches what's called the climax phase, often after hundreds of years. At that point change occurs only by disturbance: a windstorm, fire, disease outbreak, animal grazing, or human disruption. Climax habitats are not usually the most diverse of ecosystems. Without some sort of disturbance to clear out old vegetation and make room for new sprouts, diversity decreases.

An important question to answer is whether you will manage your farm's habitat to keep it in its current phase or manage for eventual climax conditions. In a natural state, fire, windstorm, disease, and pests create openings in mature systems; the pioneer species move in and restart succession. This process results in a patchwork of habitats representing various stages of maturity. To set back succession without the inconvenience of a wildfire or windstorm requires human management, such as a timber harvest. For example, aspens will sprout only in full sun and are characteristic of a young stage of succession. An aspen forest is also one of the richest animal habitats in northern states. To maintain aspen it's necessary to clear-cut it every several decades, to keep maple and other shade-tolerant trees from moving in, and the grouse and other critters that rely on aspen for food and cover from moving out.

On the other hand, a thoughtful timber harvest can accelerate succession as well, by removing crowded and unhealthy trees. These trees would eventually die out on their own as the bigger, healthier trees crowded them out. Harvesting them nets the woodland owner some income and relieves the crowding ahead of schedule.

Identifying Gaps in Ecosystems

In addition to winter food and cover and succession, a third factor in managing habitat is identifying what characteristic species are missing from your particular habitat, in order to get some clues about what you might change to increase biodiversity.

For example, if your farm features grasslands, they may be home to a diversity of plants and animals but lack ground-nesting birds like bobolinks and meadowlarks. These birds won't nest unless a grassland is large enough, probably 30 acres (12 ha) or better, because that is their niche. If you live in a region where these birds are native but aren't seeing any on your farm, maybe you could eliminate some tree lines between fields to return the area closer to what it was originally, and in that way attract endangered grassland species.

What if you see the adults, but never any babies? You may have a "population sink," where a species is unable to reproduce successfully because of some flaw in conditions. If a species can't reproduce, it is doomed. For example, if your hayfield is big enough to attract nesting pairs of bobolinks but they aren't able to raise young, it may be because you cut hay while the babies are still in the nest, and so kill them. If they're nesting in a pasture but aren't raising young, maybe grazing pressure is too high for livestock to avoid the nests. The answer might be waiting to cut hay until later in the season, or setting aside refuge areas in the pasture (maybe a different paddock each year) for young birds to escape from livestock, or grazing your paddocks less intensively. Solutions like

When you see the young bobolinks are out of the nest and perching, it's time to cut hay.

Identification

You can't assess what you have if you don't know what it is. Being able to identify what birds, bugs, and plants are, and are not, present on your farm is an important step toward developing your biodiversity plan. Of course, only an expert can identify everything, but farmers can easily learn the major species. All it takes is a little time, observation, and some good ID books.

Regional identification guides are less confusing than guides that cover the entire country simply because there are fewer species to worry about. Guides that use illustrations rather than color photos can be easier to use since an artist's rendering can point out more distinctive features of a species than a photo; on the other hand, photos are better at showing how a species looks when you're looking at the actual plant.

The easiest organisms to learn for most people are birds and trees, because they're easy to spot and relatively easy to tell apart. Bird identification in particular is a terrific way to monitor your habitat type and health, since so many birds are quite specific in their habitat requirements. Binoculars, bird feeders, and a bird book are the basic tools. (The binoculars are for those species that don't patronize bird feeders.) In general, species are either swimmers like ducks and geese, waders like cranes and sandpipers, game birds like quail and pheasant, predators like owls, hawks, and eagles, grassland birds like bobolinks and horned larks, woodland birds like woodpeckers and many warblers, or edge birds that like a combination of woods and open areas.

Foresters will tell you that they can tell the soil type and climate by the trees, and they can also tell you whether the area has been grazed or burned, whether there are soil deficiencies, and whether the last logger committed the sin of taking only the best trees and leaving the worst to reproduce. If you live near trees, learning to identify the most characteristic species in your region will also start developing your eye for type and quality of woodland habitat.

Critters, on the other hand, aren't so easy to spot as plants and birds. Learning to read tracks, scat, and other signs allows you to identify many of the shier inhabitants of your farm. For example, small, shallow digging holes are characteristic of skunks, piles of disassembled pine cones indicate red squirrels, and snow slides mean otters are enjoying your stream or pond. Identification guides for tracks and signs are available, and if you get a chance spend an afternoon in the woods, prairie, or wetlands with an experienced hunter or trapper.

Getting acquainted with local nature centers and natural areas and their staffs is also very helpful, and a welcome break from farm chores. You may be able to attend nature walks to learn about species in your area, or even bring in plants and bugs for help in identification. Somewhere in your area there are experts on the local flora and fauna who can help you out.

these often have additional good effects, such as more organic matter in the soil from the better-developed root systems of grasses that grew taller or for a longer period than usual.

The more you know about the characteristics and needs of individual plants, birds, bugs, and animals, the better you'll be able to assess habitat type and quality and manage it for maximum biodiversity.

Making a Plan

After determining your farm's natural history and assessing its current state, you're ready to plan for what habitat could be. First, how much of your land will be left for natural habitat? On some farms, there's only room for edges around fields and some native plants in the garden, while others have more wild acres than cropped acres. More land means room for bigger species, but plenty can be done with small areas as well. Whichever it is for you, there are ways to make the most of what you have.

The next step is deciding how to proceed. You may want to do nothing at all, simply monitoring your farm's wild areas. Or you may decide to enhance or restore certain habitats in order to maximize diversity on your farm. We'll talk about these three approaches next.

Monitoring Farm Habitat

On most farms that already have woods or brushy corners a common approach is to do nothing at all — simply let the wild areas fend for themselves. Nature clearly is capable of taking care of things without human interference, so why not leave well enough alone? This is the ideal, since human interference is what has caused most habitat deterioration in the first place. Unfortunately, it's not really possible, since human activity is constantly nibbling around the edges of all wild habitat, and there are almost certainly some species already present that were introduced by humans. Even if your wild areas appear to be in good shape, it's still a good idea to monitor things.

This can be as simple and pleasant as a regular Sunday walk around the farm. There are two things to watch for: that all the plants and animals are reproducing successfully, and that no invasive species are gaining a foothold.

Reproduction is the measure of success, and also of sustainability. All species of plants and animals produce more offspring than necessary to maintain a population, with the excess going to feed other species. But enough young need to survive to adulthood to keep the species going, and if that's not happening, the problem should preferably be caught and remedied early enough to reverse the decline. Remedies are specific to a species, so research and consulting with local experts are the first step when you notice something wrong. For example, if baby turtles are all getting eaten before they can make it from egg to swamp,

you might consider either putting chicken wire over the sand where the eggs are buried, or moving the eggs to a sand bed in the yard where you can fence them in from predators. But if you don't even know you have turtles nesting on the property, you'll never know if enough babies are surviving. Monitoring is the only way to find out.

Invasive species are an enormous problem in all bioregions in the United States, from the northeastern forests being choked by buckthorn to the kudzu, "the vine that ate the South," that grows indiscriminately over vast swaths of the Southeast, to the cheatgrass and knapweed that overrun prairies, to the reed canary grass that has taken over swamps in so many areas. Each region has its own group of problem species, and none of them are easy to control. Most of them require digging by hand or poisoning individual plants one by one — all backbreaking work. In

Invasive Species

Invasive species are imports from other countries or regions that have no natural enemies in their adopted ecosystem and that aggressively crowd out native species.

Most state departments of natural resources have invasive species information available on their Web sites and at their offices. On a national scale, the U.S. Geological Survey's Biological Informatics Office maintains a Web site (see the resources) on invasive species, with maps, identification information, and control options. Either of these is a good place to start learning about species of concern in your area.

Invasive species are not limited to plants. They include earthworms in northern forests, gypsy moth in hardwood forests, zebra mussels, Africanized honeybees, fire ants, and more. In most cases there has been no effective means of stopping their spread. One exception has been purple loosestrife, which a few years ago was threatening to take over wetlands across the Northeast but has been stopped by an army of volunteers breeding and releasing *Galerucella* beetles, which feed extensively on loosestrife and defoliate the plants.

Invasive species like buckthorn will crowd out native species, reducing biodiversity to a near monoculture. The native plants can't find room to sprout their next generation, and many insects, birds, and animals are forced to move elsewhere to find food and cover.

many cases it's too late to do anything; no one has yet figured out an effective way to eradicate reed canary grass and return the native ecosystem to swamps. If you are monitoring your farm's wild areas, you should be able find and kill invasives before they get out of hand. Unless it's already too late.

Enhancing Habitat

To improve what wild corners you might already have, you can plant or you can build. Your goal is to add missing pieces, such as nesting areas, winter food sources, or water.

A good place to start is by enhancing habitat for pollinating insects, such as bumblebees, since enhancement is fairly simple and so badly needed. Many species of pollinating insects are in steep decline across the country, victims of the loss of wildflowers, the disappearance of brushy areas for nests and cover, and the use of broad-

spectrum pesticides. The government has even taken notice of the decline in pollinator species, and of their importance to agriculture, by calling for restoration of pollinator habitat to be a priority in the most recent Farm Bill. This means there may at some point be funding for pollinator habitat projects.

Building Habitat Features

Brush and rock piles require nothing but your labor and free raw materials, while nest boxes, raptor perches, and similar structures can be put together from scrap materials in most cases. Creativity substitutes well for expensive materials in this area, since critters aren't fussy about the decor as long as the dimensions and location are right.

Building for Birds

Building for birds includes birdhouses, bird feeders, and nesting platforms or boxes. There are seemingly plans for nearly every bird species out there. You can also add bat houses and bee houses. Plans are available from many conservation organizations, as well as many state departments of natural resources. Build the feeders and houses to suit the species you would like to encourage on your farm.

Following plans precisely, especially the instructions on entrance hole size and where houses should be placed, is important if you don't want to get a lot of birds killed. English sparrows, for example, are notorious for invading bluebird nests and pecking open the skulls of the babies. Keeping the nests away from buildings and the entrance hole smaller than 1½ inches (3.8 cm) in most cases will exclude these birds. Other major predators of nesting songbirds are cats and raccoons, so birdhouses should have some type of excluder, or you should wrap slick metal around the post or trunk of the tree where the birdhouse is mounted, so predators can't climb up.

Disease and parasites also kill a fair share of bird babies, and many of these will overwinter in old nests. Clean out birdhouses well ahead of nesting season each year!

Putting up perches for predator birds like owls and hawks near gardens and orchards encourages them to do their rodent hunting on your property.

Perches need to be about 12 feet high. Cutting a tall tree sapling and tying it to a fence post will do the job.

Constructing Natural Cover

The other type of building you can do for wildlife is messy and inexact, and therefore an excellent project for kids on a glorious afternoon when you'd like to do something fun outdoors that doesn't involve farmwork. Brush piles, tangles, and rock piles all provide excellent cover for small birds, game birds, and small mammals, as well as many types of insects. The trick is to place them near food and other cover if you want them to be used. A brush pile in the middle of a field or pasture won't be much patronized by wildlife and is a pain to mow around as well. Brush piles along fencerows, in or next to woods and water, or in field corners, on the other hand, will be well used by all manner of creatures.

To attract the maximum number of wild users to a brush pile, crisscross the bigger branches or logs at the bottom of the pile so there are plenty of nooks and crannies, then thatch it with the smaller stuff. The bigger the pile, the more likely it will be a home to small mammals, but really any size and any type of construction will be used by something.

Tangles of vines, shrubs, and wood debris will attract wildlife as well. These most often happen naturally, but you can encourage wild grapes or other vines to cover the top of a downed tree or shrubs to create a protected space.

Rock piles will host many ground-dwelling critters. If your property tends to be rocky, you can throw a pile together using all the stones you constantly pull from the garden and fields. Again, using the bigger stones at the bottom of the pile will allow for plenty of nooks and crannies. But note that stone piles are also considered good homes by a variety of wasps and bees, so before you start moving rocks, watch to see if there's flying-insect activity.

Planting Habitat Features

Every farm has a field, pasture, or garden, and every one of these has edges that can be planted to native flowers and shrubs. It may be necessary

Encouraging Swallows

Swallows are the one exception to the rule about cleaning out birdhouses after the birds have moved out. In a 1930s essay for *Wisconsin Agriculturist*, famed naturalist Aldo Leopold wrote that the mud swallow nests on the barn beams must be left in place to shelter early-returning "scout" birds from inclement spring weather, but then the farmer must knock down all those nests by early May, before the main flock returns and starts laying eggs, to control nest-dwelling parasites. He also recommends leaving a trickle of water, either overflow from the stock tank or in the garden, to provide mud for their nests. The swallows pay for all this care with bug-control services and lovely aerial acrobatics. It's an excellent example of how just a few simple measures can help survival of a species on your farm.

Swallows may make a small mess under their nests, but they make up for it by eating massive amounts of bugs. They are fast, graceful fliers, and many old farmers thought it was lucky to have them in the barn.

to till these edges to give your new seeding a good start, and you may have to weed out unwanted invaders for a season or three until the invited plants are well established. Many farmers plant rows of annual flowers among their garden rows, and between fruit trees and vines, to provide additional habitat and food for pollinating and predator insects, as well as for bug-eating birds. Gear your plantings to what is native to your area and what species you want to encourage, and plan a variety for season-long blooms and enough height and strength to provide perches for the dragonflies and song sparrows. As noted above, species that provide seeds, berries, or cover in the fall and winter are important.

Finding out what to plant and where you can obtain seeds takes some research. Since you're looking for plants suited to local conditions, use local resources and seed suppliers if at all possible. The Internet may help, or contact your local department of natural resources or agricultural extension agent to ask about local conservation organizations and native plant and seed suppliers.

Like the rest of the natural habitat on your farm, all plantings should be monitored and, when needed, weeded, pruned, or thinned to maintain a balance of species and young and old plants.

Fencerows

Planting fencerows to native species is an excellent way to provide corridors of cover and food that also allow wildlife to move from one area to another, an important consideration for many species, like fox, whose survival depends on having areas larger than a brush patch. The only hitch is that brush and vines are extremely hard on fences, growing over the wires and dragging them down over the years till the fence falls over. Then you have to rebuild it a decade or two sooner than if it had been kept clean. Solve this problem by mowing a strip on either side of the fence and placing your planting next to, instead of in, the fence line.

Waterways

If you are fortunate enough to have a pond or stream on your property, the banks are an excellent place to encourage or restore native vegetation. Water-bank vegetation will also prevent erosion and muddy water.

Windbreaks

Trees and shrubs are as important as grasses and forbs in attracting native wildlife. They offer the added benefit that they can be planted in the form of windbreaks to keep your house warmer in winter or to keep snow from drifting into an inconvenient place. The additional height of the planting attracts more birds and wildlife, and evergreens in particular add welcome winter shelter. Plant several different species for maximum food and cover, and follow recommendations for spacing and placement.

Fence lines are excellent spots for the brushy edges that provide food, shelter, and traveling corridors for native wildlife. Leave room for mowing a strip on either side of the fence, and your fence will last decades longer than it would if you let tall grasses and climbing plants lean on it.

Other Habitat Enhancements

Over the years, as you add birdhouses and brush piles, plants and shrubs, keep your eye on the natural features of your property, how creatures use them, and whether they might accommodate additional niches. A mudbank is handy for all sorts of bugs and for some animals' burrows. Old logs are essential for drumming grouse and home to a whole miniature ecosystem of fungi, bugs, toads, and maybe salamanders. Your own observations, along with winter evenings spent reading about local natural history, can guide your habitat planning and conservation as it develops through the years.

Woodlots

Whether you plant the trees or simply have them on your property, any woodlot can benefit from a bit of tending. Judicious trimming and cutting can accelerate the development of quality woodland habitat. If you own a woodlot, consider contacting your local woodland owners' association, the American Tree Farm System, or your county forester for information on what specific steps you can take, given the species and terrain, to create a healthy, wildlife-welcoming woods.

Restoring Habitat

If you have the acreage but not the habitat, it is possible to restore a diversity of habitat to an area, though it may take decades for a full assemblage of plants, insects, and wildlife to take up residence. With good planning and effort, however, you will see immediate improvements in biodiversity and have the satisfaction of knowing that you've helped significantly to sustain an ecosystem. Any type of restoration is a major project; fortunately there is both financial and technical assistance available in many cases.

Adding Water

Water seems almost magic in its ability to draw wildlife, greatly increasing the number of species that are able to sustain themselves in the area. The term *pond* generally refers to an open body of water with solid banks, while *wetland* refers to soft, wet ground that may or may not have open water but remains water-saturated (except in a drought) and is home to a distinct community of wet-adapted plants, insects, and wildlife. Both are important for maintaining a diversity of habitats.

Building a Pond

A new pond should be situated where it will hold water. A naturally wet spot with clay soils is the natural choice, but if you don't have this and would still like a pond, it's possible to bring in clay or a manufactured liner. A purchased liner is expensive and may or may not work. Accidents, including punctures from the hooves of drinking deer, can be major problems. Since building a pond is not a restoration, such a project won't be subsidized by government conservation funds. It is still a worthwhile endeavor, however, and a local contractor can build a small pond for a reasonable cost. Or you can build one yourself; there are several excellent guides out there. (See the resources for suggestions.) Before starting, check with state and local authorities to see whether any permits are needed. In most cases connecting a constructed pond to streams or other moving water is not permitted.

Restoring Wetlands

There is government money available for the restoration of wetlands. If your presettlement vegetation map indicates a wetland on your property, or you've found drainage tile or ditches on it, then you are eligible for cost-sharing and technical assistance from the U.S. Fish and Wildlife Service, along with whatever funds might be available from state agencies. If you accept their assistance, the USFWS staff will engineer the project and help find the excavators and any other contractors needed for the job.

Since many former wetlands are now located in the middle of fields, and since some owners might still want to farm the rest of the fields, it's a concern not to flood the entire area and make it unusable. You may be able to negotiate a water-level control structure, which will drain excess water from the restored wetland, so that it doesn't spread all over the place in wet years.

Adding Prairie

Prairie looks spectacular either from a distance or close-up, where you can really see the diversity of flowers and plants. It also can be (carefully) rotationally grazed, to its benefit and yours. This most endangered of ecosystems can be a tremendous asset to your farm in terms of biodiversity, water retention, and soil health, but the seed is expensive and soil preparation and initial maintenance are labor-intensive. The results are infinitely worth the trouble.

The expense and time involved are in direct proportion to the size of the restoration. A strip of native prairie plants alongside the garden is simple; a 20-acre (50 ha) restoration project is not. But if you have the time, other people have the money and the technical expertise and are anxious to help. Start with the U.S. Fish and Wildlife Service, which will assist you in getting started and direct you to further resources. If you're not anxious to deal with the government, start with your local chapter of the Prairie Enthusiasts, another excellent source of information.

To restore a prairie requires that the current vegetation be removed, and this is usually done with a broad-spectrum glyphosate herbicide like Roundup. Even if you aren't certifying that particular patch of ground as organic, presumably you don't want to start throwing chemicals around. Use the same organic seedbed preparation methods as you would in your fields and garden: planting smother crops, preceding the restoration with a clean-tilled row crop, or using several tillage passes to germinate and kill weeds. This means preparation may have to start as much as a year ahead of the actual restoration.

Since many prairie plants establish slowly, a restoration may require weeding the first two seasons if there is heavy pressure from aggressive annual weeds. Even if you were willing to use chemicals, this job would have to be done by hand, since sprays don't discriminate between native and nonnative species.

Once the prairie is established, maintenance is minimal except for occasional mowing, burning, or grazing to eliminate shrubs and renew the vegetation.

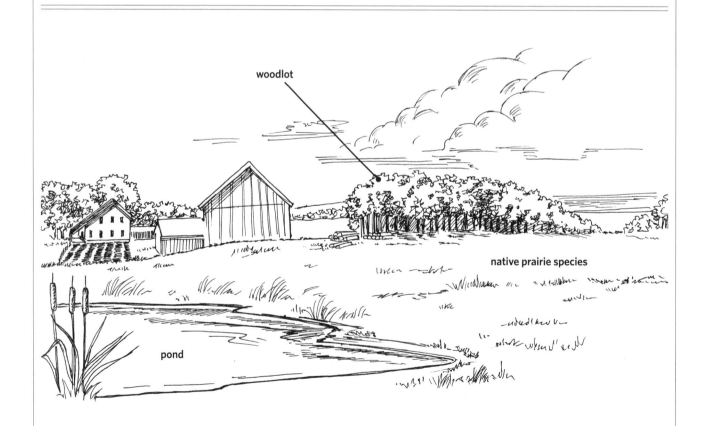

Using odd corners of your farm for native plants, restored wetlands, a pond, or a woodlot ensures greater biodiversity and plenty of good wildlife watching.

Wild Farm Resources

No matter where you're located, the Wild Farm Alliance can help you find information on local native species, how to restore and conserve them, and where to find seed for plantings. They can also instruct you about gaining access to government programs that offer financial and technical support, and that monitor the success of your activities. This nonprofit organization of farmers and conservationists promotes "a healthy, viable agriculture that helps protect and restore wild nature" and works closely with the organic farming community.

Many other organizations support habitat restoration efforts on agricultural land, including the Xerces Society, specializing in invertebrate (bug) conservation; the Prairie Enthusiasts, dedicated to the preservation of prairie lands; and Defenders of Wildlife, known for their activity on behalf of big, photogenic animals like wolves, but also working hard to conserve habitat.

Adding Woods

Planting trees removes land from agricultural production but can add so much in return: restoration of a native ecosystem, a tremendous amount of biodiversity, a renewable resource in the form of sustainably harvested timber, shade for livestock, and beauty for the farmer.

Pine trees grow the fastest and provide year-round cover, and seedlings can often be obtained very cheaply from state nurseries. On the other hand, solid plantings of pine offer few food resources for birds and animals and will shade out nearly all understory vegetation, so that biodiversity is minimal.

If you wish to plant trees, a mixed planting of species native to your region will create a more sustainable, diverse habitat than a monoculture of pines. State departments of natural resources, local woodland owner's associations, the American Tree Farm System, and other groups can assist you with information and resources. When researching what to plant and how, be careful to focus on information that promotes biodiversity and sustainability, because much of what is out there focuses on timber harvests at the expense of other concerns.

Also search for privately owned tree nurseries in your state that specialize in native species. The owner will usually have extensive knowledge of what is appropriate in your locale and how to mix and match trees and shrubs for maximum enhancement of habitat.

As with prairies, your tree-planting project can be any size you want, from a couple specimen trees in the yard to a shady border along a pasture to many fields lined with windbreaks to a solid planting of several acres or more.

Seedling trees need protection from smothering grasses, nibbling deer, and bark-chewing rodents their first several years. This is done either by planting so many trees that a few will survive no matter what, or by protecting the trees with mulch, hardware cloth around the tender trunks, and fencing or cages to keep the deer away.

Profile: **Frank Morton**

Wild Garden Seed, Philomath, Oregon

Plant breeder Frank Morton is acutely attuned to the nuances and interactions of genetic traits and environmental influences, the building blocks of biodiversity. "We've practiced plant breeding since way back in 1984, as part of our salad business. It's a way of coming up with interesting new stuff, by allowing some of our favorite things to cross-pollinate," Morton says. "To start a project we usually take two heirlooms we really like, ones that are well adapted and have good traits, and we cross them."

Frank and his wife, Karen, produce open-pollinated, certified organic seed on 3 acres (1.2 ha) of their Wild Garden Seed on the dry eastern slope of the Oregon Coast Range, at the edge of the Willamette Valley. In well over two decades of breeding work they've developed a national reputation and originated dozens of varieties. Since 1994 they've offered their seed through their own seed catalog selling both to retail customers and to other catalogs familiar to organic growers, including Fedco Seeds, High Mowing Organic Seeds, J. L. Hudson, Johnny's Select Seeds, Nichols, Seeds of Change, and Territorial Seeds. Remembering their first catalog, Morton says:

> It had all the varieties we were using in our own salad business, and a large portion of them were the result of our own breeding work. There were salad greens and wild greens like purslane and lamb's-quarter as well. We were integrating in our salad business all kinds of biodiversity, and also whole life cycles: seeds, shoots, flowering shoots, even the flower buds of parsley — they're extremely aromatic and very potent. It gave our salads a completely unique appearance.

In 1995 the Mortons began working with their neighbors John Eveland and Sally Brewer, 5 miles (8 km) down the valley at Gathering Together Farm, to produce seeds on a larger scale. "We started with amaranth and orach (a salad green), and eventually we became one company. We wanted to have a seed catalog in which all the seed was produced by one set of farmers, and that's what we do," Morton says.

Now, in addition to the 3 acres (1.2 ha) at their home farm, the Mortons manage about 7 acres (2.8 ha) for vegetable seed production at Gathering Together Farm. The rest of the 30 or so acres (12.1 ha) under tillage there are devoted to growing fresh produce.

Producing both vegetables and seeds on the same farm has had the synergistic effects of improving pest control and soil fertility. To produce seed, plants must be allowed to bolt, flower, and fully mature. Morton explains:

Flowers are the key to attracting biodiversity in the garden, whether they're flowers of weeds, wild plants, or crops. Almost all of the crop flowers attract some kind of insect, and almost all insects require flowers to complete their life cycle. In their adult stage, insects like syrphid flies and beneficial wasps have to drink nectar to get the energy they need to fly around. The pollen is essential in order for the insects to develop eggs.

Syrphid flies are some of the most widespread and beneficial insects, and if, for example, you let your early mustard greens go to flower, you encourage syrphid flies, which will lay eggs on the cabbages that you will have later on, and their young eat the aphids. So you're growing your own insect protection. You also get a lot more organic matter (to work into the soil or make compost) when you let the plants bolt and flower.

The seed crops, interspersed among the produce crops, act as insectaries for the rest of the farm. "It hardly matters what the species is," Morton says. "When you look at the rows that are flowering, you'll see this buzz of insects above the rows, getting their nectar and pollen, and then they're going to go out into the fields and lay their eggs. We think this is a basic model for bringing biodiversity to organic produce farms, which otherwise are fairly devoid of flowers."

Plant breeding and seed production offer another golden opportunity for the interested grower: the chance to develop varieties uniquely suited to the local environment and resistant to the prevalent diseases. Morton says, "Between 1900 and 1965 or so, U.S. public universities did really good plant breeding work. A lot of the breeders were pathologists, they were breeding for resistance, and all of them made use of the concept of the disease nursery, or raising a crop in the presence of its known diseases. Right around 1970 this changed . . . the land-grant universities (then) were doing work basically for the seed and chemical companies — and it would hardly make sense for those companies to develop resistant varieties."

So the Mortons created their own disease nursery, Hell's Half-Acre, to continue the work of developing resistant varieties. It began, Morton says, "as our rational way of deciding which lettuces were best."

We wanted to compare as many commercial and heirloom varieties as we could find for their resistance to downy mildew and sclerotinia (lettuce drop), a fungal omnivore that eats something like 450 different species of plants — that we know of. The reason this is so important to me is that sclerotinia used to take 50 percent of my planting every year.

We set up this disease trial, planted 40 different varieties of lettuce, replicated three times per season, for three years in a row. We inoculated the crop with sclerotinia, and this so terrified our organic inspector that he was afraid to walk in

"We think plant breeding ought to be done on farms. So one of the things we like to do is show people how to do it through workshops and demonstrations, giving them the confidence to do it themselves."
Frank Morton

> **"Flowers are the key to attracting biodiversity in the garden, whether they're flowers of weeds, wild plants, or crops."**
> Frank Morton

there. But on the other hand, our pathologist friend from Oregon State University came out and thought it was great, and she told our friends it would not increase the disease rate in their crops. We received funding from the Organic Farming Research Foundation and took data, rating all the varieties for their reactions to these diseases.

That research showed definitively that there is a genetic component to disease susceptibility in lettuces. Butter and crisp-leaf lettuces as a class are generally the most susceptible, Morton says, while loose-leafs are most resistant and romaines are in between, and there's a lot of difference between varieties in a class.

"The Merlot lettuce was the most resistant, but it doesn't taste that good," Morton says. "You have to cross it with a lot of good-tasting stuff to get flavor. That's the whole tension in plant breeding. You can come up with things that are completely disease resistant, but they don't taste like anything. You can come up with things that are fabulous tasting, but they're so fragile they won't make it into your kitchen. Disease nurseries like Hell's Half-Acre are helping agriculture move ahead without chemicals."

He continues, "We think plant breeding ought to be done on farms. So one of the things we like to do is show people how to do it through workshops and demonstrations, giving them the confidence to do it themselves."

The Mortons' interest in biodiversity extends well beyond their own crops. Besides seed crops, Shoulder to Shoulder Farm is home to 7 acres (2.8 ha) of native Oregon prairie, one of the rarest of North American ecosystems. "It's a joyous thing to see," Morton says, "but really it's just a faint shadow of what's gone away.

This is supposed to be dry land here, and anywhere we add water to the landscape, the nonnative plants take over. This is the heartbreak side of trying to manage diversity on the farm. Farming is based on fertility, and many native ecosystems are based on scarcity.

We're trying to hang on to it — we cut fir, we mow, it's a constant process. But as soon as we stop, we lose our progress. And everything our fertilizer, compost, footsteps, or water touches is going to change. That's one of the big lessons we've learned. Our feet in the garden get little bits of compost and good soil on them, and everywhere we walk it gets greener.

For the aspiring farmer, Morton suggests thinking things through carefully. "What do you want to farm? Do you want to bend over and pick lettuce? Are you an animal person or a plant person? These are personal conversations about your own nature. Have that conversation with yourself."

Then, he recommends, "take a tour, go visit organic farms, and don't be too intrusive. Get a job on a farm that practices the kind of agriculture you think you would like to do, and put in your time. I didn't do that, and I still don't know a lot of stuff that normal farmers do, like equipment. Learn about marketing — that's huge. You have to find some way in, of getting experience."

And learn how to do your own breeding research, he adds. "I want others to do this. This is important for organic agriculture."

Profile: Mike Omeg

Omeg Orchards, The Dalles, Oregon

Sometimes becoming certified organic is just not feasible, or maybe not even sensible. In such a situation, the good steward of the land still proactively experiments and innovates toward the goals of maintaining soil health and biodiversity, always conscious of how a practice impacts the entire system, not just the problem it was aimed at.

Fifth-generation cherry farmer Mike Omeg didn't immediately take over the farm after getting his master's degree in entomology. First he spent several years working for a consortium of soil and water conservation districts in northern Oregon, where he spearheaded the development of a region-wide system to forecast pest emergence and also model disease patterns, based on degree-day data from remote weather stations.

So when it comes to pest management, Omeg combines impressive education and experience with the common sense of a farmer who has grown up on the place, learning from previous generations. The result has been the evolution of a sustainable and innovative pest management system based on encouraging biodiversity and natural predators. Though not certified organic, Omeg Orchards' pest management methods would be an asset on any organic farm.

Omeg's success in utilizing owls, bats, bug-eating birds, and predatory insects for pest control and in encouraging native species of pollinator insects are inspiring neighboring growers to give his methods a try. He's received attention on a national level as well. In 2007, just two years after Omeg took over management of his family's orchards, he was given the Farmer-Rancher Pollinator Conservation Award by the National Association of Conservation Districts' North American Pollinator Protector Campaign. The Xerces Society chose Omeg Orchards for a pollinator habitat workshop, and Omeg has received grants from the USDA's Sustainable Agriculture Research and Education Program to continue his work with barn owls and from the Oregon 150 Fish, Wildlife and Habitat program to create habitat for the swallowtail butterfly.

Omeg Orchards' 400 acres (162 ha) of sweet cherry trees sit in the middle of one of the most intensive concentrations of cherry orchards in the world, the 5,000 acres (2,023 ha) around the city of The Dalles on the banks of the Columbia River in Oregon. In other words, the area is a cherry monoculture. Few things in nature are black and white, and while monocultures have some serious drawbacks, there are actually some advantages in his situation, Omeg points out.

"The decisions I've made on our farm are based on my technical training, my experience, and common sense. The best place to be for us is not purely organic and not purely conventional."

Mike Omeg

If you're a little cherry fruit fly and you're growing on a wild chokecherry tree, and I'm an orchardist in the middle of these 5,000 acres (2,023 ha), you have to fly through a lot of areas that have been sprayed to get to my place. So we've been able to lower the resident pest populations a lot. If you're spread out, there are a lot more areas around for pests to breed in, and you don't get that group protection.

On the other hand, disease control is a bigger problem, particularly since all of Omeg's cherries are sold for fresh eating, and appearance and keeping quality are critical. He says, "In cherries, mildew is one of the primary diseases, and there are very few effective products that are OMRI registered. I don't think going purely organic would be the best way to be a good steward of my land." The story might be different if mildew-resistant varieties of sweet cherry were commercially available, but plant breeders and researchers have told him that prospect is at least ten years away, Omeg says.

The other area in which he does not follow the Final Rule, Omeg says, is with plant nutrition. "The economic returns aren't there. We make the same price per pound or better being not organic, and our costs for organic would be much greater. I would have to bring in vast amounts of compost and manure to have proper nutrition for my trees, and that means a fleet of big trucks hauling from maybe hundreds of miles away — and how is that helping out the environment?"

To make a decent living as a full-time farmer on the farm where he grew up, becoming certified organic is not feasible, Omeg says. "We're a family farm, but the family farm is very different now from the mythology. It's very difficult to compete, whether you're organic or not. It's hard to make it unless you have economies of scale, because you're competing with large, professional growers. When I was in elementary school, you could do really well if you had 100 acres (41 ha). Now you need at least 300 acres (121 ha) of cherries to do well, so you don't have to have a second job and can send your kids to college."

He explains, "The decisions I've made on our farm are based on my technical training, my experience, and common sense. The best place to be for us is not purely organic and not purely conventional."

And though organic certification isn't in the cards for Omeg Orchards, he seems as close as he can get given his circumstances.

I want to be a good steward of the land that my family farms. We view it as a resource for future generations. Grandpa was always very careful about erosion, and my perspective on conservation focuses on pesticides, on making sure we use products that aren't harmful to native creatures and to the people who work on our farm. We apply very "soft" sprays — so I don't have any product that's effective against adult moths — and we use low-volume sprayers so deposition is much more efficient. With a conventional sprayer most of the material you put on ends up dripping onto the ground. We also use scouting, to ensure we are applying sprays only at damage levels above the economic threshold.

Scouting for pests and disease is an important part of Omeg's extensive integrated pest management system, which is just "brick-and-mortar good practice," Omeg says. But it's some of the other proactive management tools he's developed that have really interested other farmers, he says. "Gophers eat the roots, and voles girdle the cherry trees. Ground squirrels eat the crop and kill young trees by stripping the bark. We were spending about $43 per acre (0.4 ha), every year, using rodenticides, traps, and explosives to control these vertebrate pests, and we just weren't being very successful. Conventional notions weren't working. So I started looking at using barn owls."

In many regions barn owl populations are decreasing, and Omeg believes one reason for the decline is the loss of shelter. Barn owls like to live in barns and sheds, which are themselves in decline as agricultural lands and small farms disappear. Fortunately, barn owls adapt readily to living in nest boxes. Omeg says, "I've put up 35 barn owl boxes on our property, and I think we're at about 200 boxes around The Dalles here. We had a workshop (on the subject) last spring and about 90 growers came."

The boxes must be made to specifications, Omeg warns, or the owls won't use them (see the resources for more information on buying or building nest boxes and perches for owls, bats, and kestrels). But doing it right is well worth the effort: an adult barn owl eats at least one gopher per night, and a family during the brooding period will consume 10 to 12, for a total of 3,000 gophers per year.

As another management measure, instead of spraying to control the adult moths whose larval offspring can cause high levels of damage to cherries, Omeg puts up bat boxes. "Bat boxes are just as easy as barn owl boxes, though there is a whole science around building and locating the boxes. The limiting factor for these predators isn't the availability of food, it's not having housing. Bats are a wonderful way to augment your pest management program for adult insects," Omeg says. A nursing bat will eat 300 or more codling moths in a night, which is a lot of pest control for the minimal price of a bat box.

Starlings and sparrows can eat or damage a lot of cherries in an orchard. "A lot of times we go out in the orchards with noisemakers or shoot pyrotechnics at those birds, and we do a lot of trapping, but trapping is labor intensive and noise is irritating," Omeg says. So Omeg's newest project is putting up boxes for kestrels, which prey on starlings and sparrows. Fortunately kestrels only rarely prey on other birds that eat beneficial insects like bluebirds and swallows, which Omeg is also encouraging with nest boxes. "You never see bluebirds in orchards because the trees we grow don't have cavities for them to nest in," he says. "They need to have homes."

Omeg also is turning odd areas of brush and bare ground into insectaries, planting native plants to attract native insects by creating what he calls "high-density pollinator habitat for predatory insects, parasitoids, and pollinators — all three groups are important in our orchard system."

Certainly such a rich habitat will attract its share of pest insects, too, Omeg agrees. "Nature's a complex thing. You put out resources, and good and bad things are going to utilize the resources. But the benefits far outweigh the pests."

Organic Certification

Preparation, Application, and Inspection

Organic certification is an entirely separate process from farming organically. Farming is what you do in the field, while certification is doing the paperwork that documents what you've done in the field. The word you'll see over and over again in the certification process is *documentation*. The whole point of organic certification is to guarantee to buyers of your farm products that you are following the National Organic Program Final Rule.

You can farm organically without being certified, but you can't be certified if you're not farming according to the Final Rule. That's simple enough.

Becoming certified organic is a three-year, three-step process:

1. **Preparation.** You learn the rules, find a certification agency, set up your record-keeping system, and begin the required three-year transition period. During the transition period you must follow the Final Rule.

2. **Application for certification.** When you've begun your third year of transition, contact your certification agency and request its application packet (the packet may cost up to $60, depending on the agency). Fill it out and submit it with the application fee in the late winter, if not before.

3. **Inspection.** Once your certifier has received your application, you will be assigned an independent inspector, who will contact you to set up an inspection during the growing season. The inspector will spend anywhere from a couple of hours to an entire day at your farm and then file a report. Your certifier will review the inspector's report and your application to decide whether to give your farm organic certification.

The application and inspection are repeated annually to stay certified, but if you stick with the same certifying agency there is less paperwork, since you are renewing rather than initiating certification.

The rest is just details, and there are a lot of them. We'll cover the most important ones in the rest of this chapter. If you have additional questions about specific inputs or practices or paperwork — and you will — then call your certifier. But don't fret too much: If you understand and follow the principles and rules of organic farming, and you remember to get approval from your certifier of any new inputs or practices *before* you try them, you'll be fine. Your certifying agency will help you understand and comply with the details of the Final Rule that apply to what you are producing. You don't need to read the entire Final Rule (though it wouldn't hurt), but you should read the relevant definitions and those sections of the rule that apply to your type of operation. Look at the section headings to see which these are. If there are parts you don't understand or that are not clear — and the rule's written in "bureaucratese," so that's quite likely — your certifying agency will be able to explain them.

Why Be Certified?

Certification costs a fair amount, at this writing anywhere from about $400 to $1,000 per year, depending on what and how much you are producing. For a small-scale farmer just getting started, that's a pretty good chunk of your profit margin. So, the question you need to ask yourself before you start the whole process is this: "Why should I be certified?"

Certification is basically a marketing tool. If certification makes your product more valuable or easier to market, then it is worthwhile. If

The Final Rule on:
The Certified Organic Label

Only a certified organic farm may use the distinctive "USDA Certified Organic" label on its products.

Farms selling less than $5,000 worth of products per year and following all stipulations of the Final Rule, including record-keeping, may call their products "organic" without being certified, but they may not use the USDA label. Products from these farms may not be sold to processors as organic ingredients.

If you process any of your certified organic products to preserve them or add value, then you can label those products as "USDA Certified Organic" only if you are certified as a processor, which means, of course, that you do not add any prohibited ingredients or use any prohibited processes.

If you have your certified organic products (most commonly meat and milk) processed by someone else, the processing facility must be certified organic, and the processor must conform to the Final Rule's regulations for organic processing, in order for the product to still be labeled certified organic.

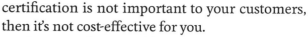

certification is not important to your customers, then it's not cost-effective for you.

For example, vegetable growers selling to local customers often don't become certified because they know their customers personally, while dairy and grain producers generally deal with buyers who require certification. That's just a rule of thumb, though, since larger market gardeners often certify, while some dairies with no access to an organic processor may wait to certify until they can get on a milk truck route. And some new farmers choose to go through the certification process simply to develop their knowledge base and as a point of pride, rather than as a marketing aid. Certification also gives a farmer the benefit of oversight by an experienced person. Though inspectors can't consult, they can discuss organic principles and management systems and refer farmers to other resources.

Remember, if you sell less than $5,000 worth of farm products each year, then the Final Rule allows you to label your products as organic even if you are not certified, as long as you follow all the rules, including the three-year transition period, and maintain the required records. Even if you're a part-time operation selling only for a few weeks each year at the farmers' market, if you're telling your customers that you're organic, you need to have the records to prove it. Many small-scale farmers tell their customers that they follow organic practices but don't keep records, which is not organic, since organic practices, by definition, include record-keeping. Even if you do keep records, you cannot use the "USDA Certified Organic" label, and you cannot sell your products to a processor or another farmer to use in a certified organic operation, unless you are actually certified by an independent certification agency.

Step One: Preparation

If you have decided to become certified, there's some homework and supplies to take care of first. You will be writing down on a near daily basis what you have done on the farm, so get a spiral notebook, a pocket notebook, or a wall calendar with lots of space — whatever is handiest for you to jot down a running record. You'll also want to start researching the rules in more depth, especially the fine points of how they are applied for your type of operation and which certifiers are active in your area and can certify your type of operation.

Fitting the Rules to the Farm

First, learn and understand the organic rules. The Final Rule is the legal standard, and certifiers may not use a higher or lower standard or make exceptions. We've discussed the Final Rule bit by bit in this book as it applies to different farm topics; any changes to the Final Rule since this printing can be found at the National Organic Program (NOP) Web site or by contacting your certifier.

If you are planning to become organically certified, you'll find you'll want more information on the details of organic methods and other aspects of farming for your particular type of operation. Farmer networks and conferences, field days, organic farming organizations, and the Midwest Organic and Sustainable Education Service (MOSES) are all excellent resources.

As you learn the rules and figure out just how you're going to apply them on your farm, start writing it all down as your farm plan. The Final Rule requires certified farmers to have and follow a comprehensive farm plan, officially called the Organic Production Plan, which details everything from crop rotations and the soil-building program to livestock preventive health care and planned inputs. The farm plan ties together all the receipts you collect and records you keep to verify what you do on your farm. Your farm plan is also the fundamental and largest part of your application

The Final Rule on:
The Farm Plan

The Final Rule requires that the Organic Production Plan include the following:

- A description of practices and procedures to be performed and maintained, including the frequency with which they will be performed

- A list of each substance to be used as a production or handling input, indicating its composition, source, location(s) where it will be used, and documentation of commercial availability, as applicable

- A description of the monitoring practices and procedures to be performed and maintained, including the frequency with which they will be performed, to verify that the plan is effectively implemented

- A description of the record-keeping system implemented to comply with the requirements established in section 205.103 (these requirements cover record-keeping)

- A description of the management practices and physical barriers established to prevent commingling of organic and nonorganic products on a split operation and to prevent contact of organic production and handling operations and products with prohibited substances

- Additional information deemed necessary by the certifying agent to evaluate compliance with the regulations

In plain terms, this means your farm plan has to describe what you're going to do, along with when and how often, and using what methods, equipment, and inputs; how you're going to check to see if it's working; and how you're going to keep a record of everything to show to the inspector. Obviously you aren't going to get all that figured out and written down overnight, or even over a single season. That's why you should get the farm plan started early in your transition period.

for certification. (A sample form for developing a farm plan can be found on the ATTRA Web site; see the resources.)

Finding a Certifier

Next, find a certifier. The United States currently has nearly 80 certifying agencies accredited by the USDA, most of them independent but some run by state governments. Some agencies operate only in their own state or region, while others will certify operations throughout the country. Agencies may also specialize in certain types of operations, such as dairy or vegetables.

To find a certifier, start with the National Organic Program, whose Web site (see the resources) lists certifying agencies by the state where they're located. Unless you're planning on selling into foreign or other unique markets your best bet is to find a certifier close by that has a specialty in your area of production. If you already have a prospective buyer for your product (which is usual for dairy and grain producers, and becoming more common with livestock), then ask the buyer which certifying agency he or she would prefer you to use.

Once you've identified possible certifiers, contact them to ask about their fee structure and application process, and whether they are a good fit for what you plan to produce and how you're going to market it. MOSES recommends asking other organic farmers in your area what agencies they use, and whether they're happy with them. You'll be working pretty closely with your certifier, so it's important to have one that is responsive and informative. Also, since you are charged for the farm inspector's mileage and costs, if you can share those costs with other farmers using the same agency it will save all of you money.

If you are interested in international markets, ask any certifiers you are considering whether they offer overseas certification. Also consult any potential overseas buyers as to what type of certification they want you to have.

When you decide on a certifier, obtain their information packet to guide you as you set up your record-keeping. This packet is different from the application and is usually free or sent for a nominal charge.

You could wait until your last year of transition before finding a certifier, but this tends to rush things and increase the possibility of mistakes. It's

Certification Agencies

Independent, privately run certification agencies began springing up around the United States during the 1970s and 1980s as the organic movement took off, as a means of assuring buyers of organic products that they were getting the real thing and no corners were being cut in production. With the implementation of the Final Rule in 2002, each independent agency was required to be approved by the USDA. In other words, the USDA now accredits the certifiers, making sure that each is complying with the federal standard in the types of operations they choose to certify. The USDA certifies agencies in four different areas: organic farms, livestock, wild crops, and processing. The certifying agency you choose should be USDA-certified for the type of farming that you do.

The certifying agencies are independent of the federal government. They often specialize in a particular field, such as vegetable, dairy, or meat production, and you should ask about this when choosing a certifier. The agencies support themselves through the fees they charge for their services. They don't rely on government money, except for those agencies that are run by state governments.

Organic inspectors may be independent or employees of a certifier. Different inspectors may have slightly different interpretations of the Final Rule. Inspectors may be accredited by the International Organic Inspectors Association, which provides training and continuing education. The IOIA also works to ensure consistency and high standards in organic inspections.

no fun going through two years of transition only to find out you've done something that disqualifies you so that all that time was wasted.

Also worthwhile is inquiring at your state department of agriculture about the availability of cost-sharing for the expense of becoming organically certified. You might also inquire about the availability of EQIP (Environmental Quality Incentives Program) cost-sharing for the expense of transitioning land to organic production.

Record-Keeping

After learning the rules and finding a certifier, the next step is setting up record-keeping. This is the part that everybody complains about. There really are a lot of records to keep, but if you make it a part of your routine, just like all the other daily farm chores, it's quite doable.

Incoming Products

The first thing you need is a file folder (or box, or envelope, or whatever) in a handy spot. In it you put every receipt and every tag for everything you buy to use on the farm, including seeds, animals, soil amendments, feed, vaccines, animal health care products, foliar feeds — in short, everything. Every time you buy an input that comes in a bag or a box, tear off the ingredient list, staple it to the receipt, and make sure there's a date on it before putting it in the file. If you're buying in bulk (like hay, lime, or bulk grain), then you need a receipt and a copy of the seller's documents that verify it's certified organic if it is something that is covered under the Final Rule. For those inputs that aren't required by the Final Rule to be certified organic, you still need documentation that the input had no additives or treatments that are prohibited. For example, lime is a mined product and so isn't required to be certified organic, but it can't be hydrated, slaked, or otherwise processed, so you'll need a document stating what sort of lime it is. This could be merely noted on the bill in most cases.

You'll need all these receipts for your income taxes, too, so you can kill two birds with one stone and use the same file also for all those purchases, like equipment, that don't have anything to do with your organic status but are deductible. Then you have all that necessary paper in one handy

The Final Rule on:
Record-Keeping

The Final Rule requires that certified organic farms maintain records "concerning the production, harvesting, and handling of agricultural products that are or that are intended to be sold, labeled, or represented as '100 percent organic,' 'organic,' or 'made with organic.'"

Such records must:

- Be adapted to the particular business that the certified operation is conducting

- Fully disclose all activities and transactions of the certified operation in sufficient detail as to be readily understood and audited

- Be maintained for not less than five years beyond their creation

- Be sufficient to demonstrate compliance with the Final Rule

These records are in addition to the farm plan. The easiest way to make sure you have every record you need for certification is to keep files for all receipts and tags and also to keep a daily diary of farm activities. Files can be old shoe boxes, big envelopes, or whatever works; just be sure to keep each year's records separately.

spot. If you want to get fancy, subdivide the file for the various categories of purchases: services, feed, seed, general supplies, and so on.

Get in the habit of checking your pockets every night before you throw your clothes in the wash. Pockets are where farmers usually stuff receipts until they remember to stick them in the file, and if you forget, well, receipts are hard to read if they've been washed.

Outgoing Products

You'll need a second file, right next to the first, for receipts or records of every product that *leaves* the farm, whether it's organic or not. This is so amounts match up. If you market ten steers but only eight are eligible for sale as organic meat, then your sales slips need to show that eight were sold as organic and two were not. Crops grown in buffer zones have to be harvested, stored, and sold separately from crops grown on certified land, and you'll need the records to show that this was done.

Production Records

If you're selling both organic and nonorganic corn, you'll need records showing that the organic crop wasn't stored in bins that had been used for the nonorganic crop. Which leads us to another point: to make your records worth the paper they are written on, you need to put numbers on storage bins, have identification on each individual animal, and make a map for each field, including a marked buffer zone. The only exception is poultry, which can be identified by flock, if every bird in it is being managed the same. Sheep, goats, cattle, and any other livestock to be used for meat or dairy need individual identification, whether by ear tag, collar, or photos for those animals with distinctive enough markings.

Then you'll need a paper record for each animal, flock, field, and storage bin that is to be certified organic. A three-ring binder with dividers is handy for this, and you can add pages as needed, or you can keep track of it all electronically. (Be sure to back up your files!)

ANIMAL RECORDS

Animal records should cover parents, date of birth, when organic management began, vaccinations, production records, and any health events like mastitis or pinkeye, and what treatment was used

Have a place to put paper! Then you can find it when you need it for farm planning and for organic inspection.

From Cradle to Grave

An explicit goal of the Final Rule is for all certified organic products to have a paper trail that runs from their origins to the final sale to the consumer. For example, if you issue a receipt for a sale of 1,000 bushels of soybeans, then these 1,000 bushels must be traceable back to the seed they grew from. That means a receipt for the soybean seed showing it was certified organic, farm records that tell when and where it was planted and what practices (cultivation, foliar feeding, and so on) were used while it was growing, a record of how harvesting equipment was cleaned or not used on nonorganic crops to guarantee the purity of the harvest, and a record of how many bushels were stored in which numbered bin during what time period.

(if it was a purchased treatment, date the label and stick it in your file). Individual animal identification and good records are standard for reputable livestock operations, so this part really shouldn't be an additional burden for the organic producer. Including growth rates, information on how quickly animals breed back, and milk production records is not required for organic certification but is an excellent idea. Those sorts of records are essential to a good breeding program, when selecting animals to cull, keep, or buy.

All these records also help you track the economic viability of the different areas of your farm business, making it clear where your profit is coming from, where you are losing money, and so where to change, contract, or expand.

FIELD RECORDS

Field records, like animal records, should record everything that happens to that field. This includes the amounts, dates, and types of soil amendments applied; planting dates and plow-down dates for green manures; what crops were planted at what rate on what dates; and how much was harvested and when. Going back to the example of our organic and nonorganic corn, field records are where you'll document the date and the

Thomas Jefferson and Farm Record-Keeping

Consider Thomas Jefferson, author of the Declaration of Independence, our third president, and an avid farmer. For nearly sixty years he kept detailed records of when and what was planted and harvested on his farm, what work was done, what the weather was, what livestock were produced, and other details. He was constantly experimenting with different varieties and methods, and he corresponded and exchanged plants with other farmers. Every day he took a few minutes to record what had happened in the fields and gardens of Monticello, building up an invaluable record of what worked, what didn't, and what might be worth further experimentation. His garden and farm books are still of interest for the insight they give into farming more than 200 years ago, and for those parts that are still relevant to how we farm today.

July 27, 1769: Millar's . . . says that 50 hills of cucumbers will yield 400 cucumbers a week during the time they are in season, which he says is five weeks, so that 50 hills will yield 2000, or one hill 40 cucumbers.

March 12, 1773: Sowed a patch of early peas, and another of Marrow fats. April 1: Both patches of peas up. May 22: First patch of peas come to table. Note this spring is remarkably forward.

March 10, 1779: This spring the weather set in remarkably mild and indeed hot and so continued until the middle of March, which had brought forward the vegetation more than was ever remembered at so early a period. Then it set in cold; the blue ridge covered with snow, and the thermometer below freezing. This killed all the fruits which had blossomed forward; the very few blossoms which were backward escaped.

Sept. 14, 1810: The largest cups of the drill hold 5 cowpeas each, and 5 cups to a turn of the wheel require 25 peas to a revolution of the wheel, which is 6 feet in circumference, then 1 pint will sow 100 revolutions = 200 yards and 18 pints or $1\frac{1}{8}$ peck will drill an acre in rows 4 feet apart, but it will be better to use the cups which take up a single pea only each, and drop with 6 cups to the band, or 6 peas to the 6 feet in this way $4\frac{1}{4}$ pints sow an acre, and 1 bushel sows 15 acres.

Jefferson, like any enlightened farmer, knew that knowing what happened in the past not only keeps you from repeating your mistakes, it guides you in the future. If you know when the apples blossomed in the past five years, which insects were a problem, and what did and didn't work, you'll be ready this year at the right time with pest management practices, and you'll probably have some new things to try. If you know how much hay and how many animal units you had on hand last fall, how many days you fed hay, and whether the winter was unusually long or short, then you'll find it easy to plan for winter feeding this year. If you know how much compost you applied and when, and can compare the impact on vegetable production with other years, taking into account differences in weather patterns, then you can figure out whether it's worthwhile to put on more or less compost this year.

Keeping a daily farm diary is not only the basis for all your certification paperwork, it will make you a better farmer. Put a notebook in a handy place and get in the habit of jotting a few notes each evening.

Should You Go Organic?

Many conventional farmers hesitate to make the transition to organic because they're afraid their crop yields will drop. They probably will. It takes time for soil health to be restored after years of synthetic fertilizers and pesticides, and organic crop production depends absolutely on soil health. Weed control is completely different from conventional systems, and it may take a few seasons to learn how to achieve good control with organic methods. Just how long all this will take depends on how well you understand and apply organic principles on your land. For this reason, transitioning from conventional to organic production is *not* a good idea if you are in poor financial shape.

procedure used to clean the combine between the two harvests.

If you also record rainfall, jot notes on the weather, what pests were problematic this year, what parts of the field are poor, and so on, you'll build an invaluable record of what works and what doesn't under what conditions on your farm.

The Daily Diary

Keeping a daily diary of farm activities is one of the simplest methods for keeping track of everything. If you jot just a few lines each evening about what you did that day, it can be transferred to the livestock and field records some cold winter day or rainy evening. A daily diary also gives precision to your record-keeping, and is a valuable resource when planning for the next season. You won't go bald from scratching your head all the time to try and remember if you cut first crop hay in the second or third week of June last summer, or just how many tomato plants you started last spring.

Making the Transition to Organic Production

By the time you've read the rules, found a certifier, and set up record-keeping you'll probably be well into the three-year transition period

from conventional to organic production that is required to certify land. The rules for transitioning livestock to organic production are a little different, but you can't have organic livestock without organic land, so we'll start with the land first.

Transitioning Land

To transition land to certified organic production requires 36 months before your first harvest, during which two things should happen:

- No prohibited inputs can be applied to the land.
- If the land is being actively farmed, during this period it must be actively managed to improve soil life, soil structure, and soil fertility and to control erosion.

Under some circumstances, if the land hasn't had any prohibited substances applied and you can prove it, your certifier may accept a shorter transition period. This can occur when land has lain fallow under a previous owner or has been in a government conservation program. Don't assume that you will automatically be granted a shorter transition period if you're buying fallow land. You must be able to document that no prohibited substances have been applied and okay it with your certifier.

Also, do not assume that fallow land will produce a big crop. Fallow doesn't mean fertile, and you should test soil before putting the land into production to find out what it needs to build fertility to acceptable levels.

If you are so fortunate as to have purchased land that is currently certified organic, you still have to get it certified in your own name before you harvest your first crop, though you'll be able to skip the transition period.

Managing land for soil health, as described in chapter 3, should include such practices as maintaining soil cover, using tillage methods that minimize soil structure disturbance and erosion, and incorporating green manures, animal manures, compost, and natural amendments like lime to increase and balance soil nutrients.

You don't have to do all of these things, nor do them all at once. And leaving land fallow with a good vegetative cover while you're getting other work done is usually perfectly acceptable, since it is nature's original method for soil-building. You will of course first check with your certifier before going this route. You do have to develop, as part of your farm plan, a plan for soil management, using those practices most appropriate to your farm, and being specific about when and how they will happen.

If you make a mistake and use a nonapproved practice or substance on your land, such as planting chemically treated seed or applying hydrated lime, then you're back to square one and have to start the three-year transition period over again. So even if the question seems stupid, ask your certifier if what you're planning to do is approved before you do it.

Transitioning Livestock

Transitioning livestock to organic production is different from land, and slightly different between dairy animals, meat animals, and poultry.

MEAT ANIMALS

To be eligible for organic certification, meat animals must have been following all of the Final Rule's stipulations for livestock since their mother's last trimester of pregnancy. In other words, being certifiably organic starts in the womb for meat animals. If you buy some conventional pregnant ewes or cows to start your operation, and you bring them onto your certified land and give them certified feed before their last trimester of pregnancy, then their offspring can be marketed as certified organic. But they (the mothers) cannot, no matter how many years they are on your farm.

Prohibited versus Allowed Substances

As a rule, anything that is natural is allowed in organic production, unless it is specifically prohibited in the Final Rule. Anything that is synthetic is prohibited, unless it's specifically allowed in the Final Rule. In most cases, synthetic substances are allowed only for specific uses.

Some examples of natural substances that are prohibited are tobacco dust, ash from manure burning, and strychnine (a natural product of some soil-dwelling organisms).

As examples of synthetic substances that are allowed, for the specific purpose of disinfection and sanitization the Final Rule allows use of isopropanol (isopropyl alcohol), sodium hypochlorite (bleach), hydrogen peroxide, and some other chemicals. Ozone gas is allowed only for cleaning irrigation systems. Newspapers and other paper are allowed in compost, as long as they contain no glossy or colored inks. Ammonium carbonate can be used as bait for insect traps so long as it doesn't ever come into contact with the soil or crop. Aspirin can be given to livestock to reduce inflammation. (Just don't give aspirin to your cat, as it could be fatal.)

The lists of allowed synthetic and not-allowed natural substances are contained in the Final Rule, and are added to occasionally. The simplest, most fail-safe method for knowing whether a substance is allowed is to check with your certifying agency.

Unfortunately the Final Rule doesn't list brand-name products. When you're standing in the farm store, or talking with a salesman out in the yard, you'll be dealing with brand names rather than the generic names of substances that are listed in the Final Rule. There are two ways to find out if a brand-name product is allowed or prohibited in organic production. First, you could call your certifier and ask. Second, you could go to the OMRI (Organic Materials Review Institute) Web site and check its extensive list of approved brand-name products. This list is not complete, since manufacturers have to pay to have their products reviewed by OMRI, and not all have done this. All the same, it's a good place to start.

POULTRY

Poultry, whether for meat or eggs, has to be organically managed from the second day after hatching. Where the eggs came from is not a concern. Since chicks are shipped immediately after hatching, any purchased poultry chicks will arrive at your farm having never eaten any food. They can be sold as certified organic if you raise them on certified feed and land.

DAIRY ANIMALS

Dairy animals must be under organic management for a year before their milk can be certified organic. So it's easier to get dairy animals certified than meat animals when you're starting out. Dairy animals just need to be on your certified organic farm for a year, eating your certified organic feed, and then they can produce certifiably organic milk. But unless the dairy animal was under organic management from its mother's last trimester of pregnancy, you can't sell her meat as organic. When her years in the milking string are done and it's time for her to go, the hamburger is not going to be certified organic. Your friends and neighbors will probably be so delighted to get some really fabulous hamburger that they won't care about that.

There are some additional rules about dairy animals. The way dairy certification usually works if you're just getting started is that a farmer buys a group of animals to begin milking. If these animals are not certified organic, then the one-year transition period is necessary. During that year, the animals must be eating organic feed and pastured on organic land, which would normally mean that the farm has to have already completed the three-year transition period. The exception allowed by the Final Rule is that the last year of the land's transition period can coincide with the year needed for the herd's transition. So, if you buy your herd in the spring and put them on pasture that is at least 24 months into the 36 month transition period, and feed them purchased certified organic grain, then the whole operation will be ready for certification the following spring. But you can't feed grain or hay if it was harvested from your land less than 24 months into transition.

The flip side of this is that there's no point in buying certified organic dairy animals unless you have a certified organic farm to bring them home to. If your farm is in transition but isn't certified, then the animals lose their certification as soon as they take a bite of your pasture. And they can never regain it, and you've paid a premium price for nothing. The exception is if you are in your final year of transition (months 24 to 36 of the transition period); then you can bring certified organic animals onto your farm and they will be able to maintain their certified status.

The second question that always arises with dairy operations is when the farmer wants to purchase additional animals from a conventional operation and add them to the organic herd. Forget it. Once a milking herd (or flock) is certified organic, you are not allowed to bring in conventional animals, even if you could figure out a convenient way to keep their milk entirely separate from that of the organic herd for a year. This rule holds even for young stock. If you buy nonorganic young stock, even if you raise them under organic management for at least a year before they begin producing milk, they and their milk will not qualify as certified organic.

Organic Animals Must Stay Organic

Once any animal is under organic management, you can't move it to nonorganic management and later back to organic. An animal gets only one shot at being certified organic. For example, it's fairly common for cow dairies to send their young heifers to a different farm until they're old enough to enter the milking string. Some farmers make their living offering custom heifer raising services. But if an organic dairy farm sends its heifers to a conventional heifer farm, those heifers can never again be certified as organic.

Step Two: The Application

Pour yourself a cup of coffee and get comfortable, because filling out the application takes a while. And don't put the job off, either, since getting it done early ensures that you'll be certified in time to sell this season's crop.

In order for you to be able to sell your crops as certified organic at the end of the three-year transition period, your farm must be inspected during the growing season of the year you want to sell an organic crop. To get the inspection done on time, you have to get your application filled out and in to your certifying agency by the end of the previous winter. So get it mailed by early March at the latest. Agencies don't have huge staffs and they have a lot of work to do in the spring, so get your papers in early to give them enough time to look them over and get an inspector lined up for your farm. You'll also want to have the inspection done early enough for the inspector to get a report filed and for the agency to review the application so you can get your certification before harvest. If you're a market gardener and you waited until March to send in your application, you may not get certified until midway through the season. But if your sales are local, your customers should understand.

If you're a dairy operation, you should have your inspection done at least four months before you plan to begin selling certified organic milk. If you are a meat producer and feeding your own hay, the hay must be certified a year before the animals are due to drop their first certifiably organic offspring.

There is no retroactive certification, so if you don't submit the paperwork and schedule the inspection in a timely manner, you may have to wait another year to sell certified organic products. For this reason, when you submit your application to the certifier, note when you plan to sell your first organic products, so the certifier knows when to schedule your inspection.

Look over the entire application before getting started. It has 20 or more pages to fill out. In addition, you'll attach field maps and histories when you send it in to the certifier.

Most of the pages are just answering questions: name, address, crops, livestock, and when is a good time to be contacted by an inspector. If you are raising field crops and you've numbered your fields and bins and kept records for each, then filling out the sections requesting information on locations, acreage, buffer zones, seed varieties, and projected yields is mostly a matter of transferring the information to the application from your records. The same applies for livestock, vegetables,

Common Mistakes Made by Organic Certification Applicants

The Midwest Organic and Sustainable Education Service (MOSES) offers numerous resources and fact sheets on organic certification, including a fact sheet on common mistakes made by organic certification applicants. That list includes:

- Missing paperwork deadlines

- Not submitting requested documentation

- Using an approved substance in a prohibited way

- Failure to pay fees

- Mistaken use of prohibited substances

- Failure to document that seeds and other products are GMO-free

- Incorrectly calculating the transition period

- Poorly kept farm records, lacking details, dates, receipts, or other information

- Failure to follow the farm plan

- Failing to let your certifier know when there have been significant changes to the farm plan

- Failure to segregate crops harvested from buffer zones or to control other commingling and contamination problems

and permanent crops: the better your records, the easier it is to fill out the application.

Once you've finished those parts of the application that cover the specifics of your farm, you move on to the general questions. These cover how you monitor your soil fertility and what the overall plan is for improving and maintaining it. You'll be asked for details of how you manage for biodiversity, practice soil conservation (erosion prevention), protect water quality, rotate crops, manage weeds, disease, and pests, and avoid chemical and GMO contamination from the neighbors. Fortunately, most of these questions are multiple choice, so you just have to check the appropriate box and move on.

Last, you may be asked to describe your record-keeping system and to attach a drawing of fields, buildings, crop storage areas, livestock facilities, and other relevant details.

Though this may sound like a lot, filling out the application is actually easier and more interesting than doing your taxes. The actual form you fill out will vary considerably depending on your certification agency and the type of operation you have, but the information requested will be pretty much the same.

Step Three: The Inspection

Inspection is usually a pleasant experience. It's your opportunity to show off what you've done and discuss your plans for the future. If you've followed the rules, kept your records neatly, and developed a comprehensive farm plan, there won't be any problems.

Organic inspectors are nothing if not thorough. Plan on walking fields, pastures, buffer zones, and any creeks or ponds you might have. You'll go through every building, and all the seed, feed, and inputs you have on hand will be examined. Any on-farm processing or storage facilities will be looked at carefully, along with compost piles, manure storage, and equipment. During the inspection be ready to answer questions about everything, from how you control thistles in a hillside pasture, to what your compost components and application rates are, to how diligent you are about replacing lost ear tags on your sheep. Before or after the tour the inspector will want to go over your records closely, so have these ready.

You won't get much farmwork done on inspection day, so schedule it during a slack time or find someone else to cover your chores.

Inspectors are not allowed to give advice or correction. They're there to inspect, not direct. They may or may not identify problem areas for you, but they certainly will for the certification agency. Once they've seen everything they'll thank you and depart to complete their report, which they'll submit to the certification agency. You may be asked to sign an affidavit confirming the inspection, and you will be billed for the inspector's time and costs (including mileage). You will receive a copy of the inspector's report from your certifier.

If you continue certification from year to year, you will complete renewal paperwork and be inspected each year, though the timing of the inspection and the individual inspector will vary.

Certification

Now you wait. That was a lot of paperwork you sent in, and your inspector sent in another big folder, and it takes a while for the certifier to work through all of it. If you got it all done early, then by the time you're picking those first crops or sending your first livestock off to the processor, you'll be able to label it "USDA Certified Organic."

Congratulations!

Profile: Harriet Behar

Organic specialist, Midwest Organic and Sustainable Education Service, Spring Valley, Wisconsin

Being inspected for organic certification takes a few hours to half a day, but there's no need to be stressed about it, says organic certification inspector Harriet Behar.

People get nervous when somebody comes to check them out, but really the inspector is just there to verify what they've written on their application. Look at your application and all the questions on it. That's what you'll need to document. Gather up all your papers and have them at the table; don't disappear to the attic for 25 minutes looking for a receipt. The inspector and you would much rather be outside, looking at the fields and livestock.

A certified independent inspector since 1991, Behar has been on the job since before the Organic Foods Production Act was passed, and long before the Final Rule standardized the requirements and changed how inspectors do their job.

"Since the government came in, inspectors aren't allowed to consult and really tell farmers anything, like 'You really need to use cover crops,'" Behar says. "Inspectors can repeat the regulations to them and tell them where to find resources, but they're supposed to remain completely objective, and there's a good point to that. What if an inspector goes back the next year and the farmer is mad about a recommendation he or she made?"

These days Behar has retired from inspection to work as the organic specialist for the Midwest Organic and Sustainable Education Service (MOSES), which gives her more time with her and her husband's own small grains and market garden operation. Her primary job now is helping people transition to organic farming, but she fondly recalls the thousands of farms she inspected during her career. "I enjoyed it," she says. "You get to go out on organic farms and meet all these interesting people and see all these interesting things. Every organic farm is a totally different thing."

But the downside of the job was the amount of time she had to spend away from her own farm and family. "As a full-time inspector I did between 125 and 175 inspections per year, from May to October, because you have to see a farm while the crops are growing. It was one week away, and one week at home to do the laundry and write the reports. We mostly raised squash, because the harvest wasn't until September, but I never really had a chance to enjoy a weed-free garden. It was a very hectic schedule."

Behar could inspect three crop farms or two dairy farms in a day. And these days, she says, inspectors are making $150 to $200 for each farm they inspect and for which

> **"It's really helpful to have an outside person who has seen many, many farms come on your farm and help you assess your organic system. The questions an inspector asks can help you think about what you're doing."**
>
> Harriet Behar

they write the report, plus expenses for mileage, motels, and food. Anyone who's interested can be an inspector, she says, if they complete the training and come to the job with farm experience. "If you're going to inspect crops, you need to know what a rotary hoe is and how it's used. Inspectors need to know what they're looking at."

She herself was working as the vegetable field person for Organic Valley (the oldest and largest farmer-owned organic cooperative in the United States, headquartered in La Farge, Wisconsin) when she attended an inspector training in 1989. "We thought it would be a good idea for the field person to know what organic inspectors were looking for," she says. "At that point there were no federal rules, and private agencies were doing certification. The training was put on by one of those agencies. I enjoyed it so much that I went to a few more, and I started doing inspections on the side in 1991. In 1996 I left Organic Valley on good terms and became a full-time organic inspector."

She also began training inspectors for the International Organic Inspectors Association (IOIA), even traveling to other countries such as Mexico and Japan to conduct trainings and inspect farms. Behar has maintained her inspector's certification and still assists occasionally by apprenticing inspectors, while her husband, Aaron Brin, is an inspector as well, which makes their own farm's yearly inspection a lively affair. The couple has been certified organic since 1989, and "a lot of times the certifiers bring apprentices here, so we'll have four inspectors," she says. "We're fairly complicated. We have a diversity of stuff and we do some on-farm processing — vegetables, small grains, some dehydration of vegetables and herbs, and a big greenhouse. That takes some time to review."

Though inspectors are not allowed to advise or make recommendations, they can still be very helpful to farmers, Behar says.

If you have questions about the regulations, they can answer them. It's really helpful to have an outside person who has seen many, many farms come on your farm and help you assess your organic system. The questions an inspector asks can help you think about what you're doing. You don't have to be afraid of the inspector, and after the first or second time inspection usually becomes routine.

The biggest sources of errors are incomplete paperwork, Behar says. "Failure to document, such as failure to keep an equipment cleaning log, or to write down the day you applied your manure." Those types of errors, if they're minor and were honest mistakes, don't generally cost a farmer his or her certification. Errors like not rotating crops or using a prohibited material do prevent a farmer from being certified.

But identifying prohibited materials "is not always that easy," Behar says. "The Final Rule says everything natural is allowed, but how are you going to know if it's natural or not? Miracle Gro says it's all natural, but it's not. We've gotten the word out to always check with your certifier."

If you are trying to cheat on the Final Rule, Behar says, "you should be nervous. We are trained, and we have our little ways to check you out and make sure you're not lying."

In reality, farmers attempting to circumvent the rules is only a minor problem, Behar says. "I went to thousands of organic farms, and I found only a couple of people cheating in all those years."

Organic farming, Behar concludes, "is a tremendous opportunity for people who want to grow good food. It's incredibly exciting. If you like to learn and like to experiment, and you like feeling like a part of nature instead of being a controller, it's the way to go. We've noticed this at the organic conference (sponsored each year by MOSES): Organic farmers tend to be optimistic and happy. Even if they have a big problem, they know what they can change, and what they can try, instead of relying on outside input. You're more in control — you make the decisions. And when things don't work, it's an opportunity for learning."

Going to Market

Turning Farm Products into Profit

There are two things you should never forget if you want to be successful at marketing a farm product: you need a great product, and you need to create a convenient, pleasant buying experience for the customer.

The food you're selling has to taste good. No matter how organic you are, no matter how sincere you are and how interesting your story is, people don't buy a second time if your food doesn't taste good. Woody carrots, stringy beef, off-flavored cheese, wormy apples, grain full of weed seeds, one bad egg in a carton — who would pay a premium price for food they don't like?

So, before you sell, make sure you can produce a consistently good-tasting product. Eat it yourself, and try it on friends who will be honest. You'll begin to get a handle on what timing, methods, and inputs it takes to raise fabulous food on your farm in salable quantities, and you'll have gained some very necessary experience before jumping into big-time production. Too often organic food sold in the past didn't look good and didn't taste good, and that's what kills future sales.

Also make sure you're producing something that people want and that is in relatively short supply. If everyone else at the farmers' market is selling tomatoes, you probably aren't going to sell too many of your own. But if your tomatoes are three weeks earlier than everyone else's, or you have heirloom varieties, or you're certified organic, then you have less competition and a premium product that folks will seek out and buy.

When organic food is good, it's superb, and that is a huge part of your marketing advantage. Quality organic grains, vegetables, milk, meat, eggs, and fruit are nutrient-dense and have an extra sweetness and fullness of flavor that is rarely, if ever, found in conventional products. Once you have a functioning organic system that naturally produces that quality, you have the potential to

develop a loyal customer base, and that's the key to succeeding as a farm business.

Also important is that your product is easy to buy. If you're wholesaling grain or milk, then you will do best if you're punctual, honest, and meticulous about product quality. If you're selling direct to consumers, profit margins are generally higher, but there's a lot more effort involved. Having a great product that promotes good health and local independence isn't enough to make most customers buy twice, and repeat customers are the core of any successful direct marketing business. Customers return for a convenient, pleasant shopping experience and a reasonable price, so that's what you'll need to deliver.

You may have heard differently. For many years now the standard advice for direct marketers has emphasized having a story behind your product to attract buyers frantic for an opportunity to support small-scale farms and local, organic food production. Sure, there will always be a small percentage of buyers who will jump on the support-your-local-

The Organic Label Advantage

Putting the "organic" label on your products guarantees to your customers that you care for your land and that your food is healthier for them than conventional alternatives. This is the organic advantage, and consumers are seeking it out in increasing numbers every year. In many cases organic products command a higher price, but there are no guarantees.

If you put the "USDA Certified Organic" logo on your products, the label must also say, "Certified organic by (your certifying agency)," and it must identify the processor and distributor of the product. In many cases, this will be you, the farmer.

If you are selling directly to consumers, you may also want to create signs or a farm brochure explaining the basics of organic production and the story of your farm to promote a clearer understanding of organic agriculture, and of the health benefits of eating organic foods.

farmers bandwagon, no matter how mediocre the food, how high the price, or how inconvenient it is to obtain. But the reality, painfully learned by too many once hopeful farmers, is that most of the people most of the time will opt for taste and convenience. An organic label and a good farm story will bring customers in the first time, but these will not carry your farm business in the long run. Good taste and a good experience are the things that will. So make it tasty, and make it convenient. Then make it attractive.

Finding the Market

Another piece of advice that is a constant at direct-marketing workshops is to find your market before you produce your product. That is usually good advice, but it contradicts what I just said about learning to produce a quality product before trying to sell it. If you start small, you can do both at the same time. As you figure out your production system, you should spend time figuring out the possibilities for future sales and make some contacts with potential buyers.

If you're planning on direct marketing, start your market research by estimating how many people live within a half hour of your farm. Are

there enough to support your enterprise? Are enough of them interested in organic foods? If not, how far will you have to drive to reach a larger community, or will you need to focus on mail-order and Web-based sales?

Check out local farmers' markets, food co-ops, grocery stores, restaurants, schools, and hospitals, and see who is now buying local organic food or might be interested in doing so in the future. Ask customers at the co-op and farmers' market, restaurant chefs, and store buyers what they're interested in buying that's not available now. Find out what is for sale and for how much, and who your competition is. Get on the Internet and check producer directories (see page 415 for some suggestions) to get an idea of what organic products are being sold, and by whom, in your area.

If you do your research, then next season, when you have the production end of things worked out, you will also know where you want to sell. If all goes well, by the time you're ready to expand production and are confident of your product quality, you'll have potential markets and buyers lined up.

Starting small and carefully in this way tends to keep the inevitable mistakes small as well, and to minimize any collateral damage from disappointed customers and a poor reputation. Starting big is exciting and fun, but the potential for major burnout and expensive mistakes is high.

ORGANIC FARMING AT WORK

Flavorful, nutrient-dense organic produce provides consumers with superb food that has been raised to the benefit of, not at the cost of, the environment.

Wholesale Marketing

Grains and milk are more often wholesaled than retailed by farmers, since they most often are produced in volumes too large for a local market to absorb. Produce and meats may also be wholesaled in some areas. Even if you want to focus on direct marketing, you may want to line up a wholesale alternative for leftovers. If you intend to wholesale those leftovers to an organic processor, you will need to be certified organic even if you qualify for the $5,000 exemption. If you are going to wholesale to a conventional processor, certification doesn't matter. Though you'll receive a lower price for wholesaled products, you also save the time it takes to process and retail them yourself.

Fortunately options for wholesaling organic grains and dairy products have greatly expanded in recent years, and the opportunities for wholesale organic meat are finally starting to take off. On the other hand, downward pressure on prices has increased as large corporations jumping on to the organic bandwagon relentlessly try to cut their costs. Doing your research before making a sale is key to making a profit.

Organic Grain

Let's start with grains, the best developed of organic wholesale markets. Organic grain dealer Prescott Bergh summed it up nicely at the Organic Farming Conference in March of 2003 when he said, "If all you want is for the sale to happen without doing the work, you will get what farmers have always gotten: bottom prices."

Bergh says farmers wholesaling grains should use the Internet to track national and international prices, shipping costs, and overall market conditions, in order to know what price to expect, as well as costs for getting a crop to a buyer. Processors and buyers have definite preferences for particular varieties for particular uses, and producers should identify potential buyers and inquire about what they want before planting, he says.

Smaller producers should seek smaller-scale buyers, such as local and regional millers specializing in organic grains, or organic livestock operations looking for feed grain. Once you've identified potential buyers, make sure they're reputable and don't have a habit of breaking contracts. You, as a producer, need to keep your end of the bargain too, by delivering a clean, high-quality crop on time and for the contracted amount.

Very small grain producers also should consider selling the grain themselves. "Why doesn't every farmers' market offer on-the-spot milling of fresh wheat?" Bergh asks. "Why not a farmstead tofu plant to deliver tofu to area CSAs? Or a fresh-fried corn chip stand at fairs and festivals?"

OFARM and Cooperative Marketing

Organic grain farmers have pioneered cooperative marketing in the organic sector, with well-established groups in many areas of the country. In 1999 the Organic Farmers Agency for Relationship Marketing (OFARM) was formed as a cooperative of cooperatives to track and share information on inventories and market conditions and to negotiate prices with buyers. In most business circumstances such widespread collaboration in setting prices would be illegal, but under the Federal Capper-Volstad Act (aka the Farmers Cooperative Act) farmers are legally allowed to communicate with each other and act jointly in pricing their products. OFARM provides a mechanism to prevent buyers from playing organic producers against each other to get a lower price, a common tactic in any sales situation in which there are multiple sellers. For more information on OFARM, visit its Web site (see the resources).

Organic Milk

Organic cow's milk markets have taken off in the past few years, enough so that buyers compete for milk in some areas. But this certainly isn't true everywhere, and even in organic hot spots milk demand will fluctuate, so it's still absolutely essential for a dairy operation to start working with buyers before production starts.

Find potential buyers on the Internet or by checking with other organic dairy farms in the region, and contact them early. Beware of brokers, who buy milk to sell to processors, taking a cut of the profit for doing nothing more than getting your milk to the plant. Some are honest and provide a real service, but some aren't, and if you can keep their percentage of the profits for yourself your bottom line will look better.

Cow dairying is the exception to the rule of producing for a season before hitting the marketing end really hard. Unless you're starting with a single cow, once the milk starts flowing you have to have a buyer because you won't be able to afford dumping it down the drain. Yet dairying demands

Why Small Farms Fail

About half of all small-scale farms (sales under $50,000 per year) fail in their first five years, according to the U.S. Department of Agriculture's Economic Research Service (ERS).

"The total number of U.S. farms has changed little in recent years," states a June 2006 article in the ERS magazine *Amber Waves*.

> *High exit rates are offset by high rates of entry into farming. There is no apparent shortage of people willing to try farming, but the challenge is creating a viable farm business.*
>
> *Many preharvest crop decisions (such as the timing and extent of soil preparation, seeding, and pest management) vary with local soil and weather conditions, and operators often learn through trial and error as much as through training, extension services, and suppliers. Similarly, successful livestock enterprises require breeding, feeding, and culling savvy that improves with experience. Marketing decisions — when to sell, how much, to whom, and under what kind of arrangement — also benefit from experience and new information. Moreover, the relevant experience is specific to a particular farm business.*

In other words, don't quit your day job until you know what you're doing on the farm.

The North Carolina University's *North Carolina State Economist* noted in a September/October 2006 article that "business failures are especially catastrophic for farmers, since their primary residence and land are generally tied up as a part of their capital investment. When failure occurs, it can be costly."

The article states that failures frequently occur 12 to 18 months after the onset of operations, at the time when most farms are increasing production and expanding their customer base. Problems tend to occur in each of the three major areas of a business: management, marketing, and finance.

> *Most individuals are incapable of being successful in all three of the major areas . . . because of this, few growing small businesses can continue to thrive with only one person handling all aspects of the business.*
>
> *Successful family farms traditionally have divided responsibilities between spouses or between parent and an adult child or between siblings. . . . One-person businesses tend to burn out because they cannot balance all of the responsibilities of their operations and perform equally well in all tasks.*

a higher level of expertise than any other type of farming, as was pointed out in chapter 12, so having experience is crucial. The message here? If you didn't grow up on a cow dairy, then you should consider working on a cow dairy before launching your own operation.

Sheep and goat dairies tend either to form their own producer cooperatives or to process and sell their milk on-farm, since commodity markets have not developed for these products. A few processing plants do exist, but the reality is that if you're producing niche milk, unless you're in one of the handful of areas that have viable networks of producers and processors, you'll likely be processing it yourself. However, demand for goat and sheep cheeses is currently on the increase, so more milk-processing plants may open up in the near future.

To find a producer cooperative or processor in your area, search the Internet, contact your state extension service, and check in with local veterinarians and farmer networks. If you do find a processor, be warned: in the past plants have opened and then closed just a short time later, leaving producers holding a lot of milk. Proceed with caution, and get a good contract!

In marketing, the article states that two common mistakes are failing to identify a distinct competitive edge and not understanding the product from the customers' perspective. "Successful agricultural business owners use their knowledge and expertise as a tool to cut through customers' information overload," the article continues. "Many value-added producers have been successful in using customer relationships, on-farm visits, and internet sites that facilitate direct-to-the-consumer communication as ways of attracting and retaining customers."

In terms of financial management, says the article, the key mistakes are poor record-keeping and poor cash-flow management.

Inexperienced farm owners often co-mingle their business funds with their household accounts. This means that they do not have a clear picture of their income and associated expenses, and, as a result, cannot get a good handle on how cash is flowing through their operations. . . . Without historical records and cash flow statements to keep things in perspective, producers may forget that money needs to be set aside to pay taxes, repay operating loans, or build savings to cover fixed expenses incurred during the months when no income is generated.

Organic Meat

Until 1999 the USDA refused to allow the "organic" label on meat. So many other labels, like "natural" and "grass-fed," were used instead — and are still used today — that consumers were left confused or with other loyalties. Fortunately, that problem is quickly disappearing, and demand for organic meats has taken off to the point that wholesalers are importing it from overseas, since there aren't enough certified organic livestock producers in North America to fill the demand.

Regional and national organic meat marketing companies and cooperatives are usually looking for more producers, though you need to be in the right location and adhere to standards and protocols. Individual producers in more remote locations have had some success with internet sales, shipping frozen products to customers across the country.

A few small meat processors are beginning to develop local markets for organic products, so also look for small slaughtering plants that are certified organic. Also consider buyers that are not certified organic but are selling their products under other labels such as "all natural" and "grass-fed"; an organic grower can certainly sell to them if the animals fit the buyers' criteria. All such buyers will have specific production and carcass quality requirements that producers have to fulfill, and many will deal with individual smaller-scale producers. On the whole, though, until you are producing more than a few animals a year, your best marketing bet is probably direct marketing to end users.

Backup Markets

Backup wholesale options do exist for some products. Cattle and sheep are commodity items and can always be sent to an auction barn for sale. You won't get any premium for being organic, but you do have an avenue to get rid of excess animals. The same goes for cow's milk and commodity grains (wheat, corn, and soybeans), which you can usually sell on the conventional market if an organic sale falls through, or at least to a local feed mill.

For other products, backup options are spottier. You may be able to sell excess products to a canning factory in your area, for example. During the growing season some areas have weekly produce auctions, where produce is auctioned off to buyers from grocery store chains and other food businesses. Again, you may not get any organic premium, but something is better than nothing.

For a niche producer with excess eggs, mushrooms, goats, specialty vegetables, and similar items, there are even fewer options. There's not often an auction barn or processor to take any excess off your hands. On the other hand, a little extra production can be turned into a terrific marketing opportunity. Consider distributing samples to local restaurant chefs, donating to a church fund-raiser, putting up a taste-testing booth at your local fair, or giving to a food shelf. Giving people a free taste of some really good food is one of the most effective ways to increase future sales.

Direct Marketing

Producers of organic poultry, eggs, fruits, vegetables, and niche items most commonly sell direct to consumers, either from the farm or at farmers' markets. There's no shortage of organic meat producers in this sector either, along with the occasional small-scale grain grower and those dairy farms doing on-farm processing.

Selling to businesses that sell to consumers (as is the case for restaurants, food cooperatives, grocery stores, schools, hospitals, and the like) is becoming more common but demands more consistent quality and volume on the part of the producer than direct sales. Other direct-marketing options include cooperating with other local growers to organize delivery routes or a storefront. There are pros and cons to each marketing option, so it's important to consider each one carefully in terms of your inclinations, resources, and any future plans for expansion.

On-Farm Sales

Let's start at home. Selling direct from the farm is handy because you don't have to load everything into a truck and drive it to the market or deliver it to customers. This saves a lot of time and gas. But if you are inviting the public onto your farm to buy, your farm should be tidy and clean at all times. Most customers will be dismayed to see pieces of dead equipment lying around the yard and a sea of mud around the barn. Keep the weeds trimmed and plant some flowerbeds, put a rocking chair on the porch, and make sure the dog is friendly to strangers. The more curb appeal your farm has, the more likely people will stop by.

Farms that are good at direct sales usually have a small stand or store with convenient parking,

since most customers will be a little hesitant about knocking on the kitchen door to see if you'll sell them some eggs. Parking is important, and it needs to be obvious. Farmers aren't shy about parking on the grass, but suburbanites tend to be, so make clear where they should put their cars.

Signs are important too. They need to be easy to read for cars whizzing by at highway speed, and you'll need several: one big one in front of the stand and a few more far enough away in either direction that people have time to slow down and make the turn.

Minding the Store

The more items you have for sale at your stand or store, the longer the sales season and the bigger the incentive for people to stop and buy. A big variety of vegetables along with some flowers and in-season fruit can make a really attractive display,

The Marketing Personality

On farms that are in business for the long term, marketing and production tend to quickly become two separate, full-time jobs done by two separate people with two different personality types.

For introverted types, one of the attractions of farming is the opportunity to work alone much of the time. There's time to think, plan, and listen to birds. Often these folks aren't interested in or good at sales.

Sales require an outgoing personality and a genuine liking for people. Oftentimes, a spouse or sibling who's not too keen on getting dirty and sweaty every day will enjoy stocking and tidying sales areas or stands, managing books and inventory, and keeping customers happy.

Fortunately, it's fairly common for an introvert to marry an extrovert, and when they start a direct-marketing farm business there's a career niche for each of them. If you're a solo farmer with no time for selling, consider pairing up with another producer or a group of producers for marketing purposes, or hiring help. Trying to do everything yourself can be a short road to burnout.

Signs

Sign makers recommend that signs be a minimum of 4 feet by 4 feet (1.2 x 1.2 m), with letters a minimum of 4 inches (10 cm) in height. Red or black letters on a white or yellow background are easy to read; green letters aren't. Use print instead of script, and always fewer words rather than smaller letters. Make sure the sign won't blow over, and since you've spent some time making it look professional, use durable materials and paints.

Local laws about sign placement and size vary widely, so check with your local and county zoning offices and your state department of transportation about regulations for your area. Generally you can't place a sign in a road right-of-way (which will vary by type of road) or attach it to a traffic sign, and some municipalities don't allow any signs that are not on your own property. (Of course, you should never place a sign on someone else's property without first getting that person's permission.)

In your store, stand, or booth, have prices clearly posted in large print for the convenience of buyers. Handouts with recipes or the story of your farm and production techniques are good sales aids, too. Above all, be sure to display your "organic" label front and center. This is exactly what many buyers are looking for, so let them know you have it!

with something for all of your customers. As your customer base grows, you might consider purchasing additional organic products to expand the inventory. Fair-trade coffee, organic spices, canned goods, and other items can really round out a grocery list for the customers and increase sales for you. Another option is to invite other organic growers in the area with different products to sell them through your store. Or you could sell your products at someone else's store. This expands the inventory, and maybe participating farmers will help staff the store so you don't have to be there every second.

On the other hand, if you end up selling more of what other people are producing and processing and less of your own production, your state or local government may decide that you are a grocery store and not a farm store and require you to have various permits and licenses in order to operate. They might even close you down if you don't conform to local zoning rules.

One of the biggest drawbacks to on-farm sales is that every time customers pull into the yard, you have to drop what you're doing and assist them. This can be a real pain if you're in the middle of a busy season. There are a few ways to deal with this problem. The first is to put a price list and cash box in the stand and trust that your customers won't steal or cheat. In many rural areas this is a time-honored tradition, and if you trust your neighbors this can work quite well. A second option is to have the stand or store handy to the house and install some kind of alert system, such as a wire in the driveway or even a doorbell customers can ring in the stand, so that when someone drives in, whoever's in the house can stroll out there to lend a hand. That way no one has to sit idly in the store when there are no customers around and other work needs to be done. A third option, and one that

Attractive, readable signs quickly let people know where to find good food.

becomes necessary if the store starts doing a fair amount of business, is to hire someone to staff it.

On-Farm Meat and Poultry Sales

On-farm meat and poultry products may require a bit of a different approach. With fresh fruits and vegetables, production is more or less continuous through the season and so are sales. With livestock, butchering occurs just once or twice a year (maybe three or four times with chickens), so either customers preorder and then come on butchering or delivery day to pick up their meat, or you'll need to have some big freezers on hand to store the meat.

With pork, beef, lamb, and cabrito (goat) sold by the whole or half carcass, it usually works best to have customers pick up direct from the processing plant rather than the farm. Poultry, on the other hand, is commonly processed on the farm, so you'll need to set up times for delivery or pickup. Even if you're one of those fortunate people with a small-scale poultry processing plant within a reasonable distance, you'll still often have to bag and deliver the chickens yourself, since many of these plants will not have the freezer space to hold your birds for customer pickup.

For whole, half, and quarter carcasses of the larger animals, marketing should start before the animals are processed. On the other hand, if you have a farm store and freezer space you don't have to have all the meat preordered; instead you can sell it by the cut. Freezers take a lot of space and are expensive to run, but they greatly increase your potential customer base, because they allow you to sell to all the folks who don't have their own chest freezer for large quantities.

Many processors can prepare meat as sausages, burger patties, jerky, and other specialty meat items for farm-store sale. These are popular with consumers and command a higher price, though of course you have to pay more to have the processing done.

The CSA Option: Community-Supported Agriculture

One of the best options for on- or off-farm sales of vegetables may be community-supported agriculture (CSA). In a CSA operation, farmers sell shares in their annual production rather than directly exchanging cash for food. Customers pay for their share or partial share in the spring and then pick up or receive delivery of a weekly box of vegetables throughout the growing season. For example, I might pay you several hundred dollars in March for a share in your CSA. You would start my weekly deliveries in, say, mid-June with greens and peas, and each week after that I'd get a share box packed with a selection of fresh, delicious, in-season vegetables. The best part would be the heirloom tomatoes from July through September, and we'd wind to a finish with root crops and winter squash in early November.

Many CSAs have expanded their share offerings to include fruits, flowers, herbs, and eggs, and CSAs can be a sales conduit for other local organic producers of meat, honey, mushrooms, and other items. To make it truly a community farm, many CSAs distribute recipes, newsletters, and information on freezing and canning the vegetables their customers can't eat in a week, as well as hosting open houses each year and sometimes requiring members to contribute a few hours of work per season. In this way CSAs build up loyal customer bases, and many hold annual appreciation events that include farm tours and great food.

With a CSA farmers know what their income will be for the year and can plan for expenditures accordingly. The customers share in the risk of losing a crop, but also in the pleasure of eating with the season, discovering new varieties and types of vegetables, and having access to a real working

ORGANIC FARMING AT WORK

A CSA offers consumers a rare opportunity — the chance to participate in producing healthy food and to be a part of the local farming industry.

farm. Best of all, a CSA farm is truly a community. You can't buy that at Wal-Mart, not even in the organic aisle!

If you're not interested in having your customers involved with your farm to that degree, then call your operation a "subscription" farm instead, and you can focus on simply selling enough shares and getting the vegetables delivered.

Farmers' Markets

The farmers' market has enjoyed an amazing renaissance over the last 20 years. In 2014 the USDA Agricultural Marketing Service listed 8,144 farmers' markets in the United States, which is 6,500 more than in 1994, when the AMS first began keeping count. In 2013 the estimated total sales volume for all local food sales was $7 billion, and average sales per vendor was $7,108, but there is enormous variation between markets and vendors. About 25 percent of vendors rely on the markets as their sole source of farm income.

In many larger cities farmers' markets have their own dedicated facilities and are tourist attractions, full of strolling musicians, sandwich and coffee vendors, and jugglers and clowns for the kids. Wares go far beyond fruits and vegetables to cut flowers, frozen meats, fresh breads and pastries, honey, shiitake mushrooms, and potted herbs. In many smaller towns the farmers' market is just a few local growers in a roped-off parking lot on a Saturday morning. Though the offerings may be a bit plainer, they're still a wonderful alternative to the supermarket's produce aisle for people who appreciate the taste of real food.

The downside of selling at a farmers' market is the time commitment and the variability in sales. If you're any distance from the market, typically you'll rise in the wee hours to pack the truck, make the drive, and set up your stand before opening. Unless the market is a major one that's always crowded, your sales volume will vary greatly depending on the temperature, whether it's raining, whether tomatoes and sweet corn are in season, and the other unpredictable whims of a smaller customer base.

The upside of a farmers' market is the delightful experience of talking to people who are really interested in what you're selling and how you produce it. For gregarious types farming can be a little lonesome at times, and a day at the market is reenergizing. It's also a great way to build a customer base for on-farm sales, a CSA, a delivery route, or other marketing avenues that are more convenient or profitable for the farmer. Some producers sell at a market only when they have excess product, while others rely solely on farmers' markets, sometimes selling at two or three different ones in a week.

As with an on-farm store, an attractive display, easy-to-read signs with prices clearly stated, and a friendly farmer greatly help sales volume. Though it's tempting to skip shaving and not to bother changing out of your barn clothes, if you look like

An attractive stand with a diversity of products and prices clearly marked makes customers want to come back.

a grunge many people will suspect your products are a little grungy as well.

Most markets have seller rules and require a payment for a sales space. Many require that you reserve a space, and some have waiting lists for vendors, so that you'll have to wait until another grower quits before you can get space. If you are hoping to sell at a farmers' market, locate the market manager to obtain information on market rules and fees. The most common rule is that only local producers can participate.

Some areas host winter markets, where customers can purchase dried, canned, or frozen goods from local producers, as well as fresh eggs, baked goods, decorative items such as wreaths and garlands, and those vegetables and fruits that keep through the year: squash, apples, carrots, cabbages, and so on. This is a good option for producers interested in year-round cash flow who have the creativity and processing facilities to create these types of products.

Restaurants, Food Co-ops, and Grocery Stores

These types of businesses are excellent marketing opportunities, especially as demand for locally grown, organic food continues to grow. But don't expect to show up out of the blue at the back door and sell a few vegetables at a premium price. Do expect to drop off product samples and be ready to discuss what you will have available in what quantities at what times.

Restaurant chefs have to plan their menus and calculate amounts ahead of time, and though many independent restaurants are delighted to feature organic foods on their menus, they have to have a reliable supplier. Consistent quality at the promised volume is essential in this market.

What's best for both parties is building a long-term relationship, where producer and chef plan together what rare and delectable varieties of heirloom vegetables and exquisite berries to grow for the coming season's menu. In fact, it's fairly common nowadays for such restaurants to credit on the menu the local farms that provide their organic pork, beef, poultry, and eggs. The restaurant gains a marketing edge and a reputation for great food, while the producer has a reliable buyer for a steadier cash flow.

One drawback to working with restaurants is that in many cases sales volume is fairly low and deliveries (of necessity) are quite frequent, so that a producer is making a lot of trips for not a lot of sales. For a small-scale producer, this might be okay. In a larger city, it may be possible to work with several restaurants, or bigger restaurants, or to expand your marketing to other urban outlets.

Food cooperatives often will deal with local producers, and so will some independently owned grocery stores. The same rules hold as for restaurants: a producer has to demonstrate superior quality, a unique product, and the ability to deliver consistent volume, on time, to get and keep these sorts of accounts. Volumes tend to be higher than with restaurants, but it's still important to inform your buyer of what is coming or not coming. You'll also want to cooperate with the buyer to publicize the unique qualities of the product (organic, local, heirloom, grass-fed, and so on) through advertising that will bring customers into the store.

Hospitals, Colleges, and Schools

Hospitals, schools, and colleges all feed a lot of people, and many are interested in using locally grown, organic food. Nevertheless, this is a much tougher market to break into, since food buyers working for these institutions often are constrained by contracts with corporate food suppliers. Unless an institution has formally adopted a policy to encourage use of local foods, a producer may be out of luck. But only for the moment — in communities around the country producers have developed some innovative strategies for getting their food into local schools and hospitals, often working as a group.

Field days at farms, special meals featuring local organic foods, and meetings with food-service staff to offer taste tests are all good ways to establish working relationships with institutions. Other ideas and resources for developing these sorts of markets are available from a number of groups, including the Community Food Security Coalition, which runs the National Farm-School Network, tracking colleges that are buying local foods and assisting producers in establishing a relationship with colleges in their communities.

Internet and Mail-Order Sales

The jury is still out on whether Internet sales are worth the trouble for small-scale farmers, and the situation is evolving constantly. Some producers find that their Web sites generate a good volume of sales; others find the opposite to be true. Much depends on the demand for your product and how easily your Web site can be found via search engines and online directories of producers. Website quality and customer service also play a role. But the overall trend is unquestionably that more and more consumers are comfortable shopping online. Even if your Web site doesn't directly generate sales, it certainly increases your ability to advertise your product, and it can direct potential buyers to your stand at a farmers' market or to your on-farm store. For that reason, it's becoming a necessity to have a presence online if you intend to be in business for the long term.

You don't need to have your own Web site to have an Internet presence; you can get on some directories instead. The granddaddy of Web-based local producer directories is Local Harvest (see the resources). Listings are free, and the site is well known, attractive, easy for buyers to navigate, and full of other information and resources.

Be sure to check your state department of agriculture Web site, too, to see if it offers a similar service for state producers, and search for private regional organizations that create directories as well. In marketing, getting your name out there is important, and the Internet has become a good tool for doing so.

Agri-tourism and You-Pick

If you enjoy talking with folks and showing off your farm, an agri-tourism business can be an enjoyable moneymaker and a terrific opportunity to create enthusiasm for your products and the idea of supporting local organic growers.

How you attract folks to your farm is limited only by your imagination. Farmers have done everything from setting up bed-and-breakfasts and retreat centers to organizing events centered on everything from apple picking and cider making to corn mazes, benefit dinners, summer farm camps for kids, farm museums, centers for local artists, barn dances, baby-animal-visiting days, catch your own fish — the list goes on and on.

The original agri-tourism attraction was the "you-pick" operation, and this is still a reliable marketing method, if you're convenient enough to a large potential customer base. The most common attractions are pumpkin patches and fruits, from apples to peaches, blueberries, and strawberries.

Many state extension services furnish information on developing agri-tourism attractions, covering all aspects from business planning and insurance to advertising and hiring help. Federal and private resources are available as well. See the resources for more information.

Other Marketing Ideas

Organic farmers are a creative bunch, their production skills honed by the job of tailoring available tools and methods to their unique farm situations. Many of them carry this attitude into their marketing as well, so that marketing methods seem to be as varied as the farms the food comes from. To get your creative juices flowing as you begin making your own marketing strategy,

Kids are a natural addition to a pumpkin patch, and a "pick your own" operation is an excellent option for the organic farmer.

here's a short list of some unique ideas now being used by organic producers:

- Remodeling trailers into attractive, portable stands for selling vegetables and fruits in shopping-mall parking lots and similar venues on regularly scheduled days (with permission of the management, of course)

- Organizing cooperative processing and sale of meat goats to ethnic communities in nearby cities

- Establishing a group home for the developmentally disabled at a CSA farm to provide those adults with meaningful work and a healthy environment

- Building a wood-fired oven and small commercial kitchen to make and sell pizzas, featuring vegetables from the market garden, two nights a week

- Using a former frozen-pizza delivery truck to sell locally produced dairy and meat products to customers along a scheduled route

- Organizing a local-foods tasting dinner as a benefit for a local nonprofit organization, such as an arts program, youth program, church, or other community group

Value-Adding Ventures

As long as you're doing your own marketing, why not expand your product line and extend your sales season by doing a little preserving, processing, or crafting? If you can turn your extra tomatoes into salsa, you've added value to the tomatoes, meaning your tomatoes are worth more. They can also now be sold year-round, which helps even out the cash flow.

Value adding has been the battle cry of small-scale producers for many years now as they launch themselves into the business of making a living in the food sector. Corporations and middlemen have the huge advantages of economies of scale and massive marketing budgets to create relentless price pressure on small farmers. The goal of

direct marketing and value adding is to take back those jobs of marketing and processing to regain a bigger share of the sales dollar. In this way, a small-scale producer might make enough money to have a decent lifestyle. The producer's big advantages are the quality of the food and the goodwill of consumers dedicated to supporting small-scale, local agriculture. The producer's disadvantages are the labor of doing three different jobs and having to compete against corporate control over retail shelf space.

A Word of Caution

Processors, brokers, wholesalers, and retailers exist for a very good reason: all are big jobs that require time and expertise. If you're interested in adding processing to your production, then do yourself a favor by thinking it through long and carefully before you start.

On-farm value-adding enterprises generally require considerable additional equipment and labor, and in most cases they involve state regulations as well. Anything beyond washing and packaging fresh fruits and vegetables is probably going to concern your local state inspector, so the first step is to find out what rules you'll need to follow.

These rules fall into three categories, governing meat and poultry, dairy, and processed fruits and vegetables.

Value-Added Meat Products

Meat regulations are the most strict; slaughtering and processing animals on your farm for sale is illegal in most cases. The exception is poultry, but only if you stay under the maximum number of birds that your state will allow you to process on your own farm.

In most cases this isn't an issue, since butchering large animals in numbers large enough for sale is a bigger and messier task than most producers would ever think of taking on. But livestock producers can work with a processor to develop value-added products like sausages, wieners, jerky sticks, specialty cuts like boneless chicken breasts, and specialty packaging such as preformed hamburger patties, or chicken wings for barbecue dips. To develop your own premium specialty meat

Restrictions on Direct Meat Sales

Meat processing plants are either state or federally inspected. Only meat processed in federally inspected plants can be sold across state lines, while meat processed in state-inspected plants must be sold within the state. (The exceptions to this law are Indiana, Ohio, North Dakota, and Wisconsin, which participate in the USDA's Cooperative Interstate Shipment Program.) Since federally inspected plants generally deal almost exclusively with large-scale producers and feedlots, it's difficult for a small-scale producer to get processing that allows them to sell across state lines. This has been a major impediment to sales for many small-scale producers who happen to farm just across the state line from a large urban area.

If you can persuade your out-of-state buyers to drive to your farm to purchase their meat, there's no problem. If you want to deliver to customers in another state, tough luck. You can't.

State-inspected plants often have standards equal to or more stringent than federal standards, and small processors are their bread and butter. Many small processors will work with you so that you can be quite fussy about carcass hanging time, how cuts are done and packaged, and developing value-added products like sausage and jerky. But you cannot label your meat products as "Certified Organic" unless both you and the processor are certified.

products you'll need to find a small-scale processor willing to work with you in developing products. In some states such as Wisconsin this is fairly simple, as the state is littered with small meat plants. But in Colorado and some other states, it may be nearly impossible because big feedlot operations have pretty much driven all the small plants out of business. One answer might be a big livestock trailer and truck for making some long hauls to a processor.

Value-Added Dairy Products

Dairy value adding requires (just like dairy production) a fair amount of up-front investment in equipment, and a close relationship with the state inspector. The reward is yogurt, flavored milks, ice cream, fresh soft cheeses, and (if you're really ambitious) your own cured cheeses. This will involve, at a minimum, a lot of experimentation on your part, and better yet some cheesemaking courses and possibly an apprenticeship with an artisan cheesemaker. Some universities and state extension services offer cheesemaking courses, and there are now artisan cheesemakers around the United States who offer classes or may take on an apprentice.

Other Value-Added Food Products

For value adding that involves drying, canning, baking, cooking, and similar activities, you will need a licensed commercial kitchen in order to be able to sell your products. Materials used, dimensions, and required temperatures for various processes are regulated by the government, so once again it's smart to contact your inspector before you get started.

Buying all-new stainless-steel counters, sinks, and other required parts is too expensive for most start-up operations, but fortunately you can find a fair amount of used stuff on the market, though you may have to search for it and drive pretty far to pick it up.

You will also need a separate facility, a building or room that is quite separate from the rest of the house. You can't just remodel your kitchen.

Those are the basics. For more information on creating value-added enterprises many additional resources are available through your state extension service, ATTRA, and similar groups.

But, as with anything connected with small-scale organic agriculture, the best way to learn about value adding is to visit operations that are doing the sort of thing that you're interested in doing.

Just to get your creative juices flowing, here's an incomplete list of some more value-added products now being successfully made and marketed by small-scale producers around the country:

- Herb-infused oils and vinegars
- Wine and beer from local fruits and grains
- Jellies and jams
- Salsa and pickles and chutneys
- Dry soup mixes
- Mixes for pancakes, bread, and muffins
- Teas
- Salves and ointments
- Soaps
- Dried herbs, fruits, and veggies
- Wreaths and garlands of dried flowers and herbs
- Custom-spun and dyed yarns
- Specialty sheep and goat cheeses
- Pies made with in-season fruit

Packaging and Processing

It would be so much easier if you could just leave the lamb tethered to your customer's back door, or toss the carrots directly into a bucket and drop them at the grocer's. But customers generally like their food clean and in nice bags, boxes, or packages.

No matter what you're selling, if you're organically certified, then the "USDA Certified Organic" label will attract customers. If you're entitled to use it, be sure to put that label on everything you sell. This assures your customers of your integrity, commitment, and quality. As the "organic" label continues to gain recognition and respect in the marketplace, it will continue to benefit you by increasing your sales potential.

A farm logo and label complete with the farm name, address, and phone number or e-mail address is good marketing strategy as well, but be careful here. You can spend more money developing a logo and buying all the business cards, labels, stationery, and other geegaws than is justified by your sales level. When you begin selling to customers who aren't your friends or neighbors, that's the time to worry about creating brand identification for your product.

Vegetables and fruits should be cleaned and bundled, bagged, or boxed. This means a washing facility with counters, sinks, and good water flow. Doing it all by hand is difficult once you've built up some sales volume. Some growers, as an example, have reworked old washing machines to rinse greens. Visit some other operations, use your imagination, and see what will work on your farm.

Most meat processors use vacuum (often called "Cryovac") packaging now, though here and there you'll still find white butcher's paper. Vacuum packaging has a longer freezer life. Either way, packages should be clearly labeled by your processor with cut, weight, and date. If you're doing further processing into wieners, jerky, or other products, the package will need a complete ingredients label.

On-farm dairy processors will have to invest in tubs, jugs, and cheese-wrapping equipment, and the state inspector should be able to inform you of any labeling requirements.

Setting a Price

Pricing is one of the trickiest parts of marketing. There are three things that influence the price you set: what a consumer is willing to pay, what the competition is charging, and how much you need to make per unit to cover your labor and input costs for that unit, whether it's a chicken, an egg, a head of broccoli, or a gallon of milk.

If you can't cover your cost of production, you're going to be out of business fairly quickly. But you can't expect 100 chickens or a row of pumpkins to pay for the entire farm, either. That's why cost of production is figured per unit, rather than by what it costs to run the entire farm.

For example, a day-old chick may cost $1.25, plus 8 pounds (3.6 kg) of feed to grow it to a light butchering weight of 8 pounds (3.6 kg). Let's say certified organic feed that season costs $2 per pound, so the total cash costs are $13.25 per bird. Then there's the cost of a waterer, feeder, pasture pen, and brooder construction. These are all

Consumer Groups

You aren't alone out there in the world of direct marketing. Aside from any organic farmers and farmer networks in your region, a plethora of non-profit consumer organizations and groups promote local and organic foods and, though not directly concerned with helping farmers, are a source of marketing ideas, information, and resources. Familiarize yourself with their names and activities in the interest of tracking consumer trends and how those might present marketing opportunities. Consider getting involved with any groups in your area, too. That's a source of potential customers, and it's a good feeling to meet some of the many folks that are out there rooting for you in your dedication to producing great organic food.

Some of the larger and better known of these consumer groups are Slow Food International, which began in Italy to promote home cooking from scratch from fresh, local foods as a way to reclaim the flavors of the past and the joy of taking family time to prepare and eat good food. The Food Routes Network works with local partners to establish state and regional "Buy Fresh Buy Local" chapters that promote local foods. Search, too, for groups working exclusively in your region. Many of these emanate from colleges and universities.

sufficient for 50 birds and should last at least 5 years, for a total of 250 birds. Let's say that all cost you $100 (just to make the math easy), or 40 cents per bird. There's also the cost of running the heat lamp for the first three weeks, let's call it 35 cents per bird. The total is now $14 per bird, and these are the costs you have to recover to not lose money.

Next there are your labor costs, or how much you want to get paid for what you do. Most beginning farmers don't care so much about this the first year or two, but then reality hits and you realize you can't work for free or even for minimum wage, since you have your own living expenses to pay. If your household's annual budget requires you to make $15 per hour, and it takes you half an hour a day to fill the waterer and feeder and move the pens, plus a full day of butchering when the chickens are ready, then you divide your total time input by the number of birds to find how much time you spent per bird. Let's call it an hour, so add $15 to your costs, and you have to realize a return of $29 per bird to recover your costs and make a decent living. If the carcass weighs 5 pounds, then your ideal price is $4.85 per pound.

This brings us to the second influence: what other producers are charging. If the other vendors in the market are charging $4 per pound, you aren't going to sell many chickens at your higher price. But if other vendors are charging $5 per pound, then you're going to sell everything you have long before they do, and earn their dislike. Nobody likes a price spoiler. Adjust your price to be at or very near the price of similar products, unless you can demonstrate a unique quality that makes your product worth more, or you can announce a short-term introductory sale in order to (briefly) charge a little less to get rid of excess product or attract new customers.

That unique selling point brings us to the third influence on price: what consumers are willing to pay. Let's face it, nobody is going to pay $4.85 a pound for regular old chicken. But if it's organic, it's locally produced, and, above all, it tastes fabulous, then a certain number of people will pay. That's why it is so important to have "organic" on your label, along with your farm name, and to take the time to explain to people who you are and what are the unique qualities of your products.

Finding Out What Everyone Else Is Charging

At a farmers' market or grocery store, it's fairly simple to find out the prevailing price for different items: you just walk around and look. But finding out what other people are charging for organic wheat, beef, goat's milk, and many other products is not so obvious. Fortunately there are Internet sites that track prices of organic, direct-marketed foods on an ongoing basis. The New Farm Organic Price Index tracks fruit, vegetable, grain, and herb prices in markets across the country and compares them to prices of conventionally produced foods. The index uses data gathered by the Agricultural

Marketing Service, which also tracks wholesale prices for organic milk. (Both indexes are listed in the resources section.) For products not covered by these indexes, such as goat's milk or organic mushrooms, you may simply have to find some producer Web sites or make some phone calls to find out what price the market will bear.

Advertising

Folks don't come looking for your products; you have to shout out that you're there. The first part of effective advertising is how you present your products, and the second part is what media you use. In both cases, budget will be a primary consideration. Professional logo development, labels, business cards, brochures, signs, and paid advertising can cost quite a bit of money. When you're just starting out, the cost may not only be more than you can afford, it might be unnecessary as well. If, for example, you're concentrating on farmers' market sales, homemade neatly lettered signs are probably all you need, with a pad and pen handy to jot down phone numbers and e-mails of customers interested in receiving your newsletter. Most home computers and printers are capable of turning out rudimentary brochures and cards. They don't have to be fancy; they do need to be easy to read and contain the essential information. If you don't feel creative, you might be able to barter some vegetables with a young writer/designer in the area to put together some basic materials for your business.

Once your production has expanded and you're seeking additional markets, then might be the time to redevelop a farm logo and label into something more colorful and fancy, have professional signs made, and put a little more money into getting the word out through paid ads.

Do-It-Yourself Advertising

But before you start spending money on paid ads, consider all the free and frugal ways you can promote yourself. In fact, the best advertising in the world you can't buy: word of mouth is hands down the most effective method of bringing in new customers, and word of mouth depends entirely on satisfied current customers. Treat your customers right, and it will repay you many times over.

This really isn't hard to do. The basis of your relationship will always be tasty, high-quality food. Your job is to use that base to build customer loyalty to your farm and your product. Farmers who genuinely like their customers and have their best interests at heart find this easy, and that description fortunately fits most small-scale organic farmers. Taking time to talk to people to find out what they're looking for and tell them how you operate starts a lot of beautiful friendships. Many farmers pass out brochures, send out newsletters, or maintain blogs to keep their steady customers informed of what is happening on the farm, for the benefit of those many customers who are seeking not just food, but a farm experience.

Samples are another low-cost and very effective means of creating sales. Offer samples at the farmers' market, or talk to your local food co-op about setting up in one of their aisles. Maybe you can organize a taste-testing event with other local producers to benefit your church or local arts organization. Drop off samples with restaurant chefs, food buyers for retail stores, and local schools. Giving people the experience of the superior taste of organic products gives them a strong incentive to buy some. With every sample, be sure to give out a brochure or business card as well, so the taster will know where to look to find more of that great food.

Another extremely effective and virtually free form of advertising is a story in your local paper or magazine on your operation. Writers are always looking for a good story — that paper has to be filled! — and small-scale, organic farms have all the specifications: scenery, local interest, nostalgia, and good products. But don't be pushy. Reporters detest getting harangued about doing a story. You could simply put the local farm and lifestyle reporters on your e-mail newsletter list, to let them know you're there and what's going on. Or, if you have a farm tour, field day, or similar event scheduled, invite the local paper and radio station.

Paid Advertising

Sometimes it's a good idea to buy ads. If you have got a lot of product to move in a short time — say, bedding plants in springtime or pumpkins before Halloween — then the quickest way to get the word out to the biggest possible audience is a paid ad. Generally your options for local sales are the local newspaper, weekly "shopper" type publications, and radio stations, but don't overlook Internet opportunities. Producer directories, tourism Web sites, and municipal Web sites might list your hours and products, especially if you're holding an event like a pumpkin tossing contest, or a special sale day for heirloom tomato plants.

Media ads in general (newspapers, magazines, and shoppers) can be hit-or-miss. Popular items with short seasons (sweet corn, garden tomatoes) seem to attract more attention than big-ticket items (sides of beef) or longer-season, less popular things (winter squash). Ads are priced according to size and where in the paper they're located: a small ad in the far back is cheap, while a big ad at the bottom of page three is pricey. This reflects a truth of media advertising: the bigger the ad and the farther forward it is in the publication, the more response it generally gets. Maybe your best bet is to talk to some of the other growers in your neighborhood or farmers' network and find out what their experience has been with media ads in your area. Then you can make an informed decision about how to approach your media advertising campaign.

Radio ads present a similar dilemma: prime-time ads on popular stations cost a lot more than off-peak ads on small stations, because it's

Bring samples when you go to talk to restaurant chefs — a taste of your high-quality organic food can make the sale.

assumed you'll have a better response. Once again, talk to other producers before making a decision.

A final word on paid advertising: keep it short and simple. Print ads cluttered with too many words and prices lose readers, while radio ads (and TV if you're really going big-time) are most effective when loud and brief. State what, when, and where, and leave out unneeded details. You can tell folks the rest of the story when they get to your farm, booth, or event.

Legal Considerations

Laws and rules are what keep us from running afoul of other people and having to settle disputes with fistfights. So, even though some regulations make no good sense, and plenty of them seem like nothing but a nuisance, if a farmer wants to stay in the business of direct marketing and value adding, and at peace with neighbors and customers, it's essential to seek out what laws and rules apply to you, and follow them.

The first category of regulation a direct marketer is likely to encounter is land-use laws. In very rural areas, land-use issues are usually handled by ordinances, which cover how you operate but don't usually restrict what you can do. Your farm stand, for example, may be limited in the hours it can operate, but you can put it anywhere on your property that you like and the neighbors can't legally object to increased traffic past their homes.

Zoning, which is the most common type of land-use regulation in more populated areas, will control what you can do. You may not be able to operate a retail business on your farm if there is a local ordinance against it, unless there are state regulations that exempt farms from business regulations.

Laws and regulations can change without your knowing, but if your use or business is already established you will be grandfathered in. In other words, since you were there first, the new regulation can't be applied to you, in most cases. But if you want to expand or change what you do, then you will have to comply with any changes in rules.

Legal Trouble Spots

The Legal Guide For Direct Farm Marketing by Neil D. Hamilton is an excellent information resource for legal questions and concerns, including all the ones you never thought about but might need to know. Prepared under a grant from SARE, the USDA's Sustainable Agriculture Research and Education program, the book is available from the Midwest Organic and Sustainable Education Service (MOSES) and other sellers that specialize in sustainable agriculture publications. (See the resources.)

If you're wondering whether any of this legal stuff applies to you, start by checking out Hamilton's list of "Eight Things That Will (Probably) Get You into Legal Trouble":

• Selling at your roadside stand more products produced by others than products you raised yourself

• Not carrying sufficient liability insurance for your operation

• Failing to comply with labor rules when hiring employees

• Conducting a commercial business in an area not zoned for such use

• Allowing unsafe conditions to exist on your property when customers are allowed to visit

• Selling processed foods that have been produced at an unlicensed facility

• Failing to observe farmers' market rules designed to protect the safety or quality of food

• Not complying with record-keeping and paperwork rules for tax or labor laws

For this reason many small-scale farmers around the country have become involved with their local planning commissions or equivalent bodies, so they can help guide how land use will be regulated in their community.

A second area of legal concern is liability. If a kid sprains an ankle by tripping in your corn maze, or a customer gets salmonella poisoning from your eggs, you can be held liable for their medical expenses and, who knows, maybe more for pain and suffering. Good liability insurance is essential for any farm operation, and you should find an insurance agent familiar with rural insurance issues to set it up for you. Many agents recommend an "umbrella policy" for farms, a relatively cheap means of covering your bases for a wide variety of potential liability issues.

The third area of legal concern is licensing and inspection for businesses and value-added enterprises. Any dairy operation and most value-added operations will require a license and regular inspection to make sure facilities and handling conform to state regulations. Any on-farm store that is selling more of other people's products than of its own farm products will lose the farm exemption and be required to have a business license, as well.

Last, as a business you'll need to maintain accounting records that will satisfy the Internal Revenue Service. Accounts for a straightforward farm operation can be fairly simple. They can be comingled with household accounts if you can keep track of what is farm and what is household expense and income, but it is usually better to set up separate bank accounts. You will need to retain and categorize receipts, too, but you're already doing that for your organic records, right? Get your accountant's advice on if, and when, you should set up the farm business as a separate legal entity for tax, insurance, and regulatory purposes.

Profile: Gray and Mary Schmidt

Winsted, Minnesota

Gray Schmidt last year pulled off a coup that any market gardener would envy. He became the preferred supplier for heirloom tomatoes to the most upscale grocery chain (21 stores) in the twin cities of Minneapolis and St. Paul. His wife, Mary, runs their bedding-plant business as well as her own landscaping company, Gardens Galore, which focuses on native plants and sustainable design. Mary is also a master herbalist specializing in Native American medicinal plants.

Finding and financing land and building their businesses all happened on a shoestring budget, sometimes less than that. But where money may have been lacking, creativity certainly hasn't been. And that has made all the difference.

Living in the western suburbs of Minneapolis, the family began looking for land in the late 1990s. When they saw an ad for 14 acres (5.7 ha) less than an hour to the west, Gray went out of his way to drive by one day on the way home from work. When he saw that the house was livable and the sheds weren't falling down, they arranged to tour the place. "We spent eight minutes in the house and said okay, we can live here, we don't have to fix anything," Mary says. "Then we spent three hours on the land, putting our hands in the soil."

The soil was black prairie loam, one of the finest agricultural soils in the world. "We were lucky. At the time land was pretty inexpensive," Gray says. Even so, finding financing was not easy.

Though they had money saved, both were self-employed and area banks thought them too poor a credit risk to lend the full mortgage amount. "So we created our own little finance package. We had a down payment to satisfy the bank and get part of the mortgage, and we had a contract for deed with the seller for a quarter of the price," Mary says. "Look at what your options are, get a little creative, and maybe come up with something that will work for you."

Their move from the suburbs included digging up and transplanting hundreds of perennials and herbs from the yard to the farm. Once the move was complete, the couple put up their first greenhouse — a modified hoop house with a double layer of plastic as a cover. They've since added a second greenhouse and an unheated cold frame for early plant starts in the spring. Their 4 acres (1.6 ha) of tillable land are devoted to production of landscaping perennials, medicinal and culinary herbs, a variety of vegetables, and, for the past two years, an acre (0.4 ha) of heirloom tomatoes in a palette of colors from yellow and orange to purple and black. Gray says:

> *Most of the purple ones are favorites for taste and marketability. Cherokee Purple and Brandywine are the most asked for, and Black Krim doesn't split a lot and produces well. I've tried a total of 60 varieties, and last year I grew 50. This year I had maybe 20. You weed out the ones that aren't as marketable. If you grow something that has great flavor but you can't market it, why grow it?*

Heirlooms crack easily and ripen fast, he says, and those varieties that can't be reliably prevented from cracking and are too delicate to survive packing and trucking won't sell. "Hybrid tomatoes are bred so they don't crack, but they've sacrificed a lot of flavor, too. Hybrids last forever and they handle trucking better. With heirlooms you have a couple of days," Gray says.

During the season, which, depending on the weather, can last anywhere from five to ten weeks, the Schmidts deliver tomatoes three times a week in 10-pound (4.5 kg) boxes. "We put a thin foam pad underneath, and there's a lid. Hybrids are usually delivered in 20-pound (9 kg) boxes, but you can't pile heirlooms because they bruise so easily. This year I sold 10,000 pounds (4.5 t) — 8,000 pounds (3.6 t) of heirlooms and a couple thousand pounds (0.9 t) or so of hybrids. It's an awful lot of labor. There's about an eight- or ten-day period when it's just nuts. Last year we sold half of our tomatoes in eight days, picking and packing until two in the morning."

Processing and packing is done in an airy, cement-floored pole shed. Usually the couple doesn't hire any outside labor, though their three grown children help out in the busiest periods. But it means long hours and hard work, and they aren't getting any younger.

"One of the problems," Mary says, "is that working sunup to sundown, day after day, wears on a body." Gray adds, "So this year I cut back. Last year I had 5,000 plants; this year I had 2,500. Last year I did cherry tomatoes, but they were too much work to pick, and you hardly get anything more for them per pound."

The busy season starts in late winter and early spring, when the greenhouses are filled with started vegetables, herbs, and flowers. Landscaping jobs, for which Mary hires help, start at this time as well. Mary estimated that three-quarters of their bedding plants are sold to area gardeners and landscaping clients, the remainder at local farmers' markets.

Tomatoes are sold only to the grocery store chain and a few restaurants, Gray says. "With what I'm doing, I can't be at a farmers' market. I don't have time, and it's questionable for profitability. The grocery stores don't want the tomatoes quite ready for the table, so if I have some that are more ripe, those are the ones the restaurants want. It's an outlet for the ripe ones, but they don't take much, and it's a lot of running around the city. For those I charge 50 cents a pound more."

It took some good luck to land the grocery store account, but it really came through due to persistence and a high-quality product. Gray recalls:

I was working nights at a friend's restaurant, helping him get up and running. I asked him if he'd be interested in some hybrid tomatoes. He said no, but he would like some heirlooms. I'd never heard of them, so we got out some catalogs and ordered and planted and all of a sudden I have a huge amount of tomatoes and no market. The only stores that handle this sort of product are the higher-end ones, so I started going around to those stores and talking to the produce managers. Most weren't interested, because they're supposed to buy from the chain's warehouse. I finally found a guy who said, "Show me what you have." I did, and he said it was much higher quality than what they got from the warehouse. He got me an intro-duction at the warehouse, where they said they'd take 100 boxes a week. Then they had a promotion, and one of the buyers said they could use 50 or 60 boxes the next day. I said, "How about 100?" He said, "You can do 100? Can you do 140?" This year they said they would buy everything I had.

> **"It was really about showing buyers the product, bringing it to the store and showing them what you grow."**
> Mary Schmidt

Though the couple has often heard the advice to find a market before producing a product, they strongly disagree that that is always the right thing to do. "It was really about showing buyers the product, bringing it to the store and showing them what you grow," says Mary. "They have to see that you can grow it, or they don't know who you are. It's the same way with herbs."

"I showed them I could bring them better product than other people," Gray says. "I never try to sell anything I haven't grown."

The farm does not have organic certification, and the couple has no plans to pur-sue it. The question "didn't come up," Gray says. "I'd explained that we were almost organic — we do use some chemical fertilizer in the greenhouse, and the straw we put on for mulch isn't certified organic. For all the record-keeping, and trying to find organic straw, it's not worth it. With heirloom tomatoes I get about the same price as organic growers."

The Schmidts follow all other organic rules, even though as their businesses have grown it's become difficult to maintain crop rotations. "We've been on a two-year rota-tion because of limited space, and you should really go four years," Gray says. "If you're going to grow an acre (0.4 ha) of tomatoes, you have to have 4 acres (1.6 ha) of tillable land."

Since much of their limited tillable ground is devoted to permanent plantings, this has been an obstacle. But once again, a little creative thinking has come to the rescue. "We're trading use of some land for use of some shed space," Mary says. So, Gray adds, "I'll have 4 acres (1.6 ha) in the neighborhood, so I can plant cover crops and so on. That's an alternative for someone trying to start out, too. If they're living anywhere close to where there is farmland, rent it. If you're growing vegetables, you can grow a heck of a pile on an acre. What you'd pay for rent is not near what it would cost in interest on bor-rowing money to buy the land. Water might be something you'd have to work out."

The Schmidts say that those starting out should have some capital for equip-ment, operating costs, and hiring labor until the business starts paying for itself. "When I started out I sold $20,000 worth of tomatoes, but I spent $10,000 for equipment and things," Gray says. "You need the cash up front to start."

Profile: **Kay and Wayne Craig**

Grassway Organics, New Holstein, Wisconsin

Wayne and Kay Craig were well into their business careers in Minneapolis when they decided to quit and start a dairy farm.

"You get to the point where you say, okay, now what?" says Kay. "You get sick of the commute, and you remember the farm with rose-colored glasses because you were a kid then and didn't have the responsibility." In 1993 the couple quit their jobs and bought an operating dairy farm. The plan was to have a very different operation from what that farm was then, and from their parents' old stanchion-barn operations.

"We had three goals: we wanted to graze, we wanted to be seasonal, and we wanted to be organic," says Kay. "We bought the farm in July, and there were 70 acres (28 ha) of standing corn, 20 acres (8 ha) of newly seeded oats, and the rest hayfields. We wanted to graze, not buy a corn planter, and a drill, and a combine. We had to set everything up, all the pastures and fence, and that winter we sold animals and bought animals so we could be seasonal. We had bitten off way too much to allow us to even consider organic at the time."

The couple had several things in their favor as new farmers. They knew dairying from growing up on dairy farms, they had money saved, they were (and are) avid readers and networkers, and they knew from their previous jobs how to set up and run a business. As Kay says:

There's no way you can start with nothing. The payments will eat you alive. You have to have money for business start-up costs. Remember the old saying, you have to have money to make money. And you have to be big enough, or smart enough in your marketing niche, to bring in the $40,000 a year to heat the house, put food on the table, pay the electric bill, and pay the phone bill. If you're direct-marketing you have to build up a customer base, and if you're wholesaling you have to have enough milk for the buyer so you can pay your bills. So you have to have assets: cows and land.

One of the best things we did — and we were able to because we came in with money — was to set up a revolving operating loan with the bank. We could borrow from it and then pay it back when money came in. That was so important. People need to become financially savvy on this. You're going to have to pay all the small businesses you buy things from to run your farm. You have to pay the breeder, you have to pay the gas bill. You can't let that stretch out.

Fifteen years later, the Craigs and two full-time employees are running an on-farm store, milking 100 cows, and raising beef steers for sale, plus 1,000 pastured meat chickens and a lot of turkeys, all on 247 acres (100 ha). They have all of their cows calve in the spring but do milk some cows through the winter, since their direct-sale milk customers like a year-round supply of milk. They feed some purchased organic grains, but their land is now entirely in hay and pasture. The cows graze about seven months a year and in winter are in deep-bedded loose housing. Milking is done in a swing-15 pit parlor retrofitted into an old calf shed. Kay says, "$75,000 paid for everything associated with the parlor. That was back in 1996. It's a lot of money, but it's not a lot of money for a parlor."

The couple achieved their third goal when they became certified organic in 2004. They ship most of their milk to the Organic Valley cooperative in southwestern Wisconsin and market the rest through their on-farm store. Kay says, "Raw milk drives the store. Right now it's about 300 gallons a week, at about double the price we get from Organic Valley." The store also stocks the farm's meat, eggs, and chicken, along with a selection of other organic products purchased wholesale. "We're a regular little grocery store. We have organic ice cream, frozen fruits and vegetables, spices. We can keep you fed off the store. If you're busy, it's only one stop."

The on-farm sales part of the enterprise grew slowly, Kay recalls.

It all started with the beef, those little brown boys, those Jersey babies. They won't give you anything at the sale barn for Jersey calves, so we started raising them. You start selling with family, friends, the guy at the hardware store. The first year we sold maybe two quarters of beef, the fourth year we sold eight, and then we saved a few more and sold them, too. Then we added 50 laying hens, which was way too many. The eggs started rolling in, so we put up a sign advertising them. Then we thought we could raise meat chickens in pens on pasture, and I ordered a hundred. The reasoning was that if we couldn't sell them we could eat them. So all the beef customers and egg customers were notified that we now had chickens, and for the first three years we didn't get to eat a single chicken that we raised. We sold them all.

When we opened up the store three and a half years ago, we already had more than a hundred customers buying something from us, so it wasn't quite as scary. And every week we sign up new customers. We have 40 steers right now. I have 17 presold, and we need the rest for the store. We go through about two per month.

The Craigs have set up the store as a separate business entity from the farm, allowing them to keep close track of expenses and profits from both businesses. "The farm has to make money, and the store has to make a profit, because the store has to buy all the milk and meat from the farm," says Kay. "You figure out what your labor is, what you need on the farm side to raise the animals, and then the store has a markup because the store has expenses to cover, too: freezers, electricity, labor, marketing."

Tracking expenses allows the Craigs to price their products to cover production costs, including labor. For example, pastured chickens in bottomless pens require daily labor for watering, feeding, and moving the pens. Kay explains:

"If you're direct-marketing you have to build up a customer base, and if you're wholesaling you have to have enough milk for the buyer so you can pay your bills. So you have to have assets: cows and land."
Kay Craig

> "With marketing, you can't be shy. You basically have to like to talk to people . . . To do a good job you have to give them some individual, one-on-one time."
>
> Kay Craig

I asked myself, "What do I need for a return to make me happy to go out there?" If I don't get $20 an hour, it's not worth it for me. And that's how we're pricing out the chickens.

Locally there's still a mind-set that products direct from the farm should be cheaper. We've had people who drive in, gasp at the price, and leave. You have to have a certain amount of thick skin. You're not going to sell to everybody. So don't lose sleep on it.

Kay says their farm is situated in a marketing epicenter, with seven sizable towns within 40 miles (64 km) of the farm and Milwaukee just an hour away, with Chicago beyond that. That means there are plenty of ways they could expand their marketing.

"There are so many things we could do. We've talked about a state-approved kitchen, or a little café. I'd like to work Joel Salatin's metro buying club idea into Milwaukee and Chicago. And that means we'd need more beef."

While Wayne handles the farming side, Kay does the marketing. It's a good fit. "With marketing, you can't be shy," she says. "You basically have to like to talk to people. They're coming to us for two things: first, they have a need — a health need or an interest in nutrition, and second they're looking for information. To do a good job you have to give them some individual, one-on-one time. That's why they're coming to you. Otherwise they'd go to the grocery store. Sometimes all you talk about is the weather, but the personal touch with their farmer is why they come. We try to be transparent in our operation, we try to answer questions. I've taken people out to pick eggs, and they still talk about it. With that kind of public relations we don't have to worry about competing with Whole Foods."

The Larger Context

Organic Farming's Effect on the Wider World

> *No man is an island, entire of itself; every man is a piece of the continent, a part of the main.*
>
> **JOHN DONNE**
> English poet and clergyman, 1571–1631

At the Upper Midwest Organic Farming Conference several years ago, six hundred jaws dropped in unison when keynote speaker Allan Savory, author of *Holistic Management*, noted that, of the several hundred civilizations that have disappeared in the course of human existence, all had been based on organic farming.

His point was that organic farming is not enough to build a sustainable society. Organic farms don't exist in isolation; they are part of regional and global ecosystems, and their owners are part of local, national, and global communities. To maintain ecosystems and economies and direct them toward a viable future, organic farmers have an interest, as does everyone, in thinking and acting beyond the borders of their land.

The Final Rule makes no comment on the organic farmer's responsibility toward either the larger ecosystem or the human community. But it was organic farmers who in the 1980s pushed for the USDA Final Rule, in the 1990s contributed to its writing, and then after the turn of the century protested its hijacking by corporate interests until the USDA came up with something that was pretty acceptable. And those farmers have always implicitly assumed that we all have a responsibility toward society that goes beyond growing good food. It may not be in the rule, but from the start it's been in the minds and hearts of organic farmers.

This responsibility is obvious if you follow the premises of organic farming to their natural conclusion. Your farm is part of a larger ecosystem and human community, regardless of whether you want it to be. If organic farming rests on the idea that good health and sustainability are the goal, then naturally the larger ecosystem and human community have to be healthy and sustainable if your farm is going to be also, and if it's going to be around for the next generations. Your liver isn't going to be healthy in the long run if the rest of your body is sick; your farm isn't going to be sustainable unless you and your local community support each other, and unless the local environment is reasonably clean and biodiverse.

So, even though there is nothing in the Final Rule on how organic farmers should relate to their communities, there has always been a clear impetus for organic farmers to be proactive if they want to claim that they are truly organic, in the larger sense of the word.

As individuals, most of us don't have much direct impact on global happenings, even though global happenings now impact us individuals more than ever before. Yet some organic farmers have been and are influential on the national and international levels, and maybe you will be one of them. But any farmer can influence the local level, and if you're aiming higher, usually being effective on those larger stages first requires some involvement and experience on the local scene.

Our local influence is economic, political, cultural, social, and environmental. How we, as organic farmers, choose to exercise our influence will help determine whether our local community and ecosystem will thrive in the long run. This in turn determines whether our farms and we ourselves will survive in the long run.

Communities, just like organic farms, rely for sustainability on having a functional system that enables its members to have clean water, adequate shelter, enough to eat, good education, access to health care, police protection, fairness under the law, waste disposal, and decent roads so as to be able to access these things. Electricity and running water are great assets as well, and libraries, churches, arts organizations, and help for those in need make a community a place where people actually want to stay. It's possible to survive without these things, and when you're 20 it even sounds romantic, but in the long run that is not what makes most of us happy.

Economic Influence

A small-scale organic farmer who plans to sell farm products in the local community requires a community that is interested in buying good food and has the money to do it. If your products taste good, educating people to want them should be no great obstacle. If you want folks to have the money to buy your products, you have to give them some money. This is done by shopping locally, at the small businesses in your town, and by hiring local contractors and other local help.

Though it's tempting to save a few bucks or get a better selection by driving another half hour to the big-box stores in a bigger town, remember that it's also tempting for your customers to save a few bucks and get a bigger selection than you can grow at the supermarket. If you buy equipment parts over the Internet and go to the local dealership only in an emergency, then they might not be there the next time you need them.

Dining at the local café not only helps keep open a place where you can meet your neighbors, it's a great place to find out where you can hire a good local carpenter or a neighbor who will do some custom plowing for you. And you'll find customers for your farm products.

A local economy is a two-way street. Every dollar we spend is a vote of support or nonsupport. Not everything is available locally, but when possible, shouldn't our economic influence be used to support those who support us?

Political Influence

Small-town and county politics are often boring when they aren't contentious, and the meetings are usually held at inconvenient times. But it's at this level that land-use planning is done, budgets for local services and infrastructure are set, and ordinances and zoning regulations are passed that determine who can do what and when with their land.

What usually happens is that after much discussion at several meetings, a town board votes to allow a sand mine, or a large-scale pig operation, or a dirt-bike track to be built in the town. Or the school district is consolidated and the kids have much longer bus rides, or a subdivision is approved for the lot next to your barn. The town residents, who never attend meetings or pay any attention to what is happening at them, suddenly realize this bad thing has happened. They organize citizens' groups, write letters to the editor, and show up at the next meeting en masse. Sometimes the decision can be reversed, other times not. The only sure thing that will happen is a lot of hard feelings among neighbors.

Wouldn't it be better to be aware of what's going on before any irreversible decisions are made? Reading the local paper, where public notices of meetings and agendas are published, as required by law, and attending an occasional meeting can keep a lot of bad things from happening. Or consider sitting on a citizen's committee or advisory board. In politics as in farming, being proactive is generally more effective than being reactive.

Social and Cultural Influence

Wendell Berry writes, in an essay titled "The Work of Local Culture," that "a human community, then, if it is to last long, must exert a sort of centripetal force, holding local soil and local memory in

Big versus Little: Corporate Organics and the Family Farm

In the 1970s organic farming was as much a political statement as a desire to raise good food. Organic farmers, by and large, were also protesting, among other things, what they considered to be an unresponsive and irresponsible government and the rapaciousness of many corporations. Since the implementation of the Final Rule in 2002 and the entry of large food corporations into the organic sector, this political context has been lost. Organic farming is now, both legally and in the public mind, simply a production system.

This loss has essentially split the organic community into two branches. On the one branch are large-scale organic farms and the many and increasing number of large food corporations that market organic foods on a national level. They have integrated organic foods into the current food system, which relies on fossil-fuel-intensive, large-scale production and national and international transportation and distribution networks.

On the other branch are what most people envision when they think of organic farms: small-scale, diversified farms that sell to their local communities. These farmers do not and never have believed that organic farming is simply a production system. They continue to remain apart from mainstream food distribution and marketing, building an alternative system with farmers' markets, farms based on community-supported agriculture (CSA), and other direct-marketing avenues.

In between are a lot of midsize organic farms and organic farmer cooperatives that are finding they need a bigger market than what's available locally in order to support a family farm.

There's been considerable dislike and disagreement between the two branches of organics over the issue of whether "big" organics is truly organic. If the answer to this disconnect between big and little, between organic ideals and marketplace realities, were easy or simple, the problem would have been solved by now. But it's not a black and white issue.

Food corporations and large-scale organic farms are in the process of realizing the small organic farmer's ideal of making organic foods part of the mainstream: convenient and affordable for the average consumer. But there is a lot of intense discomfort among small-scale farmers over the fact that this has happened courtesy of energy-intensive storage and distribution networks. If an organic tomato has to rely on fossil-fuel-intensive refrigeration and trucking in order to travel 1,500 miles (2,400 km) to get to your plate, is it really sustainable food? Is it really good for the environment? If your organic corn is cooked, mashed, mixed, and extruded into highly processed chips in an airtight plastic bag, is it still a natural, healthy, whole food? On the other hand, isn't it better that the tomato and the corn are organic and not conventional, even if they were trucked in from a long way off?

place. Practically speaking, human society has no work more important than this."

He goes on to say that "a good community, as we know, insures itself by trust, by good faith and good will, by mutual help. A good community, in other words, is a good local economy. It depends on itself for many of its essential needs and is thus shaped, so to speak, from the inside — unlike most modern populations that depend on distant purchases for almost everything and are thus shaped

from the outside by the purposes and the influences of salesmen."

A good local community depends on neighbors knowing each other, working together, and playing together. That's what builds trust, good faith, and good will. Not to mention good memories and a good place to live.

Sometimes new members of a community find it hard to become acquainted with neighbors or to acquire a social circle. But there are opportunities,

There's also a fear, based on what has happened in conventional agriculture, that corporations, in their endless quest to improve the bottom line by cutting costs and expanding markets, will create enough downward pressure on organic food prices to make it impossible for a small organic farmer to make a decent living. After all, just that sort of price pressure has made it impossible for conventional small-scale farmers to make a decent living. This in turn has contributed to the depopulation of rural areas, and to our current national dependence on large-scale farms that ignore both environmental and social degradation.

There's no question that some pressure is occurring already, as large-scale interests have liberally interpreted sections of the Final Rule in order to save on costs, while the USDA has looked the other way. Only the vigilance and vociferousness of other organic farmers is preventing organic production from shading into another form of conventional production. If this vigilance isn't maintained in the future, then the term *organic farming* could eventually cease to have any distinct meaning, even as a production system.

The small-scale organic community has by and large held to the ideal of locally based food systems. Such systems minimize fossil-fuel use, maximize self-reliance, and preserve so much of what is worthwhile and important in our culture. A local food system, in particular, preserves that all-important knowledge of where our food

originates and how it is produced. That understanding makes it possible to see why organic farming matters to human survival and happiness. But local foods are not easily available to many urban residents, especially those with limited income and transportation. Relying on one or two small farms for food can be precarious, since farmers fall ill, retire, quit, or change careers fairly regularly, leaving their customers scrambling. And it's a rare consumer or farmer, no matter how "green," who is willing to completely do without the luxuries of the national fossil-fuel-intensive infrastructure and way of life, brought to us courtesy of big business and government. Nearly all of us own cars, most of us drink coffee and eat things like bananas that are imported from far away, and we all use metal, plastic, and other products of industry.

In practical terms, the interplay between the big versus small, national versus local, and organic versus conventional dilemmas can create some difficult choices for farmers. Do you buy conventional feed from the struggling mill in town to help it stay in business, or do you import certified organic feed from some distant producer? Do you sell at the local farmers' market at higher-than-conventional prices but think it's okay to shop at the big-box store instead of the local independent hardware store, in order to save a few bucks? Do you buy imported organic coffee or give up drinking coffee?

ones that also serve to foster a lively local culture that both preserves the past and looks forward to the future. If you're looking for local history and farming information, why not volunteer at the retirement community? If you're pining for more art in your life, search out the local community choir, theater group, painting club, or library volunteers. Kids' sports teams are always in need of coaching help, craft sales in need of crafters, and community events in need of workers.

We all appreciate a competent police force or sheriff's department, an effective firefighting force, good schools, and decent medical care. In small towns many of these functions rely a great deal on volunteers and on local fund-raising organizations.

Though a farmer's time is limited during the growing season, taking an evening or two once in a while to sell hot dogs at the fair for the fire department, to staff a school event, or even just to attend a local fund-raising dinner, is appreciated, and it's an easy way for you to meet others in the community. At the very least, if you're going to buy a Christmas tree, see if you can buy it from a lot where the profits will go to serve local needs. Local needs are, in fact, your needs.

Organic farmers have a particular interest in getting organic, local foods served in their community's schools, retirement homes, and hospitals. It not only builds their farm business but is good for everyone's health, supports the local economy, and builds personal relationships. If you're already helping at the school, you're building the personal relationships that may make it possible to get good food on the cafeteria menu, and maybe some lessons on the benefits of organic farming into the science class's environmental unit. Helping others often makes them interested in helping you, a win-win situation.

Environmental Influence

Water, air, wildlife, and pollutants recognize no boundaries, and certainly not farm boundaries. In many areas farmers are working with each other, often under the aegis of a county, state, or federal land conservation program, to conserve natural resources in an entire watershed or bioregion. Private nonprofit organizations also work with groups of farmers and rural residents to conserve needed large blocks of wildlife habitat, or to create wildlife corridors for animals. Streams and rivers in particular require cooperation among landowners to achieve any real environmental progress.

Consider informing yourself about the watershed you live in, the environmental issues and opportunities in your area, and how you can be involved. Local papers, farm papers, community Web sites, state department of natural resources Web sites, and other sources carry notices of meetings, field days, conferences, and other items of interest for anyone who cares about sustaining the local environment.

Profile: **Bob Scowcroft**

Executive director, Organic Farming Research Foundation, Santa Cruz, California

Organic farming has changed a lot since Bob Scowcroft cut his teeth as an organizer in the late 1970s, fighting the use of pesticides. By 1980 he was working in organic agriculture, and in 1990 he cofounded the Organic Farming Research Foundation, the first and still by far the most important funding resource in the United States for scientific research on organic farming. He continues to serve as executive director of the OFRF, and in 2006 he was honored with the Steward of Sustainable Agriculture award at the Ecological Farming Conference in California, in recognition of his lifetime achievements.

Directed by working organic farmers, the OFRF has as its purpose "foster[ing] the improvement and widespread adoption of organic farming systems." To achieve this, the OFRF as of 2009 had distributed more than $2.3 million in support of 280 competitive grants for research projects related to organic farming systems. All research results are posted on the OFRF Web site, and hard copies are distributed free of charge to anyone who is interested.

The OFRF also conducts periodic surveys and studies of the state of organic agriculture, of the needs of organic farmers, and of whether the federal government is addressing those needs in proportion to the increasing importance of organic agriculture in the United States. This data has proven extremely useful when the OFRF goes to Washington, DC — as it routinely does — to advocate for organic agriculture.

As the long-term head of the OFRF, with his several decades of experience, Scowcroft has a uniquely broad perspective on organic agriculture as it has grown and changed over the past several decades, and he is enthusiastic about its future potential. "We have a very uniquely labeled product, and the sky's the limit," he says. "It's a very different world now. I'm heartened by the large numbers of young people who know they want to farm."

One important way in which the world of organics has changed since the turn of the twenty-first century is the entry of large farms and big corporations into the organic market. As the market share of organic foods has continued to accelerate, it's become clear that the 1970s organic ideal of small farms growing for local markets is too limited. Scowcroft says:

> **"We know organic farming works, and we want to fund the science behind how it works and make it better."**
>
> Bob Scowcroft

Scale is so twentieth century. There are small cooperatives collaborating with a hundred organic farmers that are selling into the largest supermarkets in the land. And there are individual large organic farmers selling into the smallest natural food stores.

The rivalry of the big guys (large food-processing corporations) has changed the market for the better. They buy all the number twos — the crooked carrots, the misshapen apples — for juice and cider and so on. That's common on the West Coast. And a lot of smaller food companies are very proud of their relationship with organic farmers. Amy's Kitchen is looking to transition small growers into organic production. For example, they're offering organic vegetable frozen dinners, and they want to buy the organic ingredients right down the street.

Another trend that has become clear in recent years is the multiple benefits that organic farms deliver to rural areas. Scowcroft continues:

Why does organic farming matter to a community? It's profitable and presents a fair wage for family farmers. We think organic farms use less water, and certainly the aerial application of suspected carcinogens does not occur. These farms build the soil, and those that have tours or farm days are introducing a new constituency to farming — we've lost two, maybe three generations to a cans-grow-on-trees mentality. The social thread is very important.

We know organic farming works, and we want to fund the science behind how it works and make it better. There's still very little information on homeopathic remedies and organic pasture management; we need more seed research, and research on organic dairy and ranching.

Our two long-standing objectives (at OFRF) are to build an endowment fund such that we can permanently support research projects, and then to get the results out to all the organic farmers we can find, free of charge. We have funded 280 projects, and one of the unintended consequences of this is that we've ended up funding a very exciting new generation of organic researchers. There are people out there who 12 years ago got an OFRF grant and now hold key, tenured positions (while studying organic farming) at some of the nation's leading land-grant universities. In the past ten years we've seen organic oases sprout up in certain land-grant universities, and we're getting more federal funding for organic research, via new legislation contained in the most recent Farm Bill.

The slogan was to get the scientists out of the lab and into the fields, and the farmers out of the field and into the lab.

Beginning organic farmers now need, more than ever, "a very straight, detailed, and profitable farm plan," Scowcroft says. "Not just what you're going to grow and your soil type, but what kind of resources you have to start up and get through the first year or two. How are you going to market, how are you going to diversify? You have to be a reader, and you have to know how to apply what you've learned to your own unique farm. You have to be a data collector, keeping your notes on weather patterns, frost dates, seed varieties, harvest dates. You have to have a peer network and if possible a couple of neighborly mentors. And you have to have a sense of humor, so you can laugh and get over it."

The new organic farmers of today, Scowcroft says, "are the heroes of this next generation. It's their charge — the ones who are reading this book — to get the market share to 40 or even 50 percent certified organic. Organic farming is not scale or size, it's honesty and truthfulness, values and passion. People can bring that in at an acre or a thousand acres, if they follow the rules and work to improve them. That's what this is all about."

Resources

The amount of resources and information now available to the aspiring organic farmer on the Internet is large and increasing daily. Since anyone who can conduct an Internet search can ferret out much of this information, and since so much of it is specific to particular regions and questions, I have included here only Web sites and organizations that are national in scope and of general interest to organic farmers.

Similarly, many, many books, pamphlets, and periodicals on topics related to sustainable and small-scale farming exist, and more are being published all the time. I have confined my list to those resources specific to *organic* farming (with a few useful exceptions) that are not region specific.

No doubt I've missed a few good resources, but this list will get you well on your way. In addition, spend some time searching out local and regional information on suppliers, farmer networks, events, and conditions. And be sure to introduce yourself to the neighbors.

Organic Farming Classics

Great reading on a winter's night! For these books I'm listing just the original publication date, since so many have been reissued by various publishers.

Albrecht, William A. *Soil Fertility and Animal Health*. Webster City, Iowa: Fred Hahne Printing Company, 1958.

Balfour, Eve. *The Living Soil*. London: Faber & Faber, 1943.

Bromfield, Louis. *From My Experience*. New York: Harper, 1955.

———. *Malabar Farm*. New York: Harper, 1948.

———. *Out of the Earth*. New York: Harper, 1950.

———. *Pleasant Valley*. New York: Harper, 1945.

Carson, Rachel. *Silent Spring*. Boston, MA: Houghton Mifflin, 1962.

Fukuoka, Masanobu. *One-Straw Revolution: An Introduction to Natural Farming*. New York: Rodale Press, 1978.

———. *The Road Back to Nature*. Tokyo: Japan Publications, 1987.

Howard, Sir Albert. *An Agricultural Testament*. New York and London: Oxford University Press, 1943.

———. *The War in the Soil*. Emmaus, PA: Rodale Press, 1946.

King, F. H. *Farmers of Forty Centuries: Permanent Agriculture in China, Korea, and Japan*. Madison, WI: Mrs. F. H. King, 1911.

Northbourne, Walter Ernest Christopher James Lord Baron. *Look to the Land*. London: Dent, 1940.

Rodale, J. I. *Encyclopedia of Organic Gardening*. Emmaus, PA: Rodale Press, 1959.

———. *How to Grow Vegetables and Fruits by the Organic Method*. Emmaus, PA: Rodale Press, 1961.

———. *Paydirt: Farming and Gardening with Composts*. New York: Devin-Adair Company, 1945.

———. *The Organic Front*. Emmaus, PA: Rodale Press, 1948.

General Interest

Books

Bradley, Fern Marshall, Barbara W. Ellis, and Ellen Phillips, eds. *Rodale's Ultimate Encyclopedia of Organic Gardening*. Emmaus, PA: Rodale Press, 2009.
The most recent edition of J. I. Rodale's encyclopedia. An excellent resource, with plenty of useful information for garden as well as field crops.

Duram, Leslie A. *Good Growing: Why Organic Farming Works*. Lincoln, NE: University of Nebraska Press, 2005.
Well-documented study of the broader effects of organic farming.

Logsdon, Gene. *The Contrary Farmer*. White River Junction, VT: Chelsea Green, 1994.
Not strictly organic, but full of practical advice for small-scale farming. Logsdon has written several other books and hundreds of articles on various aspects of agriculture and is always a good read.

Schwenke, Karl. *Successful Small-Scale Farming: An Organic Approach*. North Adams, MA: Storey Publishing, 1991.
Though dated in parts, this book offers excellent advice on many practical aspects of crop farming.

Periodicals

Acres USA
800-355-5313
www.acresusa.com
One of the oldest organic/alternative agriculture organizations in the country, Acres offers a monthly magazine for commercial organic and sustainable farmers, an annual conference, various other events, and a large bookstore.

New Farm
Rodale Institute
610-683-1400
www.rodaleinstitute.org/farm/new_farm
A free monthly online publication. Excellent information on all aspects of organic farming, including field crops.

The Organic Broadcaster
Midwest Organic and Sustainable Education Service (MOSES)
715-778-5775
www.mosesorganic.org/publications/broadcaster-newspaper
Published bimonthly. In-depth articles, events, and other items of interest to organic farmers.

Organizations

Agriculture and Land-Based Training Association (ALBA)
Salinas, California
831-758-1469
www.albafarmers.org
A nonprofit working to create opportunity among limited-resource and aspiring farmers. Education, demonstrations, planning assistance, and more.

Alternative Farming Systems Information Center
National Agricultural Library
Beltsville, Maryland
301-504-6559
www.nal.usda.gov/afsic
A division of the United States Department of Agriculture. Offers a plethora of free information on alternative/sustainable/organic farming

ATTRA — National Sustainable Agriculture Information Service
National Center for Appropriate Technology (NCAT)
Butte, Montana
800-346-9140
www.attra.org
ATTRA should be your first stop when you're looking for organic farming information from the government. Funded by the USDA, ATTRA has six regional offices nationwide. It offers a wealth

of publications for organic and sustainable farming online, all of it free, as well as a list of alternative soil testing laboratories, a list of sustainable farming internships and apprenticeships, and an online "Ask a sustainable agriculture expert" service, which allows farmers to get answers to specific questions.

Ecological Farming Association
Soquel, California
831-763-2111
www.eco-farm.org
Events, programs, lists of internships and job opportunities in sustainable/organic agriculture, and the annual Eco-Farm Conference held each January in California.

Ecology Action
Willits, California
707-459-0150
www.growbiointensive.org
International in scope and focusing on small-scale intensive organic gardening to grow the maximum amount of food in the smallest space, the nonprofit Ecology Action offers research, books, events, and internships.

Economic Research Service
United States Department of Agriculture
Washington, DC
202-694-5478
www.ers.usda.gov
Offers "resource regions maps" and related materials that show which areas produce the most of which crops and livestock, which can be helpful when you're considering where to start farming.

Holistic Management International
Albuquerque, New Mexico
505-842-5252
http://holisticmanagement.org
Dedicated to sustainable land use based on a goal-oriented approach to land management. This organization put the scientific data behind the management-intensive grazing movement and offers training, newsletter, and events.

International Federation of Organic Agricultural Movements (IFOAM)
Bonn, Germany
+49-0-228-926-50-10
www.ifoam.bio
The home of the worldwide organic movement; sets international standards and policies.

International Organic Inspectors Association
Broadus, Montana
406-436-2031
www.ioia.net
The nonprofit association of organic inspectors. Offers accreditations for members in crops, livestock, and processing.

Kerr Center for Sustainable Agriculture
Poteau, Oklahoma
918-647-9123
www.kerrcenter.com
A nonprofit devoted to researching and demonstrating sustainable farming and food systems. Education, field days, and workshops. Also issues policy reports on sustainable agriculture issues.

Land Stewardship Project
Minneapolis, Minnesota
612-722-6377
www.landstewardshipproject.org
Dedicated to promoting sustainable agriculture and developing sustainable communities. Administers the highly successful Farm Beginnings program, which offers training and mentorship to help new farmers.

Midwest Organic and Sustainable Education Service (MOSES)
Spring Valley, Wisconsin
715-778-5775
www.mosesorganic.org
Offers a wealth of free information on all aspects of organic farming and on transitioning to and becoming certified organic. Bookstore, publications, events, and training sessions. Organizes the Organic Farming Conference each year in late winter, the largest organic farming conference in the United States.

Minnesota Institute for Sustainable Agriculture (MISA)
University of Minnesota
St. Paul, Minnesota
800-909-6472
www.misa.umn.edu

National Center for Appropriate Technology (NCAT)
Butte, Montana
800-275-6228
www.ncat.org
Offers trainings, publications, technical assistance, and telephone help lines for sustainable agriculture and energy issues. Headquartered in Montana, with five additional regional offices.

National Organic Program (NOP)
United States Department of Agriculture
Washington, DC
202-720-3252
www.ams.usda.gov/nop
This is the home of the Final Rule. The Web site is not easy to navigate, however. If you're looking for a complete copy of the rule, don't bother trying to find it here unless you're really, really patient. Instead, contact your certifier, OMRI, or MOSES for a copy.

Natural Resources Conservation Service (NRCS)
United States Department of Agriculture
Washington, DC
202-720-7246
www.nrcs.usda.gov
Soil maps that label both type and slope of soils.

Northeast Organic Farming Association (NOFA)
Barre, Massachusetts
978-355-2853
www.nofa.org
A gold mine of events, publications, and other resources for organic farmers in the Northeast.

Oregon Tilth
Corvallis, Oregon
503-378-0690
www.tilth.org
Much, much more than a certification agency. Education, resources, research, and advocacy.

Don't miss this organization if you live in the Northwest.

Organic Farming Research Foundation (OFRF)
Santa Cruz, California
831-426-6606
www.ofrf.org
A nonprofit foundation that sponsors research related to organic farming practices, publishes the results free of charge, and works to educate the public and government decision makers about organic farming issues.

Organic Materials Review Institute (OMRI)
Eugene, Oregon
541-343-7600
www.omri.org
Reviews generic materials and brand-name products for the production of organic food and fiber. OMRI's lists are an essential resource for finding seeds, nursery stock, and other organic inputs.

Organic Trade Association (OTA)
Brattleboro, Vermont
802-275-3800
www.ota.com
Trade organization representing the organic industry in the United States and Canada. Offers an annual trade show and conference, FAQ sheets, books, and more.

Rodale Institute
Kutztown, Pennsylvania
610-683-1400
www.rodaleinstitute.org
Founded by organic pioneer J. I. Rodale, the nonprofit institute conducts research at its own organic farm, sells books and other publications, organizes events throughout the year, and publishes the online *New Farm* magazine. The institute also has an online, free, 15-hour course on transitioning to organic production.

Southern Sustainable Agriculture Working Group
Fayetteville, Arkansas
479-251-8310
www.ssawg.org
The best nongovernment, nonprofit source of organic information in the South. Resources, education, events, and more.

Sustainable Agriculture Research and Education (SARE)
College Park, Maryland
www.sare.org
info@sare.org
Sponsors grants that advance farming and ranching systems that are profitable, environmentally sound, and good for families and communities. SARE also publishes a series of excellent handbooks on sustainable agriculture under its Sustainable Agriculture Network program, as well as free bulletins, reports highlighting SARE projects, and resource guides. Headquartered in Washington, DC, with four additional regional offices.

Conferences and Events

Acres USA Conference
800-355-5313
www.acresusa.com
Held annually in December; location varies.

All Things Organic Conference and Trade Show
Organic Trade Association
802-275-3800
www.ota.com
A business-to-business event held annually in late spring.

Ecological Farming Conference
Ecological Farming Association
831-763-2111
www.eco-farm.org
Held each January in California.

MOSES Organic Farming Conference
Midwest Organic and Sustainable Education Services (MOSES)
La Crosse, Wisconsin
715-778-5775
www.mosesorganic.org
The largest organic farming conference in the United States, held annually in late winter. Immediately preceding the conference each year is the Organic University, which offers daylong intensive courses on specific topics in organic farming, such as dairy, soils, and so on.

Soil Management

Books

Hillel, Daniel. *Out of the Earth: Civilization and the Life of the Soil.* London: Aurum Press, 1992.

Magdoff, Fred, and Harold van Es. *Building Soils for Better Crops.* 2nd ed. Beltsville, MD: Sustainable Agriculture Network, 2000.

Nardi, James B. *Life in the Soil.* Chicago: University of Chicago Press, 2007.

Plaster, Edward J. *Soil Science and Management.* Albany, NY: Delmar Publishers Inc., 1992.
No doubt there are many other soil science textbooks available; I found this one to be very thorough and readable.

Government Assistance for Soil Conservation

The government offers a range of programs offering funding and/or assistance for soil restoration and conservation. Most of these programs are administered through the Natural Resources Conservation Service (NRCS), often in conjunction with state agencies. Note that the funding for these programs varies by year and region, and that programs may have been terminated or new programs initiated since this list was compiled. For more information, visit the NRCS Web site (www.nrcs.usda.gov) or contact one of its regional offices.

Conservation Reserve Program (CRP)
202-720-0673
www.nrcs.usda.gov/programs/crp
Retires highly erodible cropland; offers yearly rental payments for terms of 10 or 15 years for putting land into sod.

Conservation Security Program (CSP)
202-720-9232
www.nrcs.usda.gov/programs/csp
Assists with installing or maintaining conservation practices on working agricultural lands.

Environmental Quality Incentives Program (EQIP)

202-690-2621

www.nrcs.usda.gov/programs/eqip

Aimed at improving environmental quality; all types of land are eligible.

Farm and Ranch Lands Protection Program (FRPP)

www.nrcs.usda.gov/programs/frpp

Aimed at keeping prime agricultural land in production when threatened by development.

Grassland Reserve Program (GRP)

www.nrcs.usda.gov/programs/grp

Aimed at restoring cropland to grassland.

Wetlands Reserve Program (WRP)

www.nrcs.usda.gov/programs/wrp

Offers funding and technical assistance for returning drained agricultural land to wetland.

Wildlife Habitat Incentive Program (WHIP)

www.nrcs.usda.gov/programs/whip

Aimed at encouraging wildlife habitat development; all land is potentially eligible.

Tools and Equipment

Books

Baron, Sherry, et al., eds. *Simple Solutions: Ergonomics for Farm Workers*. DHHS (NIOSH) Publication No. 2001-111. Cincinnati, Ohio: U.S. Dept. of Health and Human Services, National Institute for Occupational Safety and Health, 2001.

Bowman, Greg, ed. *Steel in the Field: A Farmer's Guide to Weed Management Tools*. Beltsville, MD: Sustainable Agriculture Network, 2001.

Available online from Sustainable Agriculture Research and Education (SARE). Visit them at *www.sare.org*.

Deere, John. *The Operation, Maintenance, and Repair of Farm Machinery*. 27th ed. Moline, IL: Deere & Co., 1955.
The older editions of this book cover the types of machinery now used on many small farms.

Grubinger, Vernon. *Sustainable Vegetable Production from Start-Up to Market*. Ithaca, NY: Natural Resource, Agriculture, and Engineering Service, 1999.
Good information on cropping systems and tools.

Organizations

Healthy Farmers, Healthy Profits Project

University of Wisconsin
Madison, Wisconsin
608-262-1054
www.bse.wisc.edu/hfhp
Administered by the Biological Systems Engineering department, the project offers a multitude of free tip sheets on such helpful topics as how to build your own harvest cart, motorized lay-down work cart, rolling dibble marker, and hands-free vegetable washer, as well as on how to obtain such oddball handy tools as a nut picker-upper, a strap-on stool for field work, and mesh produce bags. There are also tip sheets on designing an efficient packing shed layout and sales area, as well as storage and transport methods.

Martin Diffley

Organic Farming Works
25498 Highview Ave.
Farmington, MN 55024
952-469-1855
http://organicfarmingworks.com
martin@organicfarmingworks.com

Garden, Field, and Orchard Crops

Books

Ashworth, Suzanne. *Seed to Seed: Seed Saving and Growing Techniques for Vegetable Gardeners.* 2nd ed. Decorah, IA: Seed Savers Exchange, 2002.

Coleman, Eliot. *The New Organic Grower.* White River Junction, VT: Chelsea Green, 1995.

Grubinger, Vernon. *Sustainable Vegetable Production from Start-Up to Market.* Ithaca, NY: Natural Resource, Agriculture, and Engineering Service, 1999.

Henderson, Elizabeth, and Robyn Van En. *Sharing the Harvest: A Citizen's Guide to Community Supported Agriculture.* White River Junction, VT: Chelsea Green, 2007.

Logsdon, Gene. *The Contrary Farmer.* White River Junction, VT: Chelsea Green, 1993.
Not strictly organic, but full of practical advice for small-scale farming, with plenty of attention to field crops.

Phillips, Michael. *The Apple Grower: A Guide for the Organic Orchardist.* White River Junction, VT: Chelsea Green, 2005.

Riotte, Louise. *Carrots Love Tomatoes* and *Roses Love Garlic: Secrets of Companion Planting for Successful Gardening.* North Adams, MA: Storey Publishing, 1998.

Rogers, Marc. *Saving Seed: The Gardener's Guide to Growing and Storing Vegetable and Flower Seeds.* North Adams, MA: Storey Publishing, 1991.

Schwenke, Karl. *Successful Small-Scale Farming: An Organic Approach.* North Adams, MA: Storey Publishing, 1991.
Though dated in parts, this book offers excellent advice on many practical aspects of crop farming. Of particular interest is the section on how to calibrate a grain drill.

Periodicals

Growing for Market
800-307-8949
www.growingformarket.com
A monthly newsletter for small organic growers. Offers additional publications for organic growers as well.

Organic Gardening
Rodale, Inc.
800-666-2206
www.organicgardening.com
Has the highest circulation of any gardening magazine in the world.

Organizations

International Seed Saving Institute
Ketchum, Idaho
208-788-4363
www.seedsave.org/issi/issi.html
Offers several books on how to save seed from your garden or fields.

Organic Fruit Growers Association
Minneapolis, Minnesota
http://organicfruitgrowers.org
Information on the production and marketing of organic tree fruits. Produces a quarterly newsletter.

Seed Savers Exchange
Decorah, Iowa
563-382-5990
www.seedsavers.org
A nonprofit organization conserving more than 25,000 varieties of seed, and offering many of them for sale through its catalog. The farm is open to the public and its gardens, orchard, and barnyard feature heirloom varieties and breeds.

Livestock

Books

Belanger, Jerry. *Storey's Guide to Raising Dairy Goats*. North Adams, MA: Storey Publishing, 2001.

Damerow, Gail. *Storey's Guide to Raising Chickens*. North Adams, MA: Storey Publishing, 1995.

Ekarius, Carol. *Small-Scale Livestock Farming: A Grass-Based Approach for Health, Sustainability, and Profit*. North Adams, MA: Storey Publishing, 1999.

Grandin, Temple. *Humane Livestock Handling*. North Adams, MA: Storey Publishing, 2008. See also Dr. Grandin's Web site (*www.grandin.com*) for articles and diagrams on livestock handling.

Karreman, Hubert J. *Treating Dairy Cows Naturally*. Austin, TX: Acres USA, 2006.

Klober, Kelly. *Storey's Guide to Raising Pigs*. North Adams, MA: Storey Publishing, 1996.

Minnesota Institute for Sustainable Agriculture. *Building a Sustainable Business: A Guide to Developing a Business Plan for Farms and Rural Businesses*. Beltsville, MD: Sustainable Agriculture Network, 2003. Case studies of successful organic, direct-marketing dairy businesses.

Murphy, Bill. *Greener Pastures on Your Side of the Fence*. 4th ed. Colchester, VT: Arriba Publishing, 1998. Lots of practical advice for management-intensive grazing.

Nation, Allan. *Grass Farmers*. Jackson, MS: Green Park Press, 1993. Case studies of farms around the country using management-intensive grazing.

Padgham, Jody ed. *Organic Dairy Farming: A Resource for Farmers*. Gays Mills, WI: Orangutan Press, 2006.

———. *Raising Poultry on Pasture: Ten Years of Success*. Boyd, WI: American Pastured Poultry Producers Association, 2006.

Robinson, Jo. *Why Grassfed Is Best!*. Vashon, WA: Vashon Island Press, 2000. For those interested in the 100 percent grass-fed approach.

Ruechel, Julius. *Grass-Fed Cattle*. North Adams, MA: Storey Publishing Publishing, 2006. A guide to grass-fed beef.

Salatin, Joel. *Pastured Poultry Profits*. Swoope, VA: Polyface Inc., 1996.

———. *Salad Bar Beef*. Swoope, VA: Polyface Inc., 1996.

Savory, Allan, and Jody Butterfield. *Holistic Management: A New Framework for Decision Making*. Washington, DC: Island Press, 1999.

Sayer, Maggie. *Storey's Guide to Raising Meat Goats*. North Adams, MA: Storey Publishing, 2007.

Simmons, Paula, and Carol Ekarius. *Storey's Guide to Raising Sheep*. North Adams, MA: Storey Publishing, 2001.

Thomas, Heather Smith. *Storey's Guide to Raising Beef Cattle*. North Adams, MA: Storey Publishing, 1998.

van Loon, Dirk. *The Family Cow*. North Adams, MA: North Adams, MA: Storey Publishing 1976.

Wells, Ann, Lance Gegner, and Richard Earles. *Sustainable Sheep Production*. Fayetteville, AR: ATTRA, 2000.

Periodicals

Graze
608-455-3311
www.grazeonline.com
Primarily for cow dairy producers using management-intensive grazing.

Handwoven
Interweave Press
866-949-1646
www.weavingtoday.com

Spin-Off
Interweave Press
866-949-1646
http://spinoffmagazine.com

The Stockman Grass Farmer
800-748-9808
www.stockmangrassfarmer.com
Covers management-intensive grazing for all livestock in all regions.

Organizations

American Pastured Poultry Producers Association (APPPA)
Hughesville, Pennsylvania
888-662-7772
www.apppa.org
Offers events, networking opportunities, and a list of suppliers of poultry and poultry products.

ATTRA - National Sustainable Agriculture Information Service
Fayetteville, Arkansas
800-346-9140
www.attra.org
Offers numerous publications covering all aspects of sustainable livestock production.

Holistic Management International
Albuquerque, New Mexico
505-842-5252
www.holisticmanagement.org
Dedicated to sustainable land use based on a goal-oriented approach to land management. This organization put the scientific data behind the management-intensive grazing movement and offers training, newsletter, and events.

The Livestock Conservancy
(formerly the American Livestock Breeds Conservancy)
Pittsboro, North Carolina
919-542-5704
www.livestockconservancy.org
Offers a plethora of information on threatened and endangered breeds of domestic livestock, as well as contact information for breed associations.

MidWest Plan Service
Iowa State University
Ames, Iowa
800-562-3618
www.mwps.org
Develops conceptual plans for all sorts of livestock and storage facilities; the older plans are most applicable to the small-scale operation.

Spalding Laboratories
Arroyo Grande, California
888-562-5696
www.spalding-labs.com
Publishes a fantastic pamphlet that gives ID and life-cycle information for species of flies that bother livestock, and tips for controlling them.

Weston A. Price Foundation
Washington, DC
202-363-4394
www.westonaprice.org
Promotes organic food and raw milk.

Maintaining Biodiversity

Books

Jackson, Dana L., and Laura L. Jackson, eds. *The Farm as Natural Habitat: Reconnecting Food Systems with Ecosystems.* Washington, DC: Island Press, 2002.

Leopold, Aldo. *For the Health of the Land.* Washington, DC: Island Press, 1999.
Also, Leopold's classic *Sand County Almanac*, which is less directly concerned with farming but nevertheless should be on every organic farmer's bookshelf.

Matson, Tim. *Earth Ponds: The Country Pond Maker's Guide to Building, Maintenance, and Restoration.* Woodstock, VT: Countryman Press, 1991.

Organizations

In addition to the organizations listed here, check out the programs listed under "Government Assistance for Soil Conservation" (page 409); some of those projects may intersect with habitat restoration projects.

American Tree Farm System
Washington, DC
202-765-3660
www.treefarmsystem.org
Can offer advice on developing or maintaining a woodlot for habitat diversity.

Biological Resources Discipline (BRD)
United States Geological Survey (USGS)
Denver, Colorado
888-275-8747
http://biology.usgs.gov
Offers a wide-ranging wealth of information on habitat conservation.

Defenders of Wildlife
Washington, DC
800-385-9712
www.defenders.org
Has a number of projects, including working with farmers to conserve wildlife habitat.

National Invasive Species Information Center
National Agricultural Library
Beltsville, Maryland
www.invasivespeciesinfo.gov
Offers information about invasive species across the country. The Web site also lists further resources at the state and regional levels.

Prairie Enthusiasts
Madison, Wisconsin
608-638-1873
www.theprairieenthusiasts.org
Dedicated to preserving and maintaining native prairie in the upper Midwest. Can offer advice on restoring prairie habitat.

United States Fish and Wildlife Service
Washington, DC
800-344-9453
www.fws.gov
The FWS is often overlooked but has money available for wetland, prairie, and woodland restoration projects that meet its criteria. Contact the office nearest you (under the government section in your phone book) for more information.

Wild Farm Alliance
Watsonville, California
831-761-8408
www.wildfarmalliance.org
Works to integrate agriculture with natural ecological systems. See especially their publications "Biodiversity Compliance Assessment in Organic Agricultural Systems" and "Biodiversity Conservation."

Xerces Society
Portland, Oregon
503-232-6639
www.xerces.org
A national nonprofit dedicated to preserving insects and other invertebrates. Offers a wealth of detailed articles on bugs and other small creatures, plus a magazine, books, guidelines, programs, and news bulletins.

Organic Certification

Organizations

ATTRA - National Sustainable Agriculture Information Service
Fayetteville, Arkansas
800-346-9140
www.attra.org
Sample application, documentation, and history forms, as well as a list of organic certification agencies.

Midwest Organic and Sustainable Education Service (MOSES)
Spring Valley, Wisconsin
715-778-5775
www.mosesorganic.org
Extensive and excellent materials on transitioning to organic production and becoming certified.

New Farm
610-683-1400
www.rodaleinstitute.org/farm/newfarm
Offers a free online course on transitioning to organic farming.

Marketing

Books

Gibson, Eric. *Sell What You Sow! The Grower's Guide to Successful Produce Marketing.* Carmichael, CA: New World Publishing, 1994.

Hamilton, Neil D. *The Legal Guide for Direct Farm Marketing.* Des Moines, Iowa: Drake University, 1999.
The updated 2011 edition of this book is available free online at *www.directmarketersforum.org*.

Macher, Ron. *Making Your Small Farm Profitable.* North Adams, MA: Storey Publishing, 1999.

Minnesota Institute for Sustainable Agriculture. *Building a Sustainable Business: A Guide to Developing a Business Plan for Farms and Rural Businesses.* Beltsville, MD: Sustainable Agriculture Network, 2003.

Whatley, Booker T. *Booker T. Whatley's Handbook on How to Make $100,000 Farming 25 Acres.* Emmaus, PA: Regenerative Agriculture Association, Rodale Press, 1987.

Periodicals

Growing for Market
800-307-8949
www.growingformarket.com
A monthly journal for direct-market farmers.

Organizations

Agricultural Marketing Service
United States Department of Agriculture
Washington, DC
www.ams.usda.gov
Tracks organic prices and offers a national listing of farmers' markets.

Food Routes Network
Millheim, Pennsylvania
814-571-8319
www.foodroutes.org
Works to establish "Buy Fresh Buy Local" chapters throughout the United States.

Local Harvest
Santa Cruz, California
831-515-5602
www.localharvest.org
A national directory of farms that direct-market products. Also maintains an events calendar and an online newsletter.

New Farm Grassroots Organic Price Index
http://newfarm.rodaleinstitute.org/opxgr
Lists current prices at the local farmers' market level, with a goal of tracking at least two markets in each state. Very helpful for market growers.

Organic Farmers' Agency for Relationship Marketing (OFARM)
920-825-1369
www.ofarm.org
Works to coordinate the efforts of producer marketing groups, to benefit and sustain organic producers.

Slow Food International
718-260-8000
www.slowfood.com
Ideas, inspiration, and resources for organic farmers interested in marketing their products at the local level.

Farmers Profiled

Please note: This information was accurate at the time this book was published in 2010.

Tony Azevedo
22368 West 2nd Ave
Stevinson, California 95374
209-634-0187
www.thedoublet.com
DoubleTacres@earthlink.net

Harriet Behar
Organic Specialist, Midwest Organic and Sustainable Education Service
P.O. Box 339
Spring Valley, Wisconsin 54767
715-778-5775
www.mosesorganic.org
harriet@mosesorganic.org

Marjorie Bender
Research and Technical Program Director
American Livestock Breeds Conservancy
P.O. Box 477
Pittsboro, North Carolina 27312
919-542-5704
www.albc-usa.org
mbender@albc-usa.org

Richard and Linda Byne
Byne Blueberry Farms
537 Jones Ave.
Waynesboro, Georgia 30830
706-554-6244
www.byneblueberries.com
dick@byneblueberries.com

Eugene Canales
Ferrari Tractor CIE
1065 Hazel Street
P.O. Box 1045
Gridley, California 95948
530-846-6401
530-846-0390 (fax)
www.ferrari-tractors.com
ferraritractors@ferrari-tractors.com

Cynthia Connolly
Ladybird Organics
Monticello Vineyards & Winery
P.O. Box 331
Monticello, Florida 32345
850-294-9463
www.monticellowinery.com
info@monticellowinery.com

Nancy Nathanya Coonridge
Coonridge Organic Goat Cheese
"from the wilds of New Mexico"
47 Coonridge Dairy
Pie Town, New Mexico 87827
505-250-8553 cell phone (does not work at dairy)
888-410-8433 voice mails only
www.organicgoatcheese.com
wildsofnewmexico@yahoo.com

Kay and Wayne Craig
Grassway Farm and Grassway Organics Farm Store LLC
N600 Plymouth Trail
New Holstein, Wisconsin 53061
920-894-4201
www.grasswayorganics.com
info@GrasswayOrganics.com

Glenn Drowns
Sand Hill Preservation Center
1878 230th Street
Calamus, Iowa 52729
563-246-2299
www.sandhillpreservation.com
sandhill@fbcom.net

Jeff and Lori Fiorovich
Crystal Bay Farm
40 Zils Road
Watsonville, California 95076
831-724-4137
www.crystalbayfarm.com
crystalbayfarm@mac.com

Donn Hewes and Maryrose Livingston
Northland Sheep Dairy
3501 Hoxie Gorge – Freetown Road
Marathon, New York 13803
607-849-4442
www.northlandsheepdairy.com
tripletree@frontiernet.net

Faye Jones
Executive Director, Midwest Organic and Sustainable Education Service
P.O. Box 339
Spring Valley, Wisconsin 54767
715-778-5775
www.mosesorganic.org
faye@mosesorganic.org

Dan Kelly
Blue Heron Orchard
32974 220th Street
Canton, Missouri 63435
573-655-4291
blueheronorchard@centurytel.net
www.blueheronorchard.com

Sam and Brooke Lucy
Bluebird Grain Farms
P.O. Box 1082
Winthrop, Washington 98862
509-996-3526
www.bluebirdgrainfarms.com
brooke@bluebirdgrainfarms.com

Stephen McDonnell
Applegate Farms
750 Route 202 South
Bridgewater, New Jersey 08807
866-587-5858
www.applegatefarms.com
stephen.mcdonnell@applegatefarms.com

Ralph Moore
Market Farm Implement, Inc.
257 Fawn Hollow Road
Friedens, Pennsylvania 15541
814-443-1931
www.marketfarm.com

Frank and Karen Morton
Wild Garden Seed
P.O. Box 1509
Philomath, Oregon 97370
541-929-4068
www.wildgardenseed.com
karen@wildgardenseed.com

Anne and Eric Nordell
Beech Grove Farm
3410 Route 184
Trout Run, Pennsylvania 17771
The Nordell's "Weed the Soil, Not the Crop" on DVD is available for $15 plus $3 shipping and handling. The booklet of the same title is available for $10 plus $3 for shipping and handling.

Mike Omeg
Omeg Orchards
2967 Dry Hollow Lane
The Dalles, Oregon 97058
541-296-4723
www.omegorchards.com
mike@omegorchards.com

Richard and Linda Parry
Fox Fire Farms
5737 County Road 321
Ignacio, Colorado 81137
970-563-3186
www.foxfirefarms.com
richard@foxfirefarms.com

Bob Quinn
Quinn Farm and Ranch
Kamut International
333 Kamut Lane
Big Sandy, Montana 59520
406-378-3105
www.kamut.com
Bob.quinn@kamut.com

Jim Riddle
University of Minnesota Organic Outreach Coordinator
31762 Wiscoy Ridge Road
Winona, Minnesota 55987
507-454-8310
www.organicecology.umn.edu
riddl003@umn.edu

Richard Rudolph
Rippling Waters Organic Farm
55 River Road
Steep Falls, Maine 04085
207-642-5161 (farm)
www.ripplingwaters.org
greengrower@ripplingwaters.org

Bob Scowcroft
Executive Director, Organic Farming Research
Foundation
303 Potrero St., Ste. 29-203
P.O. Box 440
Santa Cruz, California 95060
831-426-6606
www.ofrf.org

Sam and Elizabeth Smith
Caretaker Farm
1210 Hancock Road
Williamstown, Massachusetts 01267
413-458-4309
elizabeth@caretakerfarm.org
sam@caretakerfarm.org

Art Thicke
32979 Pier Ridge Road
La Crescent, Minnesota 55947

Don Zasada and Bridget Spann
Caretaker Farm
1210 Hancock Road
Williamstown, Massachusetts
413-458-9691
www.caretakerfarm.org
don@caretakerfarm.org
bridget@caretakerfarm.org

Gary Zimmer
Midwest Bio-Ag
10955 Blackhawk Drive
Blue Mounds, Wisconsin 53517
800-327-6012
www.midwesternbioag.com
bioag@mhtc.net

Glossary

Government Acronyms

ARS — Agricultural Research Service

ATTRA — Appropriate Technology Transfer for Rural Areas (div. of NCAT)

CSREES — Cooperative State Research, Education, and Extension Service

EPA — Environmental Protection Agency

ERS — Economic Research Service

FDA — Food and Drug Administration

FSA — Farm Service Agency

NAL — National Agricultural Library

NCAT — National Center for Appropriate Technology

NRCS — Natural Resources Conservation Service

SARE — Sustainable Agriculture Research and Education

SAN — Sustainable Agriculture Network

USDA — United States Department of Agriculture

(For a list of government conservation program acronyms, see Resources appendix under notes for chapter 3, Soil.)

Other Acronyms

IFOAM — International Federation of Organic Agriculture Movements

IOIA — International Organic Inspectors Association

NOP — National Organic Program

NOSB — National Organic Standards Board

allelopathic. When a plant emits chemicals from its roots that have the effect of suppressing the growth of other plants. For example, black walnut trees will suppress the growth of other trees nearby.

amendment. In organic farming, refers to a substance added to the soil to improve its tilth and/or fertility. Generally used to describe natural (nonsynthetic) substances, such as lime, compost, and manure.

Animal Unit (A.U.). Used for calculating amount of feed and/or pasture needed for a group of livestock. An animal unit is defined by the Natural Resources Conservation Service as one cow of about 1,000 pounds and her calf up to weaning.

bacteria. Single-celled organisms.

biodiversity. Variety of life-forms. The greater the biodiversity of an area, meaning the more species of plants, animals, insects, and other life-forms in the area, the more stable the ecosystem will be.

biointensive. Producing the maximum amount of food from the minimum amount of land.

Bt (*Bacillus thuringiensis*). A species of bacteria that has natural insecticidal effects, that is, it can kill crop insect pests.

CAFO (Concentrated Animal Feeding Operation). A large-scale livestock or poultry farm, where the animals are closely confined either in large buildings or feedlots.

Cation Exchange Capacity. A measure of the soil's ability to hold nutrients. Closely related to the percentage of clay and of organic matter in the soil.

chelate. A molecule that is structured in a particular way, with a metal atom surrounded by a complex organic (carbon-based) molecule. The effects of chelation vary. According to "Soil Science and Management" by Edward J. Plaster, some humus-metal chelates are insoluble in the soil, while others protect metals from being fixed in the soil, and so keep the metals accessible to plants. Artificial chelates are used as trace-metal fertilizers.

certified, certification. Organic certification is obtained when 1) a trained organic inspector has inspected the farm and written a report, 2) a certifying agency has reviewed the inspector's report and the farmer's application and verified that the farm is in compliance with the National Organic Program's Final Rule, and 3) the agency has issued a certificate to document that fact.

commodity. Those crops and livestock that are so abundantly produced and so universally in demand that there is always a market for them. Corn, wheat, soybeans, beef, and milk are examples of commodities.

Community Supported Agriculture (CSA). A farming system where consumers pay for a share of a farm's production at the start of the year, and then their portion of the production is picked up (or delivered) on a regular basis throughout the growing season. Most common in vegetable production but has been expanded to include fruits, grains, meats, eggs, and dairy by innovative farmers.

compaction. When either natural features or continual heavy traffic from machinery or animals crushes and destroys soil structure so that water, air, roots, and nutrients can not move in or out.

compost. Decomposed organic materials.

cover crop. A crop planted not for harvesting but to benefit the soil. There are several reasons to plant one: a cover crop can protect against wind and water erosion, hold nutrients that otherwise might leach away, and/or be tilled under to increase soil organic matter.

conservation tillage. Any type of tillage that minimizes soil erosion by leaving some or all the previous year's crop residues on the surface of the soil, so that 30 percent or more of the soil surface is covered. The next year's crop is then planted into the residue. This practice generally reduces the number of times a farmer has to work a field to prepare it for planting but involves specialized implements and usually a greater use of pesticides and herbicides.

continuous cropping. When the same annual crop is grown in the same field or garden area year after year. The most common example is continuous corn. This practice is prohibited for organic growers, since it depletes soil nutrients; increases disease, pest, and weed levels; and usually decreases organic matter and degrades soil structure. Continuous cropping does not refer to perennial crops, such as most fruits, or to permanent pastures.

contour strips/contour cropping. Running crop rows perpendicular to the slope of a field, so that rainfall is held in the crop rows until it can soak in, rather than being channeled directly down the slope. To improve erosion control further, strips of row crops can be alternated with forages.

conventional. Mainstream. In agriculture, this term is often used to refer to farming with high chemical inputs, continuous cropping, large-scale livestock operations, and related practices.

conventional tillage. Plowing (traditionally with a moldboard plow) so that plants and residues are completely buried and the soil surface is bare.

crop residue. The stalks and leaves left in the field after the grain has been removed.

crop rotation. The practice of growing a different crop in a particular field or garden area each year, so that over the course of four or more years the same field will grow a variety of crops. A rotation usually alternates crops that build the soil (even though they have little or no cash value) with crops that bring in income but are more demanding of the soil. Crop rotation is required by the Final Rule.

CSA. See Community Supported Agriculture.

cud. A wad of partially chewed forage that a ruminant has swallowed and then burped up from storage in the rumen (first stomach chamber) to chew thoroughly before re-swallowing it into the second and subsequent stomach chambers to complete digestion.

cultivar. Refers to plants and is equivalent to the term "breed" in animals. A cultivar is a breed or variety of plant that has been developed and maintained by humans.

cultivation. Both mechanical weeding, and, in a larger context, the growing of crops.

disc. An implement that uses offset gangs of metal discs to cut and turn soil.

drag. A type of harrow, a drag is an implement traditionally used as the last step in seedbed preparation, to smooth the soil and break up clods of dirt. Many different types of drags are used, from manufactured implements to chains, boards, or even tree branches dragged over a field.

double cropping. The practice of growing and harvesting more than one crop on the same ground in the same season. Common examples are planting soybeans into winter wheat stubble, or following garden peas with salad greens.

dryland farming. Farming in a low-rainfall area without irrigation. In the United States, dryland farming generally occurs in the area west of the 96th meridian, which runs from western Minnesota to eastern Texas. See also fallow.

ecofarmer, ecofarming. Refers to a farming system based on soil building that may still use limited amounts of synthetic fertilizers and pesticides. The terms were coined by Acres USA editor Charles Walter in the October 1971 issue. Walters said "eco" could stand for "economic" as well as "ecological," since "a farm practice cannot be economical in the long run unless it is also ecological."

ecology. The study of organisms in relation to their physical and biological environment, or how organisms relate to each other. Essentially the study of adaptations.

ecosystem. According to the Oxford English Dictionary, "A biological community of interacting organisms and their physical environment."

erosion. In soil, the carrying away of soil by mechanical means — wind or water.

environment. Surroundings, including both physical and biological factors.

fallow. Not growing a cash crop in a field for a season. In conventional dryland farming, fields are often "bare fallowed" every other year to conserve enough moisture for the following year's crop, but this bare soil has to be treated with weed killer to stay bare and is erosion prone. Organic farmers will "green fallow" fields, planting a cover crop during a fallow year that is tilled under before the cash crop is planted. Fallowing in order to plant cover crops is a common practice of organic farmers in all areas, not just dry regions.

fertility. A measure of the soil's ability to produce healthy, abundant plant life.

fertilizer. Materials added to the soil to supply essential elements (Edward J. Plaster).

Final Rule. Short-hand for 7 CFR PART 205 — USDA Agricultural Marketing Service National Organic Program Rules. The federal regulations defining and governing certified organic production and processing.

forages. Plants grown for pasture or hay, nearly always legumes and grasses. About 40 species of grasses are grown for forages in the United States, along with alfalfa and several other types of clovers. Forages are also excellent soil builders and are often used as green manures when they are not harvested for livestock feed.

GMO (Genetically Modified Organism). Any living organism that has been manipulated by other than natural means (cross breeding or cross pollination) to incorporate genes from another living organism.

grazer. An animal that grazes.

grazier. A human that manages grazing animals.

"green" labeling. Labels on food products that convey the idea that they were produced in a way that was good for the environment. Often these labels have no relation to how the product was actually produced.

green manure. A cover crop plowed down to increase soil organic matter and fertility.

hardpan. A compacted, impermeable layer in the soil, usually under the surface, that water and roots can't penetrate. Hardpan may be caused by excessive machinery traffic, or excessive tillage at a constant depth, or may be a natural feature of the soil profile. Hardpan can often be ameliorated by subsoil plowing.

harrow. An implement used to finish a seedbed by smoothing the soil surface and breaking up clods.

heat synchronization. Using injections of synthetic hormones to cause females in a livestock herd to come into heat simultaneously, to simplify breeding. Prohibited by the Final Rule.

herbicide. A type of pesticide used to kill weeds or unwanted plants.

hormone treatments. Feeding, injecting, or using slow-release boluses to administer artificial hormones to livestock in order to increase milk production or growth rates, or to synchronize heats. (See heat synchronization). BGH is the most well-known hormone treatment and is an acronym for bovine growth hormone.

humus. That part of organic matter in the soil that is highly resistant to decomposition. Humus makes the soil dark and is important for holding water and nutrients and improving soil structure.

hybrid. The offspring of a crossing between two species or varieties of plants or animals. Often more vigorous than either parent, and often infertile.

hydroponic. Refers to the method of growing plants without soil, by rooting them in a circulating water solution containing all necessary plant nutrients.

inoculation. Refers to the practice of coating legume seeds with a compatible rhizobium bacteria before planting to maximize nitrogen fixation by the crop during the growing season.

input. Any substance used in a farming system to improve fertility or pest control. Includes fertilizer, and soil amendments such as compost and lime.

insecticide. A pesticide formulated to kill insects

landrace. A breed of animal or variety of plant that has developed in and is especially adapted to a particular area or region, sometimes even an individual farm. Often the foundation for a new breed or variety.

legume. A family of plants that share the trait of being able to fix nitrogen in the soil through the interaction of their roots with rhizobium bacteria. Legumes are important sources of nitrogen for organic farmers and include alfalfa and the various species of beans, peas, and clovers.

lime. Pulverized limestone or chalk, applied to reduce soil acidity. High in calcium and, depending on the source, may also have significant levels of magnesium.

lister plow. A plow with moldboards mounted back to back to create high ridges and low furrows. The ridges are warmer and drier than the furrows, an advantage in wet, humid areas. This type of plow is also used to minimize wind erosion, or to prevent water runoff on sloped fields (see contour cropping).

macronutrient. A soil nutrient needed in large amounts by plants.

micronutrient. An essential plant nutrient, but necessary only in very small amounts.

microorganism. Any living organism too small to be seen with the naked eye.

MIG (Management Intensive Grazing). Also called rotational grazing, and managed grazing: The practice of subdividing pastures or range and rotating grazing livestock through the subdivisions as appropriate for the rate of plant growth.

moldboard plow. The classic plow. Uses a metal "moldboard" to cut and turn over the soil. Each moldboard is referred to as a "bottom," so that a three-bottom plow has three separate moldboards.

monoculture. Where only a single species is present, usually referring to plants. In agriculture it refers to the practice of growing enormous acreages of only one crop, usually corn, soybeans, or wheat, most often with little or no rotation of crops from year to year.

natural. A loosely defined term often used to specify that something occurs without human input or interference. As a product label it has no definite meaning.

natural processes. The way plants and animals would grow, reproduce, and be recycled if not influenced by humans.

nutrient cycling. The retention of nutrients in an ecosystem as they move through the various components — soil, plants, and animals. Cycling conserves nutrients. If not enough organisms and organic matter are present to process and hold nutrients, they are lost from the system through leaching, offgassing, or erosion.

organic matter. That part of the soil made up of once-living organisms.

paddock. A subdivision of a pasture.

pasture. An area growing grasses and/or legumes and fenced off for livestock grazing.

pesticide. A chemical, either man-made or naturally occurring, that kills other forms of life, usually targeting certain classes of plants or insects. Some pesticides are "broad-range," killing many types of insects or plants, while "narrow-range" pesticides target a specific genus or species. Other types of pesticides include algaecides, fungicides, and so on.

pH. A measure of the acidity or alkalinity of soil, important to know because plant growth is strongly influenced by soil pH.

plowing. Preparing a seedbed by working the soil with a plow, which either turns it over (moldboard plow) or cuts and stirs it (chisel plow).

rangeland. Uncultivated area usable for grazing. Rangeland is too rough, too dry, too inaccessible, too infertile, too erodible, and/or too rocky to be used for crops. Rangeland is also sensitive to environmental stress and can be rapidly degraded by poor grazing practices. On the other hand, well managed grazing can greatly improve the productivity and stability of rangeland.

rotation (crops and livestock). See crop rotation. Livestock can also be rotated by subdividing pastures and moving a herd from paddock to paddock throughout the grazing season (see MIG [Management Intensive Grazing]). Different species of livestock can follow each other through a pasture as well, such as sheep after cattle, and so diversify the rotation.

row crop. A field crop that is planted in distinct rows, with room between the rows to run equipment. Corn and soybeans are good examples.

ruminant. A cud-chewing animal. Ruminants have multichambered stomachs, with the first chamber, the rumen, serving as storage for food until the animal quits ingesting and begins burping up cuds of the stored food to chew thoroughly and swallow into the second and subsequent parts of the stomach for digestion.

salination. When too much irrigation or high water tables in alkaline soil leave excessive amounts of salt in the soil, inhibiting or preventing plant growth.

sewage sludge. The solid part of sewage, which accumulates at the bottom of treatment tanks at wastewater treatment plants. Conventional farmers sometimes spread sludge on their land as a fertilizer, a sort of human manure, but this is absolutely prohibited by the Final Rule, since the sludge often contains toxic metals such as cadmium and may contain human pathogens (disease-causing organisms) as well.

small grains. Grain crops with a small seed that are planted with a grain drill in very close rows, or broadcast (scattered) by hand, so that the growing crop looks like a field of grass. Examples are wheat, oats, rye, barley, and many more. Small grains sprout quickly and cover dirt rapidly, minimizing soil erosion. They are often used green manures, or as a nurse crop for slower-growing forage plants, such as alfalfa.

split application. Splitting a recommended amount of fertilizer or soil amendment into two or more separate applications. A good way of minimizing loss through leaching in sandy soils.

soil amendment. See amendment.

superphosphate. Natural rock phosphate that has been treated with sulfuric acid and/or phosphoric acid to increase the amount of phosphorus that will be immediately available to plant roots. Prohibited by the Final Rule.

supplements. Vitamins, minerals, and salt for livestock. Sometimes used to refer to soil inputs as well.

sustainable. A system that operates in such a way as to conserve and improve the diversity, health, and fertility of all its components (soil, plants, animals, people), so that it can continue indefinitely into the future. Often defined to include economic viability when referring to human-run systems.

synthetic. Man-made. In organic farming, the term is used (instead of "chemical") to distinguish between natural and man-made pesticides, fertilizers, and other agricultural inputs, since all types of all these substances are, in fact, chemicals.

terracing. On very steep land where contour cropping is not enough to stop erosion, step-like terraces can be formed instead. Terraces are most amenable to small-scale, human-powered farming and are traditional in many mountainous regions around the world.

tillable. Land that can be worked by machinery and can grow crops. Advertisements for farms and land frequently mention the number of tillable acres. Land that is too steep, wet, or rocky is not tillable, though it may be usable as pasture.

tillage. Working the soil.

tilth. The physical condition of the soil; soil structure.

topdressing. Refers both to applying fertilizer or other soil amendments on top of the soil or crop without working it into the soil, and to adding grain or feed supplements on top of the regular livestock ration.

trace element. Mineral necessary to healthy plant or animal growth, but only in very small amounts.

yield. The amount of production per unit of land or livestock. Examples are bushels per acre, or pounds of potatoes per feet of garden row. With livestock, refers to ratio between pounds of meat and live weight of the animal.

Index

Italicized page references indicate line art and photographs.
Bold page references indicate charts and tables.

Other Storey Books You Will Enjoy

Humane Livestock Handling
by Temple Grandin with Mark Deesing
Low-stress methods and complete construction plans for facilities
that allow small farmers to process meat efficiently and ethically.
240 pages. Paper. ISBN 978-1-60342-028-0.

Small-Scale Livestock Farming
by Carol Ekarius
A natural, organic approach to livestock management to produce healthier
animals, reduce feed and health care costs, and maximize profit.
224 pages. Paper. ISBN 978-1-58017-162-5.

Storey Basics® Series: Farming
The essential information you need to get things done. With titles such as *Managing Manure,*
Making Hay, and Finding Good Farmland, Storey Basics® are the ideal resource for any beginner
or seasoned farmer looking to expand their skills. These portable, highly accessible guides,
written by experts, provide the perfect amount of information to ensure success right from the start.
Paper. Learn more about each title by visiting www.storey.com.

Storey's Guide to Growing Organic Orchard Fruits
by Danny Barney
Jam-packed with information and a straightforward, easy-to-read blueprint
for both a successful orchard and a successful business.
544 pages. Paper. ISBN 978-1-60342-570-4. Hardcover. 978-1-60342-723-4.

Storey's Guide to Growing Organic Vegetables & Herbs
by Keith Stewart
Clear information and advice about equipment, crop mix, growing techniques,
irrigation, pests, harvesting, storage, labor, debt, customer management, sales,
accounting, and planning for next season.
560 pages. Paper. 978-1-60342-571-1. Hardcover. 978-1-61212-007-2.

These and other books from Storey Publishing are available
wherever quality books are sold or by calling 1-800-441-5700.
Visit us at *www.storey.com* or sign up for our newsletter
at *www.storey.com/signup.*